New Trends in Recycled Aggregate Concrete

New Trends in Recycled Aggregate Concrete

Special Issue Editors

Jorge de Brito
Chi Sun Poon
Baojian Zhan

MDPI • Basel • Beijing • Wuhan • Barcelona • Belgrade

MDPI

Special Issue Editors

Jorge de Brito
University of Lisbon
Portugal

Chi Sun Poon
The Hong Kong Polytechnic University
China

Baojian Zhan
The Hong Kong Polytechnic University
China

Editorial Office
MDPI
St. Alban-Anlage 66
4052 Basel, Switzerland

This is a reprint of articles from the Special Issue published online in the open access journal *Applied Sciences* (ISSN 2076-3417) from 2018 to 2019 (available at: https://www.mdpi.com/journal/applsci/special_issues/Recycled_Aggregate_Concrete)

For citation purposes, cite each article independently as indicated on the article page online and as indicated below:

LastName, A.A.; LastName, B.B.; LastName, C.C. Article Title. *Journal Name* **Year**, *Article Number*, Page Range.

ISBN 978-3-03921-140-1 (Pbk)
ISBN 978-3-03921-141-8 (PDF)

Contents

About the Special Issue Editors vii

Preface to "New Trends in Recycled Aggregate Concrete" ix

Jorge de Brito, Chi Sun Poon and Baojian Zhan
Special Issue: New Trends in Recycled Aggregate Concrete
Reprinted from: *Appl. Sci.* **2019**, *9*, 2324, doi:10.3390/app9112324   1

Fan Wang, Yong Yu, Xin-Yu Zhao, Jin-Jun Xu, Tian-Yu Xie and Simret Tesfaye Deresa
Performance Evaluation of Reinforced Recycled Aggregate Concrete Columns under
Cyclic Loadings
Reprinted from: *Appl. Sci.* **2019**, *9*, 1460, doi:10.3390/app9071460   6

So Yeong Choi, Yoon Suk Choi, Il Sun Kim and Eun Ik Yang
An Experimental Study on Flexural Behaviors of Reinforced Concrete Member Replaced
Heavyweight Waste Glass as Fine Aggregate under Cyclic Loading
Reprinted from: *Appl. Sci.* **2018**, *8*, 2208, doi:10.3390/app8112208 28

Yuexiu Wu, Wanpeng Song, Wusheng Zhao and Xianjun Tan
An Experimental Study on Dynamic Mechanical Properties of Fiber-Reinforced Concrete under
Different Strain Rates
Reprinted from: *Appl. Sci.* **2018**, *8*, 1904, doi:10.3390/app8101904 41

Thanapol Yanweerasak, Theang Meng Kea, Hiroki Ishibashi and Mitsuyoshi Akiyama
Effect of Recycled Aggregate Quality on the Bond Behavior and Shear Strength of RC Members
Reprinted from: *Appl. Sci.* **2018**, *8*, 2054, doi:10.3390/app8112054 59

Jorge de Brito, Rawaz Kurda and Pedro Raposeiro da Silva
Can We Truly Predict the Compressive Strength of Concrete without Knowing the Properties
of Aggregates?
Reprinted from: *Appl. Sci.* **2018**, *8*, 1095, doi:10.3390/app8071095 82

Jung-Ho Kim, Jong-Hyun Sung, Chan-Soo Jeon, Sae-Hyun Lee and Han-Soo Kim
A Study on the Properties of Recycled Aggregate Concrete and Its Production Facilities
Reprinted from: *Appl. Sci.* **2019**, *9*, 1935, doi:10.3390/app9091935 103

Eleftherios Anastasiou, Michail Papachristoforou, Dimitrios Anesiadis,
Konstantinos Zafeiridis and Eirini-Chrysanthi Tsardaka
Investigation of the Use of Recycled Concrete Aggregates Originating from a Single Ready-Mix
Concrete Plant
Reprinted from: *Appl. Sci.* **2018**, *8*, 2149, doi:10.3390/app8112149 124

Sungchul Yang
Effect of Different Types of Recycled Concrete Aggregates on Equivalent Concrete Strength and
Drying Shrinkage Properties
Reprinted from: *Appl. Sci.* **2018**, *8*, 2190, doi:10.3390/app8112190 140

Pang Chen, Wenzhong Zheng, Ying Wang and Wei Chang
Analysis and Modelling of Shrinkage and Creep of Reactive Powder Concrete
Reprinted from: *Appl. Sci.* **2018**, *8*, 732, doi:10.3390/app8050732 154

Zongping Chen, Rusheng Yao, Chenggui Jing and Fan Ning
Residual Properties Analysis of Steel Reinforced Recycled Aggregate Concrete Components
after Exposure to Elevated Temperature
Reprinted from: *Appl. Sci.* **2018**, *8*, 2377, doi:10.3390/app8122377 . . . . **168**

Zhenhua Duan, Shaodan Hou, Chi-Sun Poon, Jianzhuang Xiao and Yun Liu
Using Neural Networks to Determine the Significance of Aggregate Characteristics Affecting
the Mechanical Properties of Recycled Aggregate Concrete
Reprinted from: *Appl. Sci.* **2018**, *8*, 2171, doi:10.3390/app8112171 . . . . **186**

Won-Jun Park, Taehyoung Kim, Seungjun Roh and Rakhyun Kim
Analysis of Life Cycle Environmental Impact of Recycled Aggregate
Reprinted from: *Appl. Sci.* **2019**, *9*, 1021, doi:10.3390/app9051021 . . . . **200**

Dora Foti, Michela Lerna, Maria Francesca Sabbà and Vitantonio Vacca
Mechanical Characteristics and Water Absorption Properties of Blast-Furnace Slag Concretes
with Fly Ashes or Microsilica Additions
Reprinted from: *Appl. Sci.* **2019**, *9*, 1279, doi:10.3390/app9071279 . . . . **213**

Xinyi Wang, Chee Seong Chin and Jun Xia
Material Characterization for Sustainable Concrete Paving Blocks
Reprinted from: *Appl. Sci.* **2019**, *9*, 1197, doi:10.3390/app9061197 . . . . **226**

Yu Song and David Lange
Crushing Performance of Ultra-Lightweight Foam Concrete with Fine Particle Inclusions
Reprinted from: *Appl. Sci.* **2019**, *9*, 876, doi:10.3390/app9050876 . . . . **241**

Seungtae Lee
Effect of Nylon Fiber Addition on the Performance of Recycled Aggregate Concrete
Reprinted from: *Appl. Sci.* **2019**, *9*, 767, doi:10.3390/app9040767 . . . . **255**

About the Special Issue Editors

Jorge de Brito is a Full Professor of Civil Engineering in the Department of Civil Engineering, Architecture and Georesources, the former head of the CERIS Research Centre (2017–2018), and the director of the Eco-Construction and Rehabilitation Doctoral Programme at the Instituto Superior Técnico (IST), University of Lisbon, Portugal, from which he graduated and obtained his MSc and Ph.D. degrees. Although his research covers bridge management systems and construction technology, his main research area is sustainable construction, with an emphasis on the use of recycled aggregates in concrete and mortar. He has participated in 25 competitively financed research projects and supervised 40 Ph.D. and 180 MSc theses. He is the author of 7 books, 28 book chapters and 440 papers in peer-reviewed international journals and has two patents. He is the editor-in-chief of the *Journal of Building Engineering*, an associate editor of the *European Journal of Environmental and Civil Engineering* and a member of the editorial board of 44 international journals and of the following scientific/professional organisations: CIB, FIB, RILEM, IABMAS, and IABSE.

Chi Sun Poon obtained his Ph.D. from the Imperial College, London, and spent two further years as a post-doctoral Fellow at Oxford University specialising in cement and concrete research. Before he joined PolyU in 1992, he obtained professional experience both in a consulting company and in the Hong Kong civil service. Currently, he is the Chair Professor of Sustainable Construction Materials and Associate Head (Research) at the Civil and Environmental Engineering Department of The Hong Kong Polytechnic University. He was awarded the title of Changjiang Chair Professor by the Ministry of Education in 2017. He specialises in the teaching and research of eco-friendly construction materials, concrete technology, and waste management. He has published over 500 papers in international journals and conferences (including over 350 international journal papers and 8 patents). He is an editor of *Construction and Building Materials*. He has received a number of local and international awards for his research development in eco-construction materials including the State Technological Innovation Award in 2017 (2nd Class).

Baojian Zhan is currently a research fellow of the Department of Civil and Environmental Engineering at the Hong Kong Polytechnic University. He received his B.Eng and M. Eng in Material Science from Wuhan University of Technology, Wuhan, China, and his Ph.D. in Civil Engineering from the Hong Kong Polytechnic University. His current research interests include the characterisation and re-utilisation of industrial by-products and solid wastes, the design and testing of cement-based composites, and the development of low-carbon building materials. Dr. Zhan has recently published pieces in *Cement and Concrete Composites*, *Journal of Hazardous Materials*, *Journal of Cleaner Production*, *Materials and Structures*, and *Construction and Building Materials*. His research activities have been recognised by several awards, including GE Foundation TECH Award (2013), CIC INNOVATION AWARD (2015, by the Hong Kong Construction Industry Council).

Preface to "New Trends in Recycled Aggregate Concrete"

Concrete is the most used manufactured material in the world and has one of the greatest impacts on the environment. However, there is no envisaged alternative to this material within a developing world context. Therefore, there is a pressing need to promote the reduction of the environmental impacts of concrete, guaranteeing, at the same time, that its technological and economic advantages remain valid.

This objective could be achieved by replacing part or all of the natural aggregates of concrete with alternatives resulting from recycled materials or various types of waste/by-products from several industries, including construction. Many of the consequences on the technical, economic, and environmental performance of recycled aggregate concrete have been established by dynamic research on this subject. Nevertheless, there is a lag between research and industry practical applications and a number of subjects that still need to be explored, which represent new trends in the search for sustainable recycled aggregate concrete.

Jorge de Brito, Chi Sun Poon, Baojian Zhan
Special Issue Editors

applied sciences

MDPI

Editorial

Special Issue: New Trends in Recycled Aggregate Concrete

Jorge de Brito [1,*], Chi Sun Poon [2,*] and Baojian Zhan [2,*]

1   Instituto Superior Técnico, Universidade de Lisboa, Av. Rovisco Pais, 1049-001 Lisbon, Portugal
2   Department of Civil & Environmental Engineering, The Hong Kong Polytechnic University, Hong Kong M1504, China
*   Correspondence: jb@civil.ist.utl.pt (J.B.); cecspoon@polyu.edu.hk (C.S.P.); zhan.bj@connect.polyu.hk (B.Z.)

Received: 31 May 2019; Accepted: 3 June 2019; Published: 6 June 2019

1. Introduction

Concrete is the most used manufactured material in the world and certainly one of those having the most impact on the environment. However, there is no foreseen alternative to this material within a developing world context. Therefore, a pressing need exists to promote the reduction of the environmental impacts of concrete, guaranteeing at the same time that its technological and economic advantages remain valid.

One of the ways to achieve this objective is by replacing part or all of the natural aggregates with alternatives resulting from recycled materials or various types of waste/by-products from several industries, including construction. Many of the consequences on the technical, economic, and environmental performance of recycled aggregate concrete have been established by the very dynamic research on this subject. Nevertheless, there is both a lag between research and industry practical applications and a number of subjects that must still be explored, which represent new trends in the search for sustainable recycled aggregate concrete.

This Special Issue is therefore dedicated to "New Trends in Recycled Aggregate Concrete" and contributions on, but not limited to, the following subjects were encouraged: Upscaling the use of recycled aggregate concrete in structural design; Large-scale applications of recycled aggregate concrete; Long-term behavior of recycled aggregate concrete; Performance of recycled aggregate concrete in very aggressive environments; Reliability of recycled aggregate concrete structures; Life cycle assessment of recycled aggregate concrete; Mesostructure analysis of recycled aggregate concrete.

So far, 16 papers have been published in the Special Issue of a total of 25 submitted. The next sections provide a brief summary of each of the papers published.

2. Upscaling the Use of Recycled Aggregate Concrete in Structural Design

In their study, Wang et al. [1] evaluated the performance of reinforced recycled aggregate concrete (RRC) columns under cyclic loading, using a fiber-based numerical model. With this model, they performed a comprehensive parametric study to examine the effects of a range of variables on the hysteretic characteristics of RRC columns. A grey relational analysis was also conducted to establish quantifiable evidence of key variable sensitivities. They concluded that the use of the additional water method (AWM) for manufacturing recycled aggregate concrete was likely to reduce the lateral load-carrying capacity of the RRC columns (up to 10%), whereas the opposite would occur if a conventional mixing procedure was adopted. Moreover, compared with other factors such as steel area ratio, the content of natural aggregates replacement had a less remarkable effect on the seismic performance of the RRC columns. In general, the RRC columns have acceptable seismic-resistant properties, and they can be used in earthquake-prone regions with confidence.

Choi et al. [2] developed an experimental study on the flexural behavior of reinforced concrete members with heavyweight waste glass as fine aggregate under cyclic loading. They concluded that, for full replacement, the crack patterns were affected, and there was an increased possibility of sudden failure owing to concrete crushing. Additionally, the load capacity and flexural rigidity were affected by the content of heavyweight waste glass, but the flexural performance improved when mineral admixture as a binder or a low water-binder ratio were used.

Another experimental study was developed by Wu et al. [3] on the dynamic mechanical properties of fiber-reinforced concrete (FRC) under different strain rates. Dynamic uniaxial compression and splitting tensile experiments of FRC with polyvinyl alcohol (PVA) fibers were carried out for two matrix strengths. PVA fibers enhance and toughen concrete, improving its brittle properties and residual strength. With increasing strain rate, the compressive strength, splitting tensile strength, and elastic modulus increase to a certain extent, whereas the toughness index and the peak strain decrease to a certain degree. The post-peak deformation characteristics change from a brittle to a ductile failure with dense cracking. These effects depend on the matrix strength.

Yanweerasak et al. [4] studied the effect of recycled aggregate quality on the bond behavior and shear strength of reinforced concrete members. They concluded that the bond strength of both natural and recycled concrete increased with a decrease in water-to-cement ratio but not for the full spectrum of ratio values. Furthermore, the shear behavior of reinforced concrete beams with natural and recycled concrete is very similar, but the results depend on the size of the beams.

De Brito et al. [5] questioned the validity of concrete compressive strength predictions without knowing the properties of the aggregates. For the same mix composition (with similar cement paste quality), there was a significant difference between the results when natural aggregates of various geological nature were used in concrete. The same was true when different qualities of aggregates were used. However, the scatters significantly decreased when the mixes were classified based on the geological nature of the aggregates or on their quality. The influence of the aggregates on the compressive strength of concrete became much more discernible when recycled aggregates were used mainly due to their more heterogeneous characteristics.

3. Large-Scale Applications of Recycled Aggregate Concrete

In their research, Kim et al. [6] performed a study on the properties of recycled aggregate concrete and its production facilities. Equipment was developed to improve the quality of recycled aggregate to increase the use of that aggregate for environmental improvement purposes. The results showed improvements in the air volume, slump, compressive strength, freezing and thawing resistance, and drying shrinkage.

An investigation on the use of recycled concrete aggregates originating from a single ready-mix concrete plant was performed by Anastasiou et al. [7]. Crushed hardened concrete from test specimens (HR) and from returned concrete (CR) were tested for their suitability as concrete aggregates, and cement sludge fines (CSF) originating from the washing of concrete trucks were tested as filler. Both HR and CR can be considered good-quality recycled aggregates, especially when the coarse fraction is used. Furthermore, HR performs considerably better than CR both as coarse and as fine aggregate. CSF seems to be a fine material with good properties as a filler, provided that it is properly crushed and sieved through a 75 μm sieve.

4. Long-Term Behavior of Recycled Aggregate Concrete

Yang [8] analyzed the effect of different types of recycled concrete aggregates (RCAs) on the equivalent concrete strength and drying shrinkage properties. A total of six mixes were proportioned using the modified equivalent mortar volume (EMV) method with three RCAs. The test results show that the concrete with RCAs produced from concrete sleepers exhibited compressive strength, Young's modulus, and flexural strength values equivalent, within 2% variation, to those values of the companion natural aggregate concrete. In other mixes, compressive strength was found to decrease to

11–20%. For 100% replacement, the Young's modulus increased up to 10% and the drying shrinkage increased up to 8%, while for 50% replacement, the Young's modulus decreased up to 8% and the drying shrinkage dropped up to 4%.

Chen et al. [9] performed an analysis and modelling of the shrinkage and creep of reactive powder concrete (RPC) with various steel fiber contents. It was revealed that the compressive strength and modulus of elasticity increased with increasing steel fiber content, contrary to shrinkage and creep. A good linear relationship was found between the axial stress ratios and creep strain. All four existing models were unable to accurately predict the shrinkage and creep of RPC. However, a good agreement between the experimental results and the proposed shrinkage and creep numerical models was observed.

5. Performance of Recycled Aggregate Concrete in Very Aggressive Environments

Chen et al. [10] analyzed the residual properties of steel-reinforced recycled aggregate concrete (RAC) components after exposure to elevated temperatures. Significant physical phenomena occurred on the surface of RAC and steel-reinforced recycled aggregate concrete (SRRAC) components after exposure to elevated temperatures. The mechanical properties deteriorated significantly with the increase of temperature, namely, the strength of the RAC, and compressive capacity, bending capacity, shear capacity, and stiffness of the SRRAC. The ductility and energy dissipation of SRRAC components were insignificantly affected by the elevated temperatures. Mass loss ratio, peak deformation, and bearing capacity showed a slight increase trend with the increase of replacement ratio. However, the stiffness showed significant fluctuation for high replacement ratios. The ductility and energy dissipation showed significant fluctuation for intermediate replacement ratios.

6. Reliability of Recycled Aggregate Concrete Structures

Duan et al. [11] used artificial neural networks (ANNs) to determine the significance of aggregate characteristics on the mechanical properties of recycled aggregate concrete (RAC). The results show that water absorption has the most important effect on aggregate characteristics, further affecting the compressive strength of RAC, and that combined factors including concrete mixes, curing age, specific gravity, water absorption, and impurity content can reduce the prediction error of ANNs to 5.43%. Moreover, for elastic modulus, water absorption, and specific gravity, they are the most influential, and the network error with a combination of mixes, curing age, specific gravity, and water absorption is only 3.89%.

7. Life Cycle Assessment of Recycled Aggregate Concrete

Park et al. [12] performed an analysis of the life-cycle environmental impact of recycled aggregate. The environmental impact of recycled aggregate (wet) was up to 16–40% higher than that of recycled aggregate (dry), mostly due to the amount of energy used by impact crushers. The environmental impact of using recycled aggregate was found to be up to twice as high as that of using natural aggregate, largely due to the greater simplicity of production of natural aggregate requiring less energy. However, the abiotic depletion potential was approximately 20 times higher when using natural aggregate because its use depletes natural resources, whereas recycled aggregate is recycled from existing construction waste. Among the life-cycle impacts determined from the assessment of recycled aggregate, global warming potential was lower than for artificial lightweight aggregate but greater than for slag aggregate.

8. New Applications of Recycled Aggregate Concrete

Foti et al. [13] studied the mechanical characteristics and water absorption properties of blast-furnace slag concrete with fly ash or micro silica additions. The results show the following: (i) the use of fly ashes, and especially silica fume, together with slag in concrete enhances the compressive strength of concrete mixes and shows very high water/cement ratios; (ii) micro silica concrete shows a specific weight lower than slag and fly ash concrete; (iii) for both types of concrete, the splitting tensile strength is consistent with the compressive strength, whereas flexural tensile strength is rather low, especially for slag and silica fume concrete; and (iv) compared to an ordinary concrete, the types of concrete examined in this research have a lower water absorption value, especially silica fume concrete.

Wang et al. [14] presented a material characterization for sustainable concrete paving blocks. Five types of waste materials were used in this project, including recycled concrete coarse aggregate (RCCA), recycled concrete fine aggregate (RCFA), crushed glass (CG), crumb rubber (CB), and ground granulated blast-furnace slag (GGBS). Using either RCCA or RCFA can decrease the blocks' strength and increase their water absorption. The suggested incorporation levels of RCCA and RCFA are 60% and 20%, respectively. Adding CG to the concrete paving blocks as a type of coarse aggregate can improve their strength and decrease their water absorption. The addition of CR causes a significant deterioration of the blocks' properties, except for their slip resistance.

Song and Lange [15] analyzed the crushing performance of ultra-lightweight foam concrete with fine particle incorporation. The results indicate that the use of fine-graded sand particles in a small content simultaneously reduces the cement content and enhances the crushing performance. However, poor performance is observed for a high sand content. The cellular structure of the foam–sand composite, and thus its mechanical behavior, can be substantially weakened by larger sand particles, especially when the particle size is larger than the voids in the foam.

The effect of the addition of nylon fibers (NFs) on the performance of recycled aggregate concrete (RAC) was studied by Lee [16]. RAC showed a lower performance than crushed stone aggregate concrete (CAC) because of the adhered mortars in the recycled aggregates. However, it was obvious that the addition of NFs in RAC mixes was much more effective in enhancing the performance of concrete, due to the crack bridging effect from NFs. In particular, a high content of NFs (1.2 kg/m^3) led to a beneficial effect on the concrete properties compared to a low content of NFs (0.6 kg/m^3) with respect to mechanical properties and permeability, especially for RAC mixes.

Funding: This research received no external funding.

Acknowledgments: Thanks are due to all the authors and peer reviewers for their valuable contributions to this Special Issue. The MDPI management and staff are also to be congratulated for their untiring editorial support for the success of this project.

Conflicts of Interest: The authors declare no conflict of interest.

References

1. Wang, F.; Yu, Y.; Zhao, X.Y.; Xu, J.J.; Xie, T.Y.; Deresa, S.T. Performance evaluation of reinforced recycled aggregate concrete columns under cyclic loadings. *Appl. Sci.* **2019**, *9*, 1460. [CrossRef]
2. Choi, S.; Choi, Y.; Kim, I.; Yang, E. An experimental study on flexural behaviors of reinforced concrete member replaced heavyweight waste glass as fine aggregate under cyclic loading. *Appl. Sci.* **2018**, *8*, 2208. [CrossRef]
3. Wu, Y.; Song, W.; Zhao, W.; Tan, X. An experimental study on dynamic mechanical properties of fiber-reinforced concrete under different strain rates. *Appl. Sci.* **2018**, *8*, 1904. [CrossRef]
4. Yanweerasak, T.; Kea, T.; Ishibashi, H.; Akiyama, M. Effect of recycled aggregate quality on the bond behavior and shear strength of RC members. *Appl. Sci.* **2018**, *8*, 2054. [CrossRef]
5. De Brito, J.; Kurda, R.; Raposeiro da Silva, P. Can we truly predict the compressive strength of concrete without knowing the properties of aggregates? *Appl. Sci.* **2018**, *8*, 1095. [CrossRef]
6. Kim, J.H.; Sung, J.H.; Jeon, C.S.; Lee, S.H.; Kim, H.S. A study on the properties of recycled aggregate concrete and its production facilities. *Appl. Sci.* **2019**, *9*, 1935. [CrossRef]
7. Anastasiou, E.; Papachristoforou, M.; Anesiadis, D.; Zafeiridis, K.; Tsardaka, E.C. Investigation of the use of recycled concrete aggregates originating from a single ready-mix concrete plant. *Appl. Sci.* **2018**, *8*, 2149. [CrossRef]
8. Yang, S. Effect of different types of recycled concrete aggregates on equivalent concrete strength and drying shrinkage properties. *Appl. Sci.* **2018**, *8*, 2190. [CrossRef]

9.   Chen, P.; Zheng, W.; Wang, Y.; Chang, W. Analysis and modelling of shrinkage and creep of reactive powder concrete. *Appl. Sci.* **2018**, *8*, 732. [CrossRef]

10.   Chen, Z.; Yao, R.; Jing, C.; Ning, F. Residual properties analysis of steel reinforced recycled aggregate concrete components after exposure to elevated temperature. *Appl. Sci.* **2018**, *8*, 2377. [CrossRef]

11.   Duan, Z.; Hou, S.; Poon, C.S.; Xiao, J.; Liu, Y. Using neural networks to determine the significance of aggregate characteristics affecting the mechanical properties of recycled aggregate concrete. *Appl. Sci.* **2018**, *8*, 2171. [CrossRef]

12.   Park, W.J.; Kim, T.; Roh, S.; Kim, R. Analysis of life cycle environmental impact of recycled aggregate. *Appl. Sci.* **2019**, *9*, 1021. [CrossRef]

13.   Foti, D.; Lerna, M.; Sabbà, M.F.; Vacca, V. Mechanical characteristics and water absorption properties of blast-furnace slag concretes with fly ashes or microsilica additions. *Appl. Sci.* **2019**, *9*, 1279. [CrossRef]

14.   Wang, X.; Chin, C.S.; Xia, J. Material characterization for sustainable concrete paving blocks. *Appl. Sci.* **2019**, *9*, 1197. [CrossRef]

15.   Song, Y.; Lange, D. Crushing performance of ultra-lightweight foam concrete with fine particle inclusions. *Appl. Sci.* **2019**, *9*, 876. [CrossRef]

16.   Lee, S. Effect of nylon fiber addition on the performance of recycled aggregate concrete. *Appl. Sci.* **2019**, *9*, 767. [CrossRef]

applied
sciences

MDPI

Article

Performance Evaluation of Reinforced Recycled Aggregate Concrete Columns under Cyclic Loadings

Fan Wang [1], Yong Yu [1], Xin Yu Zhao [1,*], Jin Jun Xu [2,*], Tian Yu Xie [3] and Simret Tesfaye Deresa [2]

[1] State Key Laboratory of Subtropical Building Science, South China University of Technology, Guangzhou 510640, China; wangfan@scut.edu.cn (F.W.); yuyong1990@foxmail.com (Y.Y.)
[2] College of Civil Engineering, Nanjing Technology University, Nanjing 210023, China; simtes2009@gmail.com
[3] School of Civil, Environmental and Mining Engineering, University of Adelaide, Adelaide 5005, Australia; tianyu.xie@adelaide.edu.au
* Correspondence: ctzhaoxy@scut.edu.cn (X.-Y.Z.); jjxu_concrete@163.com (J.-J.X.)

Received: 26 February 2019; Accepted: 28 March 2019; Published: 8 April 2019

Abstract: Recycled concrete aggregates (RCAs) generated from construction and demolition activities have been recognized as a feasible alternative to natural aggregates (NAs). Naturally, the columns fabricated with reinforced recycled concrete (RRC) have been proposed and investigated to promote the structural use of recycled aggregate concrete (RAC). There is still, however, very limited modeling research available to reproduce, accurately and efficiently, the seismic response of RRC columns under lateral cyclic loading; proper evaluations are also lacking on addressing the columns' seismic behaviors. To fill some of those research gaps, a fiber-based numerical model is developed in this study and then validated with the experimental results published in the literature. Subsequently, the numerical model justified is applied to carry out a comprehensive parametric study to examine the effects of a range of variables on the hysteretic characteristics of RRC columns. Furthermore, a grey relational analysis is conducted to establish quantifiable evidence of key variable sensitivities. The evaluation results imply that the use of the additional water method (AWM) for manufacturing RAC is likely to reduce the lateral load-carrying capacity of the RRC columns (up to 10%), whereas the opposite would occur if a conventional mixing procedure is adopted. Moreover, compared with other factors such as steel area ratio, the content of RCA replacement has a less remarkable effect on the seismic performance of the RRC columns. In general, the RRC columns possess acceptable seismic-resistant properties, and they can be used in earthquake-prone regions with confidence.

Keywords: reinforced concrete; recycled aggregate concrete; columns; seismic performance; numerical analysis; variable sensitivity

---

1. Introduction

Recycled concrete aggregates (RCAs) generated from construction and demolition waste (CDW) have been deemed as a potential alternative to natural aggregates (NAs) with the advantage of minimizing the environmental impacts of CDW [1–4], where the resulting concrete products, termed recycled aggregate concrete (RAC), are now received as a type of "green concrete" [5–10]. Along this line, the reinforced recycled concrete (RRC) members (i.e., reinforced concrete slabs, beams, columns and shear walls manufactured with RCAs) have been developed and explored to promote the sustainable use of RAC at the structural level.

Research efforts have been dedicated to characterizing the properties of RRC beams and columns in static loading conditions. The influence of RCAs on the monotonic shear [11–23] or flexural [24–35] behaviors of RRC beams has also been extensively experimentally investigated, with the replacement ratio (i.e., r) varied from 0% to 100%. The experimental outcomes indicate that: (i) in either shear or flexural failure, the presence of various RCA replacement ratios only has a limited influence on the damage process and failure pattern of RRC beams; (ii) an increased replacement ratio appears to reduce the load-carrying capacity of RRC beams. It should be highlighted that the RCAs used in the aforementioned studies were generally not treated before concreting. For the purpose of improving the performance of RCAs, Katkhuda and Shatarat [36] adopted an acid treatment to eliminate the weak layer resulting from the adhesive mortar on RCAs so as to enhance the behavior of RRC beams.

In the case of RRC columns, similar findings have been reported in a number of compression tests using untreated RCAs [37–39]: (i) RCA replacement ratio had no remarkable influence on the failure progression of the columns, and (ii) the compressive strength and the elastic stiffness of RRC columns were generally lower than those of conventional RC columns made with NAs, where the reductions in these properties depended on the RCA content.

It is noteworthy that two thorough review studies by Silva et al. [40] and Tošic et al. [41] have clearly shown that the design code Eurocode 2 can provide a good prediction for the load-carrying capacity of RRC beams without stirrups. However, for the beams with stirrups, which are more significant in practice—the predictions by Eurocode 2 showed a remarkable gap compared to the measurements, suggesting that the underlying mechanisms still needed to be revealed towards more reliable designs.

Up to now, however, research is still scarce on the feasibility of using RAC in concrete elements designed with seismic-resistant purposes. Only a few experimental studies have been conducted aiming to clarify the effect of the incorporation of RCA on the seismic response of RRC columns. Xiao et al. [42] and Yang [43] investigated the cyclic behaviors of RRC columns subjected to axial compression and lateral load reversals. They observed that an increased RCA replacement ratio tended to reduce the lateral load resistance of RRC columns. Moreover, both ductility and energy-dissipating capacities of RRC columns declined to an extent, owing to the incorporation of RCA.

Seismic modeling is essential to generalize the experimental outcomes of RRC columns so as to gain the confidence of their structural use in seismic regions. Again, it is not surprising that only very few studies have been conducted on that important issue. Xiao et al. [44] have numerically modeled a three-dimensional RAC test frame in the OpenSees platform and validated that fiber-based finite elements can be used to approximate the seismic response of RRC frame structures. Still, the potential influences of the unique material properties of RAC on the RRC columns have not yet been numerically and systematically studied.

Thus, the main purposes of the present study are two-fold: (i) to investigate the modeling method of RRC columns under combined action of constant axial load and cyclic lateral loading, based upon a fiber-based numerical approach; (ii) to examine the effects of a range of variables (such as the replacement ratio of RCAs, the mixing method of RAC, and the yield strength and area ratio of longitudinal reinforcing bars) on the seismic performance of RRC columns. This allows an insightful seismic evaluation of those columns that has been less often addressed before.

The remainder of this paper is organized as follows: a fiber-based modeling method is firstly developed to reproduce the cyclic response of RRC columns. Upon a careful benchmarking, the model is employed to perform a comprehensive parametric investigation to examine the effects of a set of potential influencing variables on the seismic performances of RRC columns. Furthermore, a grey relational model, capable of detecting the underlying, not easily discernible tendencies, is used to study the sensitivity of key variables. The research outcomes presented in this study provides valuable insights on the seismic design of RRC columns. The outcomes can be easily extended to performance-based evaluations, hence they are beneficial to the safe and rational use of those columns as lateral-load-resisting elements.

2. Numerical Modelling

2.1. Description of the Modeling Method

A fiber beam model based primarily on the section discretization into fibers was used in this study to predict the hysteretic response of RRC columns subject to lateral cyclic loadings. The SeismoStruct software [45] was employed as the platform to implement the model. A cantilever-type column (i.e., the column's lower end is fixed while the top end is free) was modeled in order to simulate the boundary conditions commonly adopted in column cyclic tests (see Figure 1).

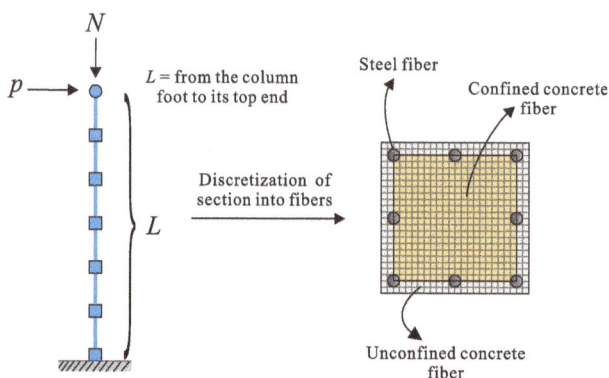

Figure 1. Loading scheme, boundary conditions, and fiber discretization for reinforced recycled concrete (RRC) columns.

The element type "inelastic displacement-based frame element" provided in SeismoStruct was chosen, which was capable of accounting for both material and geometric nonlinearities accurately. It assumed that the concrete and steel bars were rigidly connected, since Zhang et al. [46] have clearly concluded that the bond-slip effect on the flexure-controlled RC columns was generally insignificant. Only flexural bending failure was considered in the proposed model, because such a failure is typical for slender or well-reinforced RC columns [47–51]. Coping with column shear failure was, thus, out of the scope of this study.

2.1.1. Constitutive Model for Recycled Aggregate Concrete (RAC)

The "con_ma" model [52–54] available in SeismoStruct was used to represent the uniaxial constitutive behavior of RAC. This model can be completely determined by the following three key properties: the elastic modulus (E_c), the cylindrical compressive strength (f_c), and the peak strain (ε_{co}^r). Based on a thorough review and extensive experimental database of RAC, Gholampour et al. [55] have proposed an accurate stress–strain relationship for obtaining f_c and E_c of RAC:

$$f_c(\text{MPa}) = \frac{23.5 \times 0.998^r \times (w_{\text{eff}}/c + 0.09)}{w_{\text{eff}}/c^{1.7}}, \tag{1}$$

$$E_c(\text{GPa}) = 0.016 \times (6.1 - 0.015r) \times (5.3 - 1.7 w_{\text{eff}}/c)^{3.9}, \tag{2}$$

where r = the replacement ratio of RCAs ($0\% \leq r \leq 100\%$), and w_{eff}/c = the effective water-to-cement ratio ($0.3 \leq w_{\text{eff}}/c \leq 0.8$).

Xiao et al. [56] suggested an equation to determine the peak strain of RAC (ε_{co}^r) as a function of r:

$$\varepsilon_{co}^r = \varepsilon_{co}^n \left(1 + \frac{r}{65.715r^2 - 109.43r + 48.989} \right). \tag{3}$$

The value of ε_{co}^n in the above equation was determined based on the work by Lim and Ozbakkaloglu [57]:

$$\varepsilon_{co}^n = \frac{f_c^{0.225w_d}}{1000}w_s w_a; \; w_d = \left(\frac{2400}{\rho_{c,f}}\right)^{0.45}; \; w_s = \left(\frac{152}{D_c}\right)^{0.1}; \; w_a = \left(\frac{2D_c}{H_c}\right)^{0.13}, \tag{4}$$

where $\rho_{c,f}$ = the bulk density of concrete (2250 kg/m^3 $\leq \rho_{c,f} \leq$ 2550 kg/m^3); D_c and H_c = the diameter and height of cylinder concrete samples, respectively (50 mm $\leq D_c \leq$ 400 mm, 100 mm $\leq H_c \leq$ 850 mm). w_d, w_s, and w_a are, respectively, the coefficients accounting for the concrete density and the samples' aspect ratio. It should be highlighted that $w_d = w_s = w_a = 1.0$ for the common NAC samples. Consequently $\rho_{c,f}$ = 2400 kg/m^3, D_c = 152 mm, and H_c/D_c = 2.0.

2.1.2. Constitutive Model for Steel Reinforcement

The "stl_mp" model in SeismoStruct was used to describe the uniaxial tensile and compressive constitutive relationships of steel reinforcing bars. This model was initially proposed by Yassin [58], and then modified and extensively utilized by Menegotto-Pinto [59], Filippou et al. [60], and Monti [61]. For more details the reader can refer to the above articles.

2.2. Validation and Discussion

As discussed in Section 1, Xiao et al. [42] and Yang [43] have conducted cyclic tests on the seismic performance of RRC columns under constant axial load and cyclic lateral reversals. The loading process in the above experiments consisted of a load-controlled phase followed by a displacement-controlled phase. During the first phase, the lateral load was progressively exerted on the specimen with an of increment of, respectively, $0.20P_y$ [42] and 5 kN [43]. After steel yielded, the loading method was switched to the displacement-controlled mode. In both experiments the displacement increment was adopted as $1.0\Delta_y$ (Δ_y = the displacement at steel yielding). The loop at each drift level was repeated three times until the lateral load resistance dropped below 80% [42] or 70% [43] of the peak load. Figure 2 and Table 1 show the material properties and the geometries of the RRC cantilever-type test columns. Note that to compensate the larger water absorption of RAC, a so-called additional water method (AWM) (see Section 3.1 for detail) for RAC mix proportion was used in the aforementioned cyclic tests. The experimental outcomes from these tests were employed to validate the numerical model developed herein. The failure mode of all the test specimens was solely flexural-dominated failure (see Figure 3).

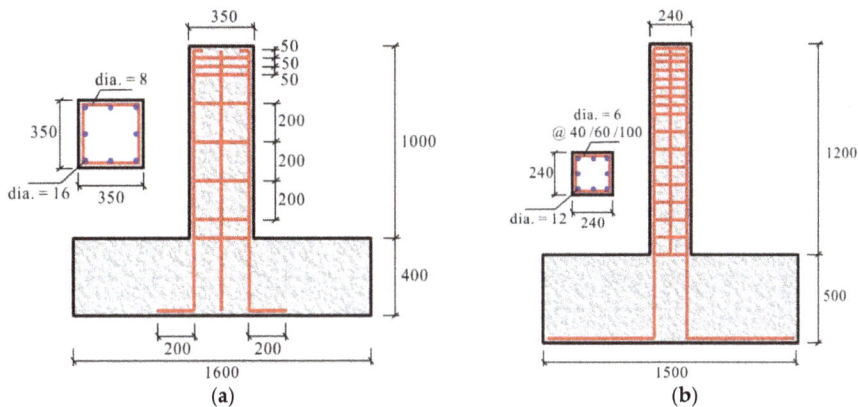

Figure 2. Reinforcement and geometry diagram of reported specimens (unit: mm). (a) Specimens in Xiao et al. [42], and (b) specimens in Yang [43].

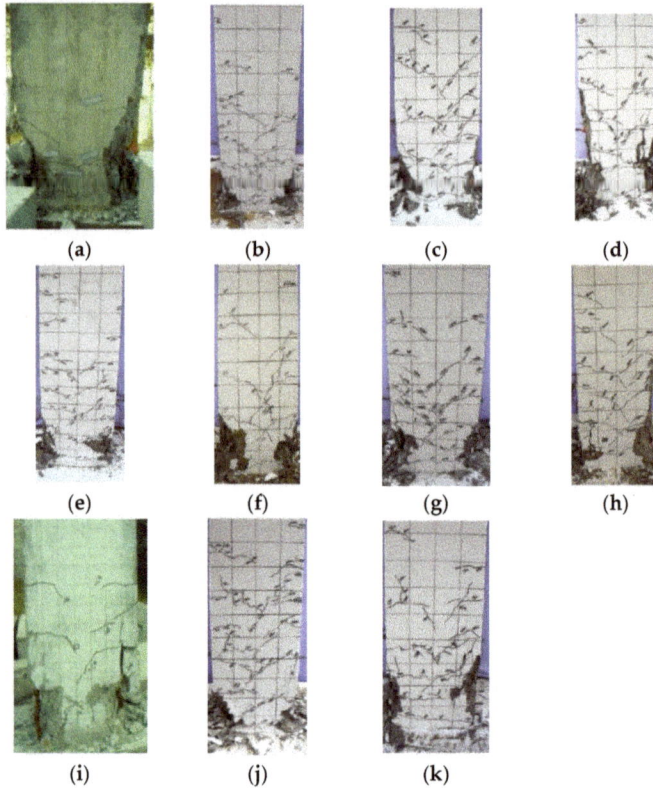

Figure 3. Flexural failure modes of test specimens. (**a**) NCCC-1 [42]; (**b**) NAC-0.30-40 [43]; (**c**) NAC-0.30-60 [43]; (**d**) NAC-0.30-100 [43]; (**e**) RAC50-0.30-40 [43]; (**f**) RAC50-0.30-60 [43]; (**g**) RAC50-0.30-100 [43]; (**h**) RAC100-0.30-60 [43]; (**i**) RCCC-2 [42]; (**j**) RAC50-0.15-60 [43]; and (**k**) RAC50-0.45-60 [43].

Following the modeling approach described previously, the global cyclic response of the specimens in [42,43] can be calculated. Figure 4 compares the lateral load-displacement hysteresis loops obtained from the experiments and the simulations.

Table 1. Experimental information of RRC specimens collected from literature.

Sample	w_{eff}/c	r [%]	f_c [MPl]	E_c [GPl]	L [mm]	h = b [mm]	n	d [mm]	f_y [MPl]	E_s [GPl]	d_{hoop} [mm]	S [mm]	f_{yv} [MPl]	$P_{u,t}$ [kN]	$P_{u,s}$ [kN]	$P_{u,t}/P_{u,s}$	μ_t	μ_s	μ_t/μ_s
Xiao et al. [42]																			
NCCC-1	0.488	0	28.98	36.3	1280	350	0.30	16	353	196	8	200	340	178	174	1.02	5.98	5.67	1.05
RCCC-2	0.43	100	26.98	25.0	1280	350	0.30	16	353	196	8	200	340	164	162	1.01	4.56	3.93	1.16
Yang [43]																			
NAC-0.30-40	0.45	0	33.52	33.0	1000	240	0.30	12	549	241	6	40	389	93	95	0.98	3.80	4.10	0.93
NAC-0.30-60	0.45	0	33.52	33.0	1000	240	0.30	12	549	241	6	60	389	92	94	0.98	3.63	3.94	0.92
NAC-0.30-100	0.45	0	33.52	33.0	1000	240	0.30	12	549	241	6	100	389	87	93	0.93	3.29	3.70	0.89
RAC50-0.30-40	0.45	50	28.88	31.6	1000	240	0.30	12	549	241	6	40	389	84	90	0.93	4.13	4.42	0.93
RAC50-0.30-60	0.45	50	28.88	31.6	1000	240	0.30	12	549	241	6	60	389	91	90	1.02	3.79	4.27	0.89
RAC50-0.30-100	0.45	50	28.88	31.6	1000	240	0.30	12	549	241	6	100	389	87	89	0.98	3.53	3.77	0.94
RAC50-0.15-60	0.45	50	28.88	31.6	1000	240	0.15	12	549	241	6	60	389	69	61	1.12	4.02	4.27	0.94
RAC50-0.45-60	0.45	50	28.88	31.6	1000	240	0.45	12	549	241	6	60	389	94	89	1.06	3.08	4.44	0.69
RAC100-0.30-60	0.45	100	28.00	31.3	1000	240	0.30	12	549	241	6	60	389	82	89	0.92	3.35	4.56	0.73
Average valve																1.00			0.92
Cov																0.06			0.13

Note: Cov—Coefficient of variation.

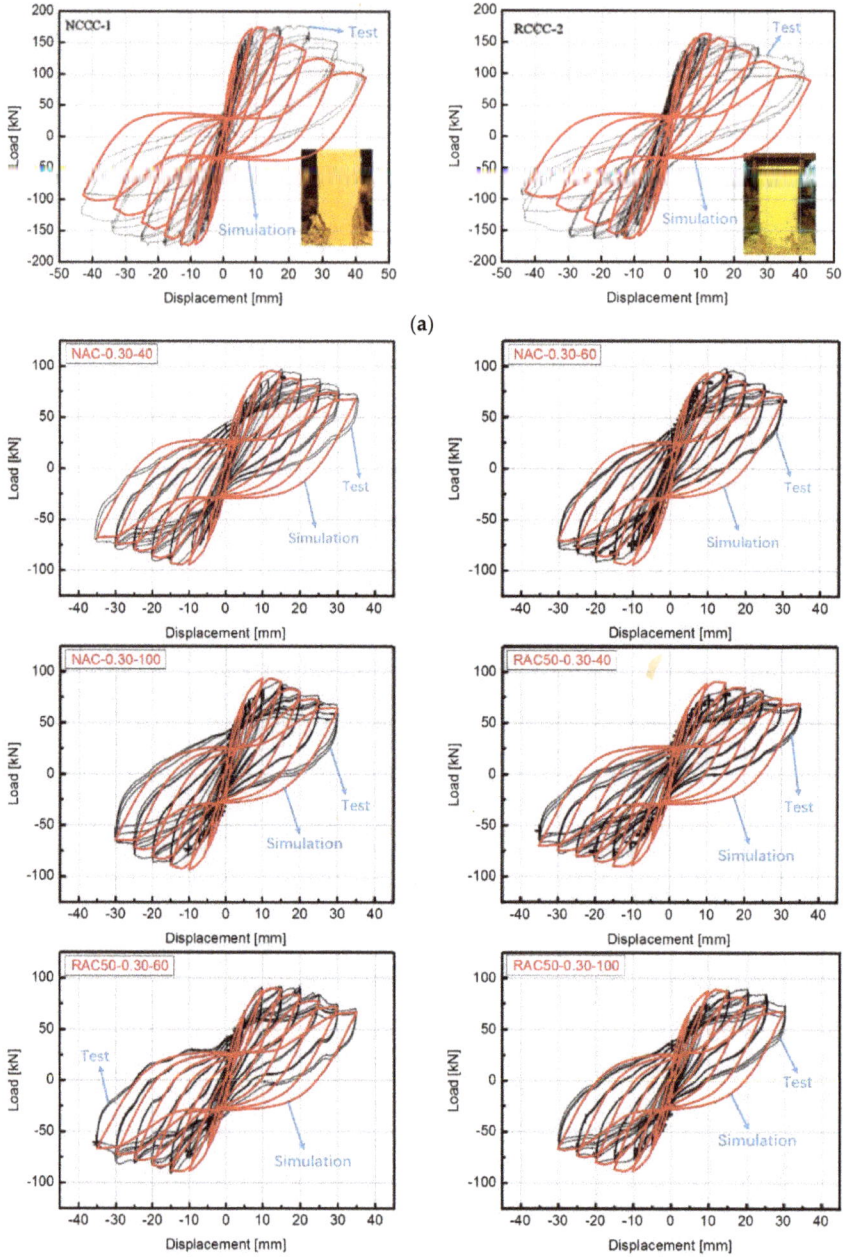

Figure 4. *Cont.*

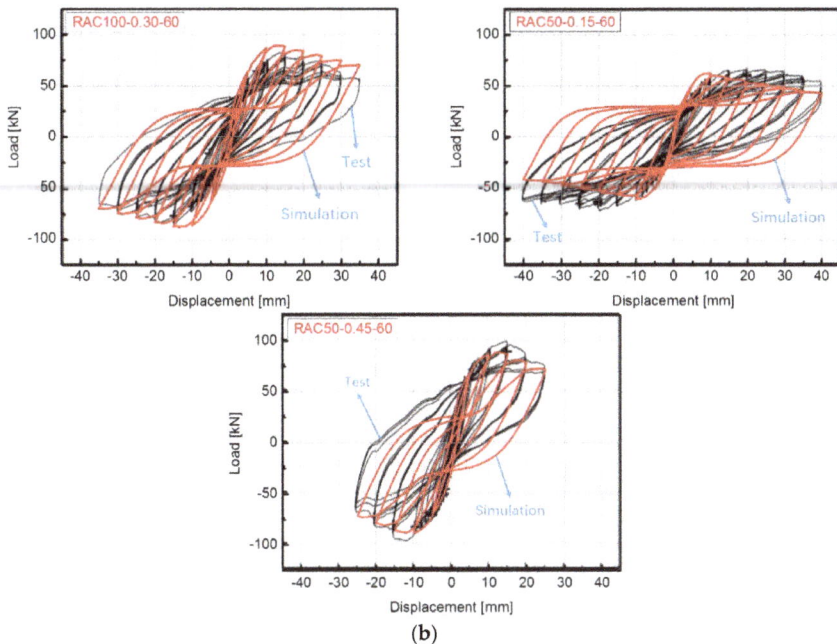

(b)

Figure 4. Comparison of hysteresis loops between experimental and numerical results. (**a**) Test specimens in Xiao et al. [42], and (**b**) test specimens in Yang [43].

The lateral load-carrying capacity and the ductility in Figure 4 were defined as the maximum lateral load (P_u) and the displacement ductility coefficient (μ), respectively. In Table 1, $P_{u,t}$ is the experimental ultimate load, $P_{u,s}$ is the corresponding numerical ultimate load, and μ_t and μ_s are the experimental and numerical displacement ductility coefficients, respectively. The value of μ is defined as the ratio of the ultimate lateral displacement (denoted as $\Delta_{0.85}$) at the load corresponding to 85% of the peak load in the descending backbone curve to the yield lateral displacement (denoted as Δ_y) determined based on the energy equivalence principle (see Figure 5).

$$\mu = \frac{\Delta_{0.85}}{\Delta_y}$$

(5)

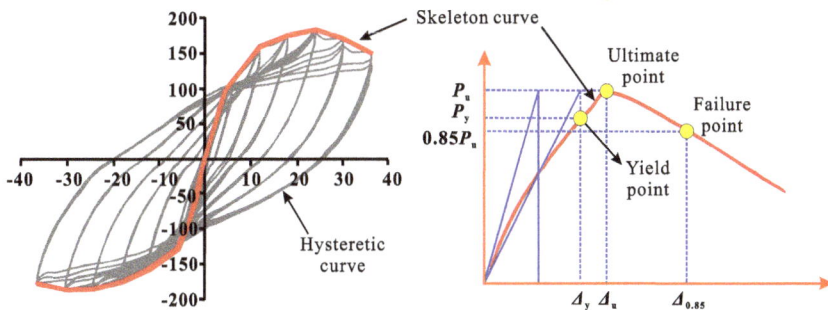

Figure 5. Load-displacement backbone curve.

From Figure 4 and Table 1, it is clear that the predicted hysteresis loops correlated well with the measured ones; the prediction-to-test ratio for the two major seismic performance indexes (P_u and μ) was close to unity ($P_{u,s}/P_{u,t} = 1.00$ and $\mu_s/\mu_t = 0.92$). It can also be seen that the coefficients of variation on the ultimate lateral load and displacement ductility were small on the whole. Hence, the finite-element model developed can provide an overall faithful tool to estimate the seismic performance of RRC columns.

3. Parametric Investigation

A parametric study was conducted so as to extrapolate the existing experimental outcomes for evaluating the seismic response of RRC columns. For convenience, the extended numerical models were established and discussed on the basis of the typical test columns in Yang [43] (i.e., NAC-0.30-60, RAC50-0.30-60, and RAC100-0.30-60; see Table 1). The numerical calculation was terminated once the lateral strength dropped to 70% of the peak strength in the descending branch, which was consistent with the loading protocol used in [43].

Four key variables were considered and varied in the parametric study: (1) the RCA percentage ($0\% \leq r \leq 100\%$), (2) the yield strength of longitudinal steel reinforcement ($300 \text{ MPa} \leq f_y \leq 500 \text{ MPa}$), (3) the area ratio of the steel ($1.57\% \leq \rho_s \leq 6.28\%$), and (4) bi-directional cyclic loading represented by the loading angle α ($0° \leq \alpha \leq 45°$).

The related properties for the numerical analyses were selected as (basically in accordance with [43]): (1) $b = h = 240$ mm, $c = 25$ mm, and $L = 1000$ mm; (2) $f_y = 549$ MPa, $f_{y,v} = 389$ MPa, and $E_s = 241$ GPa; (3) $d = 12$ mm, $d_{hoop} = 6$ mm, and $S = 60$ mm; and (4) axial load ratio $n = 0.3$.

Note that the concrete mixing method was crucial to the mechanical properties of RAC. In view of this, two widely-accepted mixing methods—the additional water method (AWM) and the equivalent total water method (ETWM)—were considered in this numerical study. For each case studied, the following values of RCA replacement percentage were used: $r = 0\%$, 50%, or 100%, which corresponded, respectively, to the RAC's target strength $f_{c,0\%} = 33.52$ MPa, $f_{c,50\%} = 28.88$ MPa, or $f_{c,100\%} = 28.0$ MPa, according to Yang [43].

It should be highlighted that the pre-saturation method may lead to inferior properties of RAC as compared to the AWM as a result of the bleeding effect. However, there are still different opinions. Ferreira et al. [62] have demonstrated that concrete mixes using the pre-saturation method exhibited slightly worse fresh and hardened state behaviors than mixes made with the AWM; however, the mechanical behavior differences observed were generally small and sometimes inconclusive. González-Taboada [63] have further clarified that this indeed depended on the water absorption of RCA: when the water absorption was low, both methods negatively affected concrete compressive strength, whereas when water absorption was high, compressive strength was not affected and both methods can be accurately used. The current authors also found that pre-saturation was, in general, similar to the AWM [64]. Therefore, regarding the AWM and pre-saturation methods, only the former was chosen to compare with the ETWM in this study.

3.1. Influences of r and of the Addition Water Method

Previous experimental investigations [62–75] have clearly reported that the mechanical properties of both RAC and the members containing RAC are predominated by the characteristics of RCAs (RCA percentage, the water absorption and density of RCAs, crushing damage in RCAs, etc.). Based on the water absorption capacity of RCAs, Xu et al. [64] suggested the underlying mechanism of strength reduction or enhancement of RAC manufactured with the AWM and ETWM mixing methods. Figure 6 illustrates the principle of strength variations on RAC determined by the two manufacturing methodologies: the AWM (i.e., [62,63,66]) and ETWM (i.e., [69–72]).

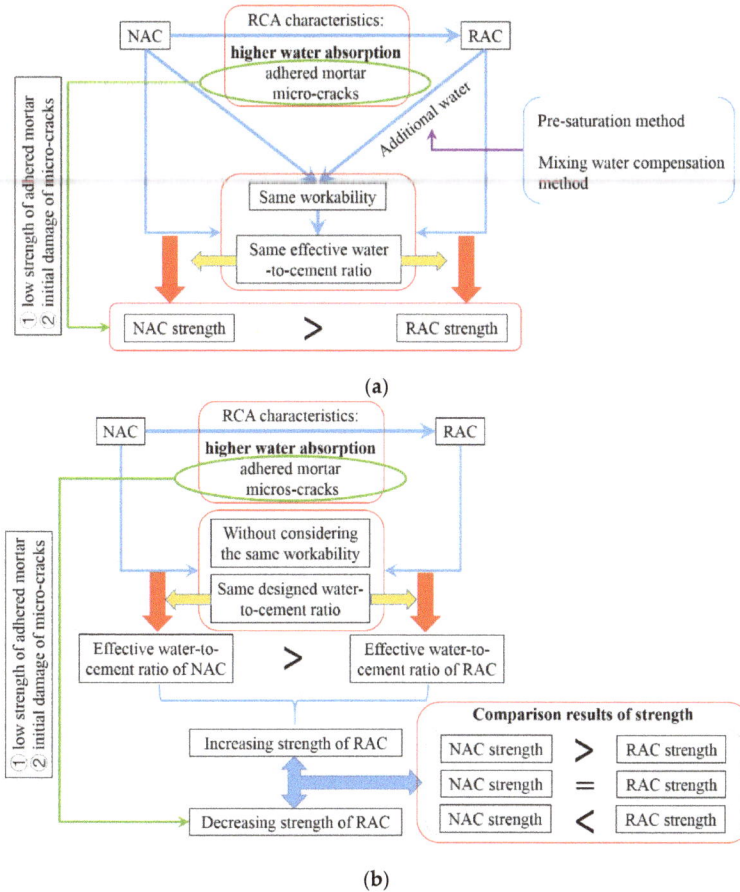

Figure 6. The relationship between recycled concrete aggregate (RCA) characteristics and concrete compressive strength using: (**a**) the additional water method, and (**b**) the equivalent total water method.

A careful inspection of compressive strength of RAC was carried out by examining the test results reported in Xiao et al. [56], Zega and Maio [76] (using AWM), as well as Chen et al. [73] (using ETWM). Figure 7 displays the compressive strength ratio of RAC with respect to that of an equivalent NAC versus the RCA content in light of the above works. It shows that an increase in the RCA percentage reduces the compressive strength ratio between RAC and NAC when the AWM is employed, whereas an increase in the RCA percentage results in an overall increase in the strength of RAC when the ETWM is otherwise used. These experimental results ascertain the principles of the equivalent effective water method and the equivalent total water method based on the way the method affects the relationship between the strengths of NAC and RAC (Figure 6).

In the present modeling, the RAC material properties were adopted from Xiao et al. [56] (in the case of the AWM) and Chen et al. [73] (the ETWM). Figures 8 and 9 show the predicted horizontal load-displacement hysteresis loops and seismic performance using different concrete mixing methods. It can be seen from Figures 8 and 9 that: (i) An increase in r resulted in an up to 10% decrease in P_u of RRC columns using the AWM, whereas the opposite occurred when the ETWM was adopted. However, (ii) the variation in P_u was generally lower than that of the f_c of RAC, consistently true for the two types of mixing methods. This was because the impacts of RCAs (i.e., low strength of adhered mortar and crushing damage in RCAs (i.e., [63–67]) could be largely reduced in RRC columns as a result of structural effects such as the confinement provided by transverse steel reinforcement. Despite the above detrimental effects, (iii) the ductility of RRC columns manufactured with the AWM increased slightly with an increase in the ratio of RCA percentage, whereas for the ETWM no such obvious dependency on the replacement ratio could be found. This can be explained as the AWM may have led to a lower f_c compared to the ETWM, which in turn resulted in a decreasing brittleness [77].

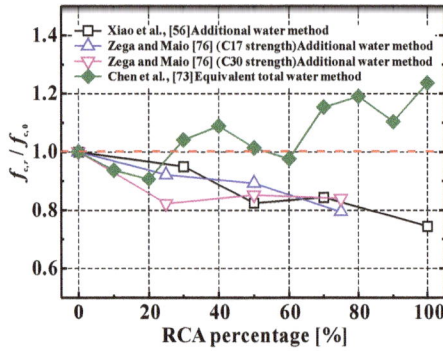

Figure 7. Effect of RCA percentage on 28-day compressive strength of recycled aggregate concrete (RAC) (Note: $f_{c,0}$ is the cylinder compressive strength when $r = 0\%$; $f_{c,r}$ is the cylinder compressive strength when $r \neq 0\%$).

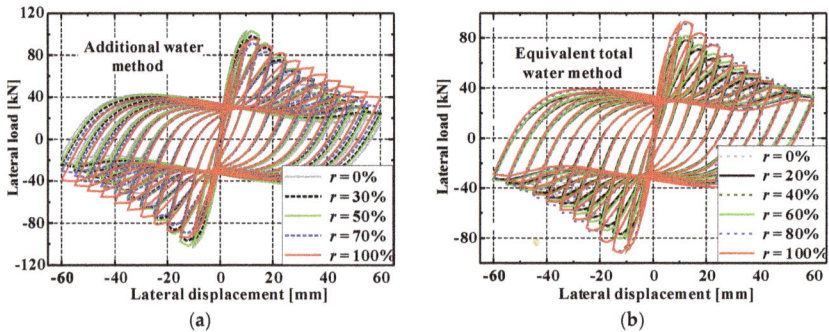

Figure 8. Effect of RCA percentage on hysteresis loops of RRC columns ($n = 0.3$). (**a**) Additional water method (Xiao et al. [56]). (**b**) Equivalent total water method (Chen et al. [72]).

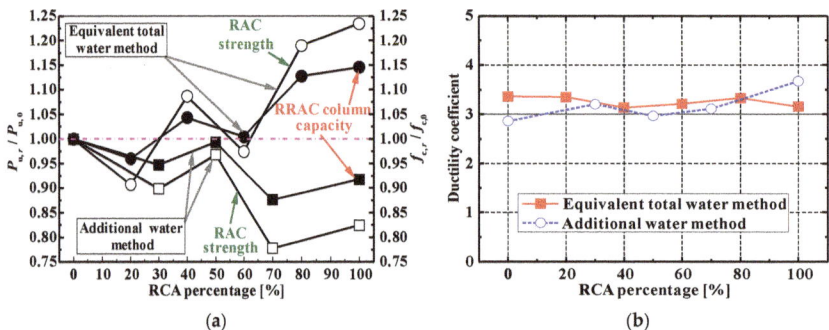

Figure 9. Effect of RCA percentage on seismic performance of RRC columns (Note: $P_{c,0}$ is the ultimate lateral load when $r = 0\%$; $P_{c,r}$ is the ultimate lateral load when $r \neq 0\%$). (**a**) Capacity. (**b**) Ductility.

Due to the relatively pronounced and interesting effects caused by using the AWM, the numerical study in the following sections was based on the specimens (NAC-0.30-60, RAC50-0.30-60, and RAC100-0.30-60) reported in Yang [43], where the AWM was adopted.

3.2. Influences of f_y and ρ_s

Four classes of steel reinforcement in accordance with the Chinese concrete design code (GB 50010-2010 [78]) were employed to conduct the parametric study. The values of steel yield strength f_y corresponding to the four classes were 300, 335, 400, and 500 MPa. The elastic modulus of the steel reinforcement E_s was taken as 200 GPa. The numerical models of RRC columns in this section were established based on the specimens reported in Yang [43] via varying the values of f_y described above.

Figure 10 illustrates the calculated hysteresis loops for the RRC columns, and Figure 11 shows the seismic performance indexes (i.e., P_u and μ) of those columns. From Figures 10 and 11 it can be concluded that improving f_y definitely resulted in a marked increase in P_u. But increasing f_y also reduced the ductility of RRC columns. This reduction can be explained by the inconsistent increase in the yield displacement (i.e., Δ_y) and the ultimate displacement (i.e., $\Delta_{0.85}$) when higher-strength steel was used. The former (Δ_y) was increased more significantly than the latter ($\Delta_{0.85}$), thus reducing the ductility ratio, as defined in Equation (5).

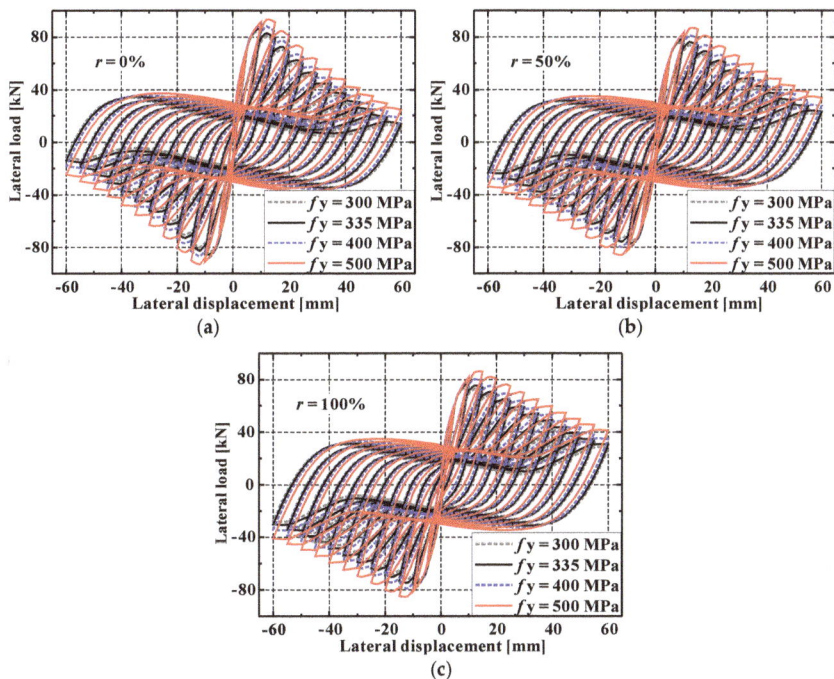

Figure 10. Effect of steel yield strength on hysteresis loops of RRC columns (n = 0.3). (a) r = 0%, (b) r = 50%, and (c) r = 100%.

From Figure 10a, it was also noticed that the value of P_u at a specific yield strength (f_y) generally decreased with an increase in r because the AWM was adopted as discussed in Section 3.1. It should be highlighted that a similar study on the axial load capacity of RRC columns manufactured using the ETWM have been reported in the previous investigation conducted by authors (i.e., [64]), showing that the ETWM generally leads to a slight increase in the load capacity as a function of r.

The influence of area ratio of steel reinforcement ($\rho_s = A_s/A$) on the seismic performance of RRC columns was also discussed. The values of $\rho_s = 1.57\%$, 2.79%, 4.36%, or 6.28% were considered by varying the diameter of the longitudinal steel reinforcement (i.e., $d = 12, 16, 20$, and 24 mm).

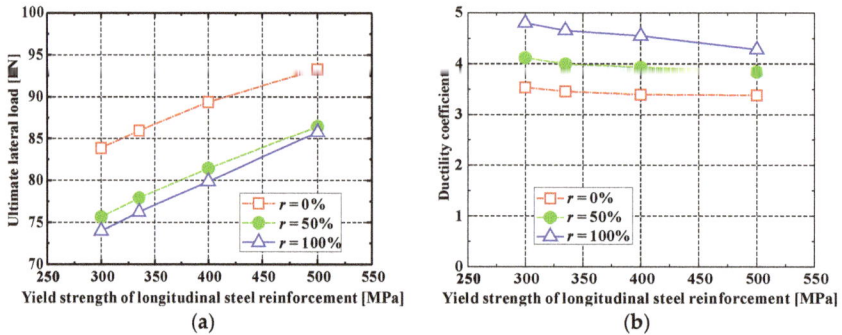

Figure 11. Effect of steel yield strength on seismic performance of RRC columns. (**a**) Capacity. (**b**) Ductility.

Figure 12 shows the hysteresis loops of RRC columns with different ρ_s. Figure 13 shows the influence of ρ_s on the seismic performance indexes. As expected, an increase in ρ_s resulted in a substantial increase in P_u. This was because the steel reinforcement ratio contributed significantly in the flexural strength of the members. On the other hand, an increase in ρ_s resulted in a reduction in μ of RRC columns. In fact, the ductility characteristic was not only affected by concrete but also related to the stress level in the longitudinal bars caused by the axial compressive load; the following derivation clearly shows this.

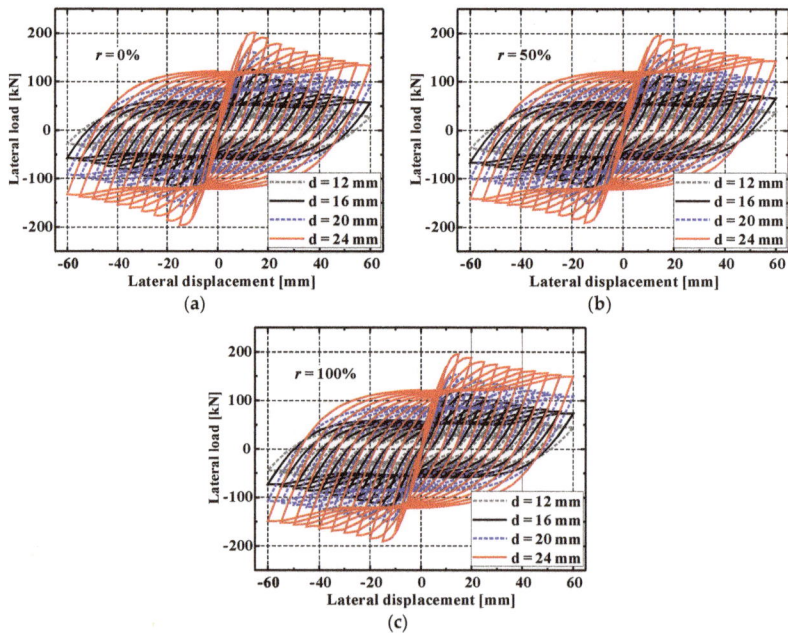

Figure 12. Effect of steel ratio on hysteresis loops of RRC columns ($n = 0.3$). (**a**) $r = 0\%$, (**b**) $r = 50\%$, and (**c**) $r = 100\%$.

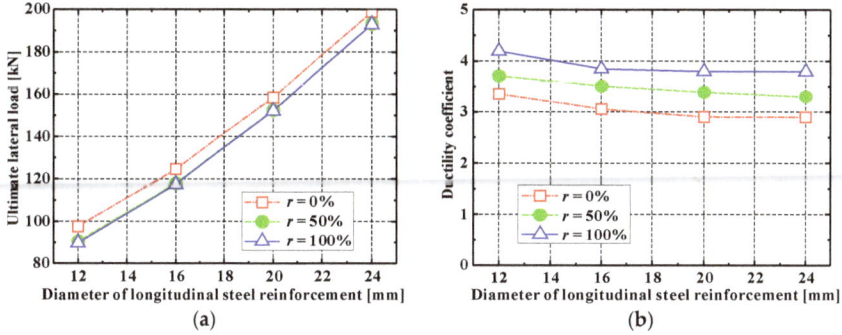

Figure 13. Effect of steel ratio on the seismic performance of RRC columns. (**a**) Capacity. (**b**) Ductility.

The strain compatibility can be assumed between concrete and longitudinal steel reinforcement, leading to:

$$\varepsilon_c = \varepsilon_s \ \rightarrow \ \frac{\sigma_c}{E_c} = \frac{\sigma_s}{E_s} \ \rightarrow \ \sigma_c = \frac{E_c}{E_s}\sigma_s = \alpha_{cs}\sigma_s, \tag{6}$$

where ε_c and ε_s are the strain of concrete and longitudinal steel reinforcement, respectively; σ_c and σ_s are the stress of concrete and longitudinal bars, respectively; E_c and E_s are the elastic modulus of concrete and steel, respectively; and α_{cs} is equal to the ratio of E_c/E_s.

A relationship between the axial load ratio and the stress level of longitudinal bars can be obtained:

$$n = \frac{N}{f_{cp}A} = \frac{\sigma_c A_c + \sigma_s A_s}{f_{cp}A} = \frac{\alpha_{cs}\sigma_s(A - A_s) + \sigma_s A_s}{f_{cp}A} = \frac{\sigma_s[\alpha_{cs}A + (1 - \alpha_{cs})A_s]}{f_{cp}A}$$

$$\sigma_s = \frac{n f_{cp}A}{\alpha_{cs}A + (1 - \alpha_{cs})A_s} \ \Rightarrow \ \frac{n f_{cp}A}{\alpha_{cs}A + (1 - \alpha_{cs})\ \boxed{A_s \uparrow}} = \boxed{\sigma_s \downarrow} \tag{7}$$

where A_c and A_s are the sectional area of concrete and longitudinal bars, respectively; and $A = A_c + A_s$.

It can be seen from Equation (7) that given a certain axial load ratio and strength class of concrete, increasing the steel ratio (ρ_s) led to a decrease in the stress level of steel reinforcement. Consequently, more stress would be transferred and sustained by the concrete, which accelerated its damage in cyclic loading. The ductility of RRC columns thus decreased with increasing the area ratio of steel reinforcement.

3.3. Influence of Bi-Directional Loading

RC structures are often subjected to multi-directional loadings under earthquake ground motions [79,80]. A bi-directional loading scheme was used herein to investigate the seismic performance of RRC columns. Figure 14 shows the definition of loading angle (α) of P to P_x, in which $P = (P_x^2 + P_y^2)^{0.5}$ is the resultant force, and P_x and P_y are the components in X and Y directions, respectively. Loading angles of $0°$ and $45°$ were employed to study the impact of bi-directional loading on the seismic performance of RRC columns.

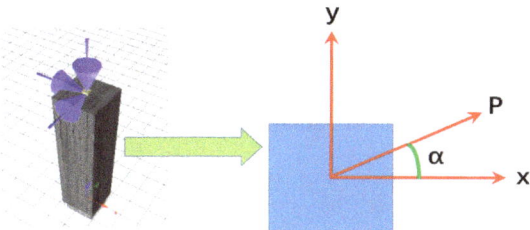

Figure 14. Lateral bi-directional loading for RRC columns.

Figure 15 shows the hysteresis loops of RRC columns with different values of α, and the influences of α on P_u and μ are plotted in Figure 16. Remarkable findings can be observed from Figures 15 and 16: (1) α had no remarkable influence on the initial stiffness of RRC columns, (2) P_u of RRC columns subjected to the loading angle of 45° was lower than that of RRC columns with a 0° loading angle, and (3) μ of RRC columns with the loading angle of 45° was larger than that of RRC columns with 0°. The reason was that the bi-directional loading scheme resulted in the coupling behavior of RC columns—that is, one direction loading can weaken the load-carrying capacity in another direction. On the contrary, one direction loading can expedite the lateral deformation in another direction.

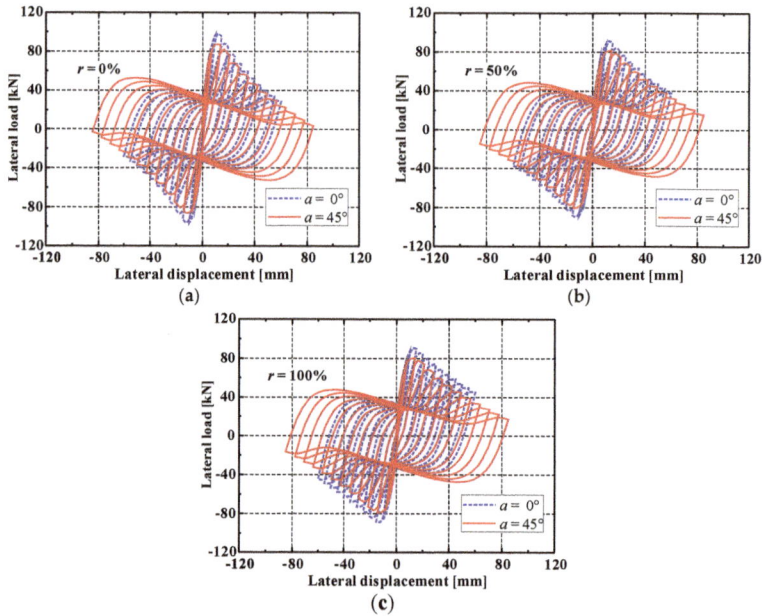

Figure 15. Effect of bi-directional loading on hysteresis loops of RRC columns ($n = 0.3$). (a) $r = 0\%$, (b) $r = 50\%$, and (c) $r = 100\%$.

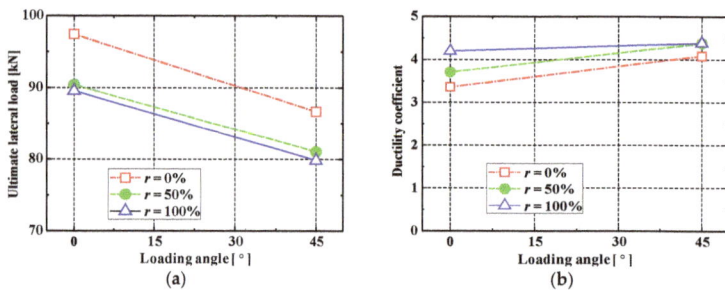

Figure 16. Effect of bi-directional loading on the seismic performance of RRC columns. (a) Capacity. (b) Ductility.

4. Grey Relational Analysis (GRA)

In addition to the numerical model, grey relational analysis (GRA) was also presented in this study to further evaluate the variable sensitivity of the seismic performance of RRC columns. As a set of system theory, grey relational model is a mathematics-based approach to compute the degrees of the correlation between the independent variables (IVs) and dependent variables (DVs) (i.e., [81,82]). Many

20

investigation efforts (i.e., [83–86]) have demonstrated that GRA was an effective solution to optimize the engineering materials and structures showing the level of influence of each variable involved in the problem. Based on the experimental dataset collected from Xiao et al. [42] and Yang [43], GRA was employed to determine the effects of material strengths, geometry dimensions, reinforcement configurations, axial load ratios, and RCA percentages on the hysteretic performance of RRC columns.

4.1. Mathematical Model of GRA

The maximum load (P_u) and the displacement ductility coefficient (μ) reported in Xiao et al. [42] and Yang [43] were determined as the reference matrix, $X_0(j)$, where, $j = 1, 2, \ldots, n$. The key experimental parametric variables, including $r, f_y, L/h, \rho_s$ and n, were selected as the comparative matrix, $X_i(j)$, where, $i = 1, 2, \ldots, m$. The following equation is a mathematical model for the construction of the reference matrix and the comparative matrix.

$$\begin{aligned} X_0 &= X_0(1), X_0(2), \ldots X_0(n) \\ X_1 &= X_1(1), X_1(2), \ldots X_1(n) \\ &\cdots \\ X_m &= X_m(1), X_m(2), \ldots X_m(n) \end{aligned} \tag{8}$$

Normalization of variables in the matrix is processed in order to eliminate their numerical fluctuation.

$$x_i(j) = \frac{X_i(j)}{\frac{1}{n}\sum_{i=1}^{n} X_i(j)}. \tag{9}$$

The grey relational coefficient ξ_i is calculated:

$$\xi_i[x_0(j), x_i(j)] = \left| \frac{\min_{i=1,n}\min_{j=1,m}\Delta_i(j) + \rho\max_{i=1,n}\max_{j=1,m}\Delta_i(j)}{\Delta_i + \rho\max_{i=1,n}\max_{j=1,m}\Delta_i(j)} \right|, \tag{10}$$

$$\Delta_i(j) = |x_0(j) - x_i(j)|, \tag{11}$$

$$\min_{i=1,n}\min_{j=1,m}\Delta_i(j) = \max_i\left(\max_j|x_0(j) - x_i(j)|\right), \tag{12}$$

$$\max_{i=1,n}\max_{j=1,m}\Delta_i(j) = \min_i\left(\min_j|x_0(j) - x_i(j)|\right), \tag{13}$$

where $0 \leq \rho \leq 1$, and its frequently-used value is equal to 0.5 [81].

In GRA, grey relational entropy density (γ) can be used to measure the degree of correlation between the reference matrix and the comparative matrix:

$$\gamma = \frac{1}{n}\sum_{i=1}^{n} \xi_i[x_0(j), x_i(j)] \leq 1.0. \tag{14}$$

It is worth noting that values of γ approaching the unit indicates a closer correlation between the IVs and the DVs; when γ is over 0.7, there is a strong correlation between the IVs and the DVs; when γ is less than 0.5, the correlation between the IVs and the DVs can be considered negligible [81].

4.2. Evaluation of Variable Sensitivity

Table 2 shows the results of the grey relational evaluation between hysteretic performance and parameters. From Table 2, it can be seen that:

(i)   for the maximum lateral load P_u, the influencing sequence was: $L/h > \rho_s = f_y = n > r$;
(ii)   for the displacement ductility μ, the influencing sequence was: $L/h > \rho_s = f_y > n > r$.

Table 2. Ranking results of grey relational evaluation.

Performance	Grey Relational Entropy Density γ_i				
	r	L/h	f_y	ρ_s	n
P_u	0.62	0.87	0.86	0.86	0.85
μ	0.58	0.84	0.83	0.83	0.80

Observations from Table 2 are summarized as:

(1) The steel strength (i.e., f_y), the geometric ratio (i.e., L/h), the reinforcement area ratio (i.e., ρ_s), and the axial load ratio (n) had the most significant influences on the seismic performance of RRC columns, with γ ranging from 0.80 to 0.87;

(2) The sensitivity of RCA percentage on the seismic performance of RRC columns was much less remarkable than that of the steel strength and the other three structural factors (i.e., L/h, ρ_s, and n); however, considering that concrete constituted a large portion of the overall resistance, the influence of RCA percentage should be considered in the seismic design and evaluation of RRC columns.

5. Conclusions

Previous studies on evaluating the seismic performance of concrete columns made with RCAs are still not often seen. These columns have many potential applications in areas with seismic design requirements. With the aim of addressing this significant gap, this research presents a simple, yet practical and efficient, numerical method implemented in the SeismoStruct software to provide an in-depth understanding of the seismic performance of RRC columns. The findings from this study support the following conclusions:

(1) The lateral load-carrying capacity of RRC columns using the AWM generally reduces (up to 10%) with an increase in the RCA replacement percentage, whilst this trend is reversed when the ETWM is used.

(2) Increasing the steel strength is advantageous to the gain in lateral strength of the RRC column, but this leads to a reduction in ductility. Similar two-edged results are also observed when increasing the area ratio of steel reinforcement.

(3) Bi-directional loading has a negative influence on the lateral load-carrying capacity, but it has a positive influence on the ductility of RRC columns.

(4) The steel strength and some well-recognized structural factors (i.e., the shear-span ratio, the area ratio of steel reinforcement, and the axial load ratio) are identified by the GRA method as the most essential parameters affecting the seismic performances of RRC columns, with the grey relational entropy density, γ, ranging from 0.80 to 0.87.

(5) The sensitivity of RCA percentage on the seismic performance of RRC columns is quite modest compared to those of the four factors listed in (4); however, quality and percentage of RAC still should be well-controlled in seismic design of RRC columns.

Overall, it can be said that the use of the RRC columns in seismic regions is generally viable. Those green columns can be properly seismically designed with confidence despite some of the deleterious aspects. However, further studies are still needed, such as determining the minimum longitudinal and transverse steel ratios used in RRC columns, to suppress any negative effects caused by the incorporation of RAC. More experimental data are also called for to investigate the variability of the seismic performance of RRC columns and, finally, to determine their seismic reliability.

Author Contributions: X.-Y.Z., J.-J.X., and F.W. performed the numerical analysis and wrote the whole paper; Y.Y., T.-Y.X., and S.T.D. revised the paper.

Funding: The authors would like to gratefully acknowledge research grants from the National Natural Science Foundation of China (Project Nos: 51708289 and 51608435), Postdoctoral Science Foundation of China (Project No: 2017M611796), Fund of State Key Laboratory of Subtropical Building Science of China (Project Nos: 2017KC18, 2018ZB34 and 2019ZB22), Special Fund for Energy Conservation of Guangzhou (Project No: J-2016-01) and Fundamental Research Fund for the Central Universities (Project No: 2018ZD43).

Conflicts of Interest: The authors declare that there is no conflict of interest regarding the publication of this paper.

Abbreviations

A	Column sectional area
A_c	Concrete sectional area
A_s	Total sectional area of longitudinal rebars
b	Column sectional width
D_c	Diameter of concrete cylinder
d	Diameter of longitudinal rebar
d_{hoop}	Diameter of transverse rebar
E_c	Elastic modulus of concrete
E_s	Elastic modulus of longitudinal rebar
f_c	Concrete cylindrical compressive strength
f_y	Yield strength of longitudinal rebar
$f_{y,v}$	Yield strength of transverse rebar
h	Column sectional height
H_c	Height of concrete cylinder
L	Column effective length
N	Axial load
n	Axial load ratio
P	Lateral load
r	Replacement ratio of RCAs [%]
S	Hoop spacing
w_{eff}/c	Effective water to cement ratio
w_d	Coefficient accounting for concrete density
w_s	Coefficient considering sample aspect ratio
w_a	Coefficient considering sample aspect ratio
σ_c	Stress of concrete
σ_s	Stress of longitudinal bar
ε_{cor}	Peak strain of RAC
ε_c	Strain of concrete
ε_s	Strain of longitudinal rebar
ε_{con}	Peak strain of NAC
$\rho_{c,f}$	Bulk density of concrete
ρ_s	Area ratio of steel rebar
ξ_i	Grey correlation coefficient
γ_i	Grey correlation entropy density
α	Bi-directional cyclic loading angle
μ	Displacement ductility coefficient
$\Delta_{0.85}$	Lateral displacement at 85% peak load
Δ_y	Yield lateral displacement

References

1. Loo, Y.H.; Tam, C.T.; Ravindrarajah, R.S. Recycled concrete as fine and coarse aggregate in concrete. *Mag. Concr. Res.* **1987**, *39*, 214–220.
2. Pacheco, J.; de Brito, J.; Chastre, C.; Evangelista, L. Uncertainty models of reinforced concrete beams in bending: Code comparison and recycled aggregate incorporation. *J. Strut. Eng.* **2019**, *145*, 04019013. [CrossRef]

3.   Pepe, M.; Filho, R.D.T.; Koenders, E.A.B.; Martinelli, E. A novel mix design methodology for recycled aggregate concrete. *Constr. Build. Mater.* **2016**, *122*, 362–372. [CrossRef]

4.   Fathifazl, G.; Abbas, A.; Razaqpur, A.G.; Isgor, O.B.; Fournier, B.; Foo, S. New mixture proportioning method for concrete made with coarse recycled concrete aggregate. *J. Mater. Civ. Eng.* **2009**, *21*, 601–611. [CrossRef]

5.   Matias, D.; de Brito, J.; Rosa, A.; Pedro, D. Mechanical properties of concrete produced with recycled coarse aggregates—Influence of the use of super plasticizers. *Constr. Build. Mater.* **2013**, *44*, 101–109. [CrossRef]

6.   Ccia, F.; Raposo, J.; Guerra, M.; Júlio, E.; de Brito, J. Shear strength of recycled aggregate concrete to natural aggregate concrete interfaces. *Constr. Build. Mater.* **2016**, *109*, 139–145. [CrossRef]

7.   Kurda, R.; de Brito, J.; Silvestre, J.D. Indirect evaluation of the compressive strength of recycled aggregate concrete with high fly ash ratios. *Mag. Concr. Res.* **2017**, *70*, 1–13. [CrossRef]

8.   Xu, J.J.; Zhao, X.Y.; Yu, Y.; Xie, T.Y.; Yang, G.S.; Xue, J.Y. Parametric sensitivity analysis and modelling of mechanical properties of normal- and high-strength recycled aggregate concrete using grey theory, multiple nonlinear regression and artificial neural networks. *Constr. Build. Mater.* **2019**, *211*, 479–491. [CrossRef]

9.   Thomas, C.; Setién, J.; Polanco, J.A.; Alaejos, P.; de Juan, M.S. Durability of recycled aggregate concrete. *Constr. Build. Mater.* **2013**, *40*, 1054–1065. [CrossRef]

10. Ozbakkaloglu, T.; Gholampour, A.; Xie, T.Y. Mechanical and durability properties of recycled aggregate concrete: Effect of recycled aggregate size and content on the behaviour. *J. Mater. Civ. Eng.* **2017**, *30*, 04017275. [CrossRef]

11. Han, B.C.; Yun, H.D.; Chung, S.Y. Shear capacity of reinforced concrete beams made with recycled-aggregate. *ACI Spec. Publ.* **2001**, *200*, 503–516.

12. Sogo, M.; Sogabe, T.; Maruyama, I.; Sato, R.; Kawai, K. Shear behaviour of reinforced recycled concrete beams. In Proceedings of the Conference on the Use of Recycled Materials in Building and Structures, Barcelona, Spain, 8–11 November 2004.

13. González-Fonteboa, B.; Martínez-Abella, F. Shear strength of recycled concrete beams. *Constr. Build. Mater.* **2007**, *21*, 887–893. [CrossRef]

14. Choi, H.B.; Yi, C.K.; Cho, H.H.; Kang, K.I. Experimental study on the shear strength of recycled aggregate concrete beams. *Mag. Concr. Res.* **2010**, *62*, 103–114. [CrossRef]

15. Fathifazl, G.; Razaqpur, A.G.; Isgor, O.B.; Abbas, A.; Fournier, B.; Foo, S. Shear strength of reinforced recycled concrete beams without stirrups. *Mag. Concr. Res.* **2009**, *61*, 387–400. [CrossRef]

16. Fathifazl, G.; Razaqpur, A.G.; Isgor, O.B.; Abbas, A.; Fournier, B.; Foo, S. Shear strength of steel reinforced recycled concrete beams with stirrups. *Mag. Concr. Res.* **2010**, *62*, 685–699. [CrossRef]

17. Fathifazl, G.; Razaqpur, A.G.; Isgor, O.B.; Abbas, A.; Fournier, B.; Foo, S. Shear capacity evaluation of steel reinforced recycled concrete (RRC) beams. *Eng. Struct.* **2011**, *33*, 1025–1033. [CrossRef]

18. Arezoumandi, M.; Smith, A.; Volz, J.S.; Khayat, K.H. An experimental study on shear strength of reinforced concrete beams with 100% recycled concrete aggregate. *Constr. Build. Mater.* **2014**, *53*, 612–620. [CrossRef]

19. Knaack, A.M.; Kurama, Y.C. Behavior of reinforced concrete beams with recycled concrete coarse aggregates. *J. Struct. Eng.* **2015**, *141*, B4014009. [CrossRef]

20. Sadati, S.; Arezoumandi, M.; Khayat, K.H.; Volz, J.S. Shear performance of reinforced concrete beams incorporating recycled concrete aggregate and high-volume fly ash. *J. Clean. Prod.* **2016**, *115*, 284–293. [CrossRef]

21. Rahal, K.N.; Alrefaei, Y.T. Shear strength of longitudinally reinforced recycled aggregate concrete beams. *Eng. Struct.* **2017**, *145*, 273–282. [CrossRef]

22. Ivan, S.I.; Snežana, B.M.; Nikola, T. Shear behaviour of recycled aggregate concrete beams with and without shear reinforcement. *Eng. Struct.* **2017**, *141*, 386–401.

23. Brandes, M.R.; Kurama, Y.C. Behavior of shear-critical prestressed concrete beams with recycled concrete aggregates under ultimate loads. *Eng. Struct.* **2018**, *165*, 237–246. [CrossRef]

24. Sogo, M.; Sogabe, T.; Sato, R.; Kawai, K. Flexural properties of reinforced recycled concrete beams. In Proceedings of the Conference on the Use of Recycled Materials in Building and Structures, Barcelona, Spain, 8–11 November 2004.

25. Sato, R.; Maruyama, I.; Sogabe, T.; Sogo, M. Flexural behavior of reinforced recycled concrete beams. *J. Adv. Concr. Technol.* **2007**, *5*, 43–61. [CrossRef]

26. Fathifazl, G.; Razaqpur, A.G.; Isgor, O.B.; Abbas, A.; Fournier, B.; Foo, S. Flexural performance of steel reinforced recycled concrete (RRC) beams. *ACI Struct. J.* **2009**, *106*, 858–867.

27. Choi, W.C.; Kim, S.W.; Yun, H.D. Flexural performance of reinforced recycled aggregate concrete beams. *Mag. Concr. Res.* **2012**, *64*, 837–848. [CrossRef]

28. Ignjatovic, I.; Marinkovic, S.; Miškovic, Z.; Savic, A. Flexural behavior of reinforced recycled aggregate concrete beams under short-term loading. *Mater. Struct.* **2013**, *46*, 1045–1059. [CrossRef]

29. Disfani, M.M.; Arulrajah, A.; Haghighi, H.; Mohammadinia, A.; Horpibulsuk, S. Flexural beam fatigue strength evaluation of crushed brick as a supplementary material in cement stabilized recycled concrete aggregates. *Constr. Build. Mater.* **2014**, *68*, 667–676. [CrossRef]

30. Kang, T.H.K.; Kim, W.; Kwak, Y.K.; Hong, S.G. Flexural testing of reinforced concrete beams with recycled concrete aggregates (with appendix). *ACI Struct. J.* **2014**, *111*, 607–616.

31. Arezoumandi, M.; Smith, A.; Volz, J.S.; Khayat, K.H. An experimental study on flexural strength of reinforced concrete beams with 100% recycled concrete aggregate. *Eng. Struct.* **2015**, *88*, 154–162. [CrossRef]

32. Evangelista, L.; de Brito, J. Flexural behaviour of reinforced concrete beams made with fine recycled concrete aggregates. *KSCE J. Civ. Eng.* **2017**, *21*, 1–11. [CrossRef]

33. Mohammeda, T.U.; Das, H.K.; Mahmood, Z.H.; Rahman, M.N.; Awal, M.A. Flexural performance of RC beams made with recycled brick aggregate. *Constr. Build. Mater.* **2017**, *134*, 67–74. [CrossRef]

34. Yang, S.; Lee, H. Structural performance of reinforced RCA concrete beams made by a modified EMV method. *Sustainability* **2017**, *9*, 131. [CrossRef]

35. Seara-Paz, S.; González-Fonteboa, B.; Martínez-Abella, F.; Eiras-López, J. Flexural performance of reinforced concrete beams made with recycled concrete coarse aggregate. *Eng. Struct.* **2018**, *156*, 32–45. [CrossRef]

36. Katkhuda, H.; Shatarat, N. Shear behavior of reinforced concrete beams using treated recycled concrete aggregate. *Constr. Build. Mater.* **2016**, *125*, 63–71. [CrossRef]

37. Ajdukiewicz, A.B.; Kliszczewicz, A.T. Comparative tests of beams and columns made of recycled aggregate concrete and natural aggregate concrete. *J. Adv. Concr. Technol.* **2007**, *5*, 259–273. [CrossRef]

38. Wang, Y.Y.; Chen, J.; Zong, B.; Geng, Y. Mechanical behavior of axially loaded recycled aggregate concrete filled steel tubular stubs and reinforced recycled aggregate concrete stubs. *J. Build. Struct.* **2011**, *32*, 170–177.

39. Choi, W.C.; Yun, H.D. Compressive behavior of reinforced concrete columns with recycled aggregate under uniaxial loading. *Eng. Struct.* **2012**, *41*, 285–293. [CrossRef]

40. Silva, R.V.; de Brito, J.; Evangelista, L.; Dhir, R.K. Design of reinforced recycled aggregate concrete elements in conformity with Eurocode 2. *Constr. Build. Mater.* **2016**, *105*, 144–156. [CrossRef]

41. Tošic, N.; Marinkovic, S.; Ignjatovic, I. A database on flexural and shear strength of reinforced recycled aggregate concrete beams and comparison to Eurocode 2 predictions. *Constr. Build. Mater.* **2016**, *127*, 932–944. [CrossRef]

42. Xiao, J.Z.; Huang, X.; Shen, L.M. Seismic behavior of semi-precast column with recycled aggregate concrete. *Constr. Build. Mater.* **2012**, *35*, 988–1001. [CrossRef]

43. Yang, Y.Q. *Experimental Study on Seismic Performance of Recycled Aggregate Concrete Columns*; Harbin Institute Technology: Harbin, China, 2016.

44. Xiao, J.Z.; Wang, C.Q.; Pham, T.L.; Yang, Z.J.; Ding, T. Nonlinear analysis and test validation on seismic performance of a recycled aggregate concrete space frame. *Struct. Des. Tall Spec.* **2015**, *23*, 1381–1405. [CrossRef]

45. SeismoSoft. SeismoStruct–A Computer Program for Static and Dynamic Nonlinear Analysis of Framed Structures. 2014. Available online: www.seismosoft.com (accessed on 20 March 2019).

46. Zhang, Y.Y.; Harries, K.A.; Yuan, W.C. Experimental and numerical investigation of the seismic performance of hollow rectangular bridge piers constructed with and without steel fiber reinforced concrete. *Eng. Struct.* **2013**, *48*, 255–265. [CrossRef]

47. Ahmad, S.H.; Weerakoon, S.L. Model for behavior of slender reinforced concrete columns under biaxial bending. *ACI Struct. J.* **1995**, *92*, 188–198.

48. Bahn, B.Y.; Hsu, C.T.T. Cyclically and biaxially loaded reinforced concrete slender columns. *ACI Struct. J.* **2000**, *97*, 444–454.

49. Barrera, A.C.; Bonet, J.L.; Romero, M.L.; Miguel, P.F. Experimental tests of slender reinforced concrete columns under combined axial load and lateral force. *Eng. Struct.* **2001**, *33*, 3676–3689. [CrossRef]

50. Barrera, A.C.; Bonet, J.L.; Romero, M.L.; Fernández, M.A. Ductility of slender reinforced concrete columns under monotonic flexural and constant axial load. *Eng. Struct.* **2012**, *40*, 398–412. [CrossRef]

51. Babazadeh, A.; Burgueño, R.; Silva, P.F. Evaluation of the critical plastic region length in slender reinforced concrete bridge columns. *Eng. Struct.* **2016**, *125*, 280–293. [CrossRef]
52. Madas, P. *Advanced Modelling of Composite Frames Subjected to Earthquake Loading*; Imperial College, University of London: London, UK, 1993.
53. Mander, J.B.; Priestley, M.J.N.; Park, R. Theoretical stress–strain model for confined concrete. *J. Struct. Eng.* **1998**, *114*, 1804–1826. [CrossRef]
54. Martinez, R.J.C.; Elnashai, A.S. Confined concrete model under cyclic load. *Mater. Struct.* **1997**, *30*, 139–147. [CrossRef]
55. Gholampour, A.; Gandomi, A.H.; Ozbakkaloglu, T. New formulations for mechanical properties of recycled aggregate concrete using gene expression programming. *Constr. Build. Mater.* **2017**, *130*, 122–145. [CrossRef]
56. Xiao, J.Z.; Li, J.B.; Zhang, C. Mechanical properties of recycled aggregate concrete under uniaxial loading. *Cement Concr. Res.* **2005**, *35*, 1187–1194. [CrossRef]
57. Lim, J.C.; Ozbakkaloglu, T. Stress-strain model for normal- and light-weight concretes under uniaxial and triaxial compression. *Constr. Build. Mater.* **2014**, *71*, 492–509. [CrossRef]
58. Yassin, M.H.M. *Nonlinear Analysis of Prestressed Concrete Structures under Monotonic and Cyclic Loads*; University of California: Berkeley, CA, USA, 1994.
59. Menegotto, M.; Pinto, P.E. Method of analysis for cyclically loaded R.C. plane frames including changes in geometry and non-elastic behaviour of elements under combined normal force and bending. In *Symposium on the Resistance and Ultimate Deformability of Structures Acted on by Well Defined Repeated Loads*; International Association for Bridge and Structural Engineering: Zurich, Switzerland, 1973.
60. Filippou, F.C.; Popov, E.P.; Bertero, V.V. *Effects of Bond Deterioration on Hysteretic Behaviour of Reinforced Concrete Joints*; Report EERC 83-19; Earthquake Engineering Research Center, University of California: Berkeley, CA, USA, 1983.
61. Monti, G.; Nuti, C.; Santini, S. *CYRUS—Cyclic Response of Upgraded Sections*; Report No. 96-2; University of Chieti: Chieti, Italy, 1996.
62. Ferreira, L.; de Brito, J.; Barra, M. Influence of the pre-saturation of recycled coarse concrete aggregates on concrete properties. *Mag. Concr. Res.* **2011**, *63*, 617–627. [CrossRef]
63. González-Taboada, I. *Self-Compacting Recycled Concrete: Basic Mechanical Properties, Rheology, Robustness and Thixotropy*; Universidade da Coruña: A Coruña, Spain, 2016.
64. Xu, J.J.; Chen, Z.P.; Ozbakkaloglu, T.; Zhao, X.Y.; Demartino, C. A critical assessment of the compressive behavior of reinforced recycled aggregate concrete columns. *Eng. Struct.* **2018**, *161*, 161–175. [CrossRef]
65. González-Fonteboa, B.; Martínez-Abella, F.; Carro, L.D.; Seara-Paz, S. Stress-strain relationship in axial compression for concrete using recycled saturated coarse aggregate. *Constr. Build. Mater.* **2011**, *25*, 2335–2342.
66. Silva, R.V.; de Brito, J.; Dhir, R.K. The influence of the use of recycled aggregates on the compressive strength of concrete: A review. *Eur. J. Environ. Civ. Eng.* **2015**, *19*, 825–849. [CrossRef]
67. Pedro, D.; de Brito, J.; Evangelista, L. Performance of concrete made with aggregates recycled from precasting industry waste: Influence of the crushing process. *Mater. Struct.* **2015**, *48*, 3965–3978. [CrossRef]
68. Xu, J.J.; Chen, Z.P.; Xiao, Y.; Demartino, C.; Wang, J.H. Recycled aggregate concrete in FRP-confined columns: A review of experimental results. *Compos. Struct.* **2017**, *174*, 277–291. [CrossRef]
69. Chen, G.M.; He, Y.H.; Jiang, T.; Lin, C.J. Behavior of CFRP-confined recycled aggregate concrete under axial compression. *Constr. Build. Mater.* **2016**, *111*, 85–97. [CrossRef]
70. Teng, J.G.; Zhao, J.L.; Yu, T.; Li, L.J.; Guo, Y.C. Behavior of FRP-confined compound concrete containing recycled concrete lumps. *J. Compos. Constr.* **2016**, *20*, 04015038. [CrossRef]
71. Xu, J.J.; Chen, Z.P.; Xue, J.Y.; Chen, Y.L.; Zhang, J.T. Simulation of seismic behavior of square recycled aggregate concrete-filled steel tubular columns. *Constr. Build. Mater.* **2017**, *149*, 553–566. [CrossRef]
72. Chen, Z.P.; Xu, J.J.; Chen, Y.L.; Lui, E.M. Recycling and reuse of construction and demolition waste in concrete-filled steel tubes: A review. *Constr. Build. Mater.* **2016**, *126*, 641–660. [CrossRef]
73. Chen, Z.P.; Xu, J.J.; Xue, J.Y.; Su, Y.S. Performance and calculations of recycled aggregate concrete-filled steel tubular (RACFST) short columns under axial compression. *Int. J. Steel Struct.* **2014**, *14*, 31–42. [CrossRef]
74. Zhao, X.Y.; Wu, B.; Wang, L. Structural response of thin-walled circular steel tubular columns filled with demolished concrete lumps and fresh concrete. *Constr. Build. Mater.* **2016**, *129*, 216–242. [CrossRef]

75. Wu, B.; Peng, C.W.; Zhao, X.Y.; Zhou, W.J. Axial loading tests of thin-walled circular steel tubes infilled with cast-in-place concrete and precast segments containing DCLs. *Thin Wall Struct.* **2018**, *127*, 275–289. [CrossRef]
76. Zega, C.J.; Maio, A.A.D. Recycled concretes made with waste ready-mix concrete as coarse aggregate. *J. Mater. Civ. Eng.* **2011**, *23*, 281–286. [CrossRef]
77. Gettu, R.; Bazant, Z.P.; Karr, M.E. Fracture properties and brittleness of high-strength concrete. *ACI Mater. J.* **1990**, *87*, 608–618.
78. China Standards Publication. *Code for Design of Concrete Structures*; GB 50010-2010; China Planning Press: Beijing, China, 2010. (In Chinese)
79. Tseng, C.C.; Hwang, S.J.; Mo, Y.L.; Yeh, Y.K.; Lee, Y.T. Experiment of torsionally-coupled RC building subjected to bi-directional cyclic loading. In Proceedings of the 3rd International Conference on Advances in Experimental Structural Engineering, San Francisco, CA, USA, 15–16 October 2009.
80. Mo, Y.L.; Luu, C.H.; Nie, X.; Tseng, C.C.; Hwang, S.J. Seismic performance of a two-story unsymmetrical reinforced concrete building under reversed cyclic bi-directional loading. *Eng. Struct.* **2017**, *145*, 333–347. [CrossRef]
81. Liu, S.F.; Yang, Y.J.; Forrest, J. *Grey Data Analysis: Methods, Models and Applications*; Springer: Singapore, 2017.
82. Patton, M.Q. *Qualitative Evaluation and Research Methods*; Sage Publications: Thousand Oaks, CA, USA, 1990.
83. Zhang, Y.J.; Zhang, X. Grey correlation analysis between strength of slag cement and particle fractions of slag powder. *Cem. Concr. Compos.* **2007**, *29*, 498–504. [CrossRef]
84. Ho, P.H.K. Forecasting construction manpower demand by gray model. *J. Constr. Eng. Mag.* **2010**, *136*, 1299–1305. [CrossRef]
85. Lai, W.C.; Chang, T.P.; Wang, J.J.; Kan, C.W.; Chen, W.W. An evaluation of mahalanobis distance and grey relational analysis for crack pattern in concrete structures. *Comp. Mater. Sci.* **2012**, *65*, 115–121. [CrossRef]
86. Wang, Z.J.; Wang, Q.; Ai, T. Comparative study on effects of binders and curing ages on properties of cement emulsified asphalt mixture using gray correlation entropy analysis. *Constr. Build. Mater.* **2014**, *54*, 615–622. [CrossRef]

*applied
sciences*

MDPI

Article

An Experimental Study on Flexural Behaviors of Reinforced Concrete Member Replaced Heavyweight Waste Glass as Fine Aggregate under Cyclic Loading

So Yeong Choi [1], Yoon Suk Choi [2], Il Sun Kim [1] and Eun Ik Yang [1,*]

[1] Department of Civil Engineering, Gangneung-Wonju National University, Jukheon-gil 7, Gangneung-si, Gangwon-do 25457, Korea; csy7510@gwnu.ac.kr (S.Y.C.); iskim@gwnu.ac.kr (I.S.K.)
[2] Convergence Technology Division, Korea Conformity Laboratories, Seoul 08503, Korea; yoons0305@kcl.re.kr
* Correspondence: eiyang@gwnu.ac.kr; Tel.: +82-33-640-2418

Received: 27 September 2018; Accepted: 7 November 2018; Published: 9 November 2018

Abstract: The development of electronic technology has accelerated in recent decades. Consequently, electronic wastes such as cathode ray tube (CRT) glass are accumulated, and hazardous wastes including heavy metals are generated. Simultaneously, natural resources are required to create concrete; however, they are already exhausted. Furthermore, heavyweight waste glass is considered to be the most suitable substitute for aggregate owing to its physical characteristics and chemical composition. However, structural results regarding the recycling of heavyweight waste glass as fine aggregate in Reinforced Concrete (RC) members are insufficient. Thus, herein, experimental study is conducted to evaluate whether RC members with heavyweight waste glass as fine aggregate can be applied for concrete structures. Flexural behavior tests of reinforced concrete members were performed. Fifteen specimens with different substitution ratios of heavyweight waste glass were prepared. The results showed that when all the fine aggregate is replaced by heavyweight waste glass in RC members, the heavyweight waste glass substitution ratio affected the crack occurrence patterns, and the possibility of a sudden failure of a member increased owing to concrete crushing in the compression zone. Additionally, the load capacity and flexural rigidity were affected by the substitution ratio of heavyweight waste glass; however, the flexural performance is improved when mineral admixture as a binder or a low water-binder ratio were used. Therefore, heavyweight waste glass is considered applicable for use as fine aggregate of concrete.

Keywords: flexural behavior; recycling; heavyweight waste glass; cyclic load; reinforced concrete member

---

1. Introduction

Concrete is the primary material in construction worldwide. Additionally, the need for concrete will increase to almost 7.5 billion m^3 (approximately 18 billion ton) a year by 2050 [1,2]. Particularly, the rapid development on large-scale infrastructures is causing the exhaustion of natural resources; several countries are facing the shortage of natural resources and the supply of aggregates is being exhausted [1]. Therefore, those countries are relying on imports to satisfy their needs. To handle this situation, a variety of studies are performed for the development of alternative resources for concrete [3–5]. Meanwhile, the recycling technologies of electric waste have been emphasized as a global issue, owing to its rapidly growing volume and complex nature [6]. Additionally, since 2012, when analog TV broadcasting ended, systems were converted to digital TV broadcasting in South Korea, and a large volume of cathode-ray tube (CRT) TVs and monitors were discarded. CRT glass products are classified into panels and funnels, wherein the panels may be reused as glass after washing; however, it is difficult to treat the funnels (heavyweight waste glass) using conventional recycling technology because they contain many heavy metals such as iron and lead [7,8]. Therefore, heavyweight waste glass has been studied extensively for many years. Conventionally, it is produced with crushed concrete materials. Using heavyweight waste glass as fine aggregate in concrete demonstrated some benefits such as improved durability [9,10].

Meanwhile, almost all other studies focused on the removal of heavy metal in heavyweight waste glass. However, in our previous studies, we demonstrated the applicability of crushed heavyweight waste glass as fine aggregate in shielding concrete [11,12]. The results of those studies in indicated that, to improve the shielding performance, it is important to increase the density of the material in the radiation shielding concrete. Therefore, we did not perform any treatment to remove the heavy metals. Furthermore, we investigated the effect of heavyweight waste glass on the volume change properties of mortar according to the substitution ratio. The result indicated that it may be feasible to utilize heavyweight waste glass as fine aggregate in mortar specimens [1].

Additionally, previous experimental studies have focused primarily on the mechanical properties or durability of concrete or mortar specimens [9,13–16]. Hitherto, structural behavior investigations are rarely conducted in RC members that use recycled heavyweight waste glass as fine aggregate. Namely, structural results regarding recycling heavyweight waste glass as fine aggregate are insufficient. Particularly, few have been investigated under the cyclic loading condition. The objective of this paper is to investigate experimentally the flexural behavior of RC members under cyclic loading, and the effect of the heavyweight waste glass substitution ratio.

2. Materials and Methods

2.1. Materials

In this study, Ordinary Portland Cement (OPC) was used in all the RC members. To investigate the effect of mineral admixtures on the flexural behavior of RC members, the cement part was replaced with a mineral admixture at a water-binder ratio of 45%. The mineral admixture type was Fly Ash (FA) and Blast Furnace Slag (BFS). The physical and chemical compositions of the binders are shown in Table 1.

Table 1. Physical and chemical compositions of the binders.

Binders Properties	Binders Type	OPC	FA	BFS
Physical	Specific gravity	3.15	2.34	2.82
	Blaine (cm^2/g)	3200	3700	4000
Chemical (%)	SiO$_2$	21.36	52.83	31.85
	Al$_2$O$_3$	5.03	18.08	14.55
	Fe$_2$O$_3$	3.31	7.74	0.59
	CaO	63.18	5.95	34.95
	MgO	2.89	1.43	5.63
	SO$_3$	2.30	0.01	2.97
	LOI	1.40	6.14	0.60

Crushed gravel was used as coarse aggregate with the maximum aggregate size, G_{max} of 20 mm. The specific gravity and absorption ratio of the coarse aggregate were 2.68% and 1.35%, respectively. River sand having a fineness modulus of 2.79 was used as a natural fine aggregate in the RC member. The specific gravity and absorption ratio of this fine aggregate were 2.6% and 1.07%, respectively. Furthermore, crushed heavyweight waste glass was used as fine aggregate.

2.2. Heavywegiht Waste Glass

The heavyweight waste glass was supplied from the TV funnel or cathode-ray tube, and it was crushed by a jaw crusher for use as fine aggregate in concrete [12]. Only crushed heavyweight waste glass that could pass through a No. 4 sieve was used. Heavyweight waste glass has a specific gravity of 3.0 and fineness modulus of 3.34, respectively. To determine the fineness modulus of heavyweight waste glass, we used sieve of No. 4, No. 8, No. 16, No. 30, No. 50, and No. 100 according to ASTM C 136 [17]. The physical and chemical compositions of the heavyweight waste glass are shown in Table 2, as determined using X-ray fluorescence. Furthermore, Type 1-4 refers to different manufactures and size of television. These are composed of heavy metals such as iron, lead, and chromium, regardless of the manufacturer and size of the television.

Table 2. Physical and chemical compositions of heavyweight waste glass.

Properties / Products		Type 1	Type 2	Type 3	Type 4	Avg.
Physical	Specific gravity			3.0		
	F.M.			3.34		
Chemical (%)	Fe_2O_3	49.9	40.3	40.3	42.0	43.1
	PbO	15.1	12.7	12.7	12.8	13.3
	Cr_2O_3	16.7	14.4	14.4	14.4	15.0
	SiO_2	9.6	20.4	20.4	18.7	17.3
	K_2O	1.8	2.8	2.7	2.6	2.5
	Other	6.8	9.4	9.5	9.5	8.8

2.3. Test Variables and Mix Proportions

To investigate the flexural behaviors with the heavyweight waste glass substitution ratio, we performed a compressive strength test on the concrete specimens, and a flexural test on the RC members. The test variables are listed in Table 3.

Table 3. Test variables.

Item	Contents
W/B ratio	35%, 45%, 55%
Mineral admixture (replacement ratio)	FA (20%), BFS (50%) (at W/B 45%)
Waste glass substitution ratio	0, 50, 100 (%)
Age (testing days)	28 (35OPC, 45OPC, 55OPC)
	91 (45FA20, 45BFS50)
Specimen size (mm)	150 × 205 × 1400 mm (RC member)
	Ø100 × 200 mm (Compressive strength)

In this study, the water-to-binder ratios of the specimen were varied at W/B 35% (i.e., 35OPC), W/B 45% (i.e., 45OPC) and W/B 55% (i.e., 55OPC) for the evaluation of the properties of the RC member. The heavyweight waste glass was used as a substitute for fine aggregate at 0%, 50% and 100% by volume. In addition, to investigate the effects of mineral admixture on the RC members, the cement part was replaced by a mineral admixture at a W/B ratio of 45%. The replacement ratios were 20%, for FA (i.e., 45FA20), and 50% for BFS (i.e., 45BFS50). The flexural test for the OPC case was conducted at 28 days; meanwhile, the FA and BFS case was conducted at 91 days obtain to the sufficient reaction of mineral admixture. The concrete mixture proportions with the heavyweight waste glass substitution ratio for all test members are listed in Table 4. The target slump and air content values of concrete were 100 ± 20 mm and 4.5 ± 1.0%, respectively.

Table 4. Concrete mix proportion.

Specimen ID	W/B (%)	H.R [1]	Unit Weight (kg/m^3)						
			W	C	G	S	H.G [2]	FA	BFS
35OPC-0	35	0	167	477	999	673	-	-	-
35OPC-50	35	50	167	477	999	337	388	-	-
35OPC-100	35	100	167	477	999	-	777	-	-
45OPC-0	45	0	170	378	1008	738	-	-	-
45OPC-50	45	50	170	378	1008	369	426	-	-
45OPC-100	45	100	170	378	1008	-	851	-	-
55OPC-0	55	0	173	315	998	792	-	-	-
55OPC-50	55	50	173	315	998	396	457	-	-
55OPC-100	55	100	173	315	998	-	914	-	-
45FA20-0	45	0	170	302	996	729	-	76	-
45FA20-50	45	50	170	302	996	364	420	76	-
45FA20-100	45	100	170	302	996	-	841	76	-
45BFS50-0	45	0	170	189	971	713	-	-	189
45BFS50-50	45	50	170	189	971	357	421	-	189
45BFS50-100	45	100	170	189	971	-	842	-	189

[1] Heavyweight waste glass substitution ratio. [2] Heavyweight waste glass.

2.4. Test Method

2.4.1. Experimental Set-Up for Compressive Strength

For the compressive strength tests, the specimens (Ø100 mm × 200 mm) were prepared based on ASTM C 39 [18]. The compressive load was supplied by a universal testing machine (UTM) with a capacity of 1000 kN. Each compressive strength value was the average of three specimens. Moreover, for the calculation of the elastic modulus, the linear variable displacement transducer (LVDT) was attached on the opposite side of the cylindrical specimen at a mid-height level (measuring distance = 100 mm) and the displacement value was obtained from the LVDT. The elastic modulus was calculated from the stress-strain relationship obtained at each step. The modulus of elasticity of ASTM C 469 [19] was calculated.

2.4.2. Experimental Set-Up for Flexural Behavior

RC members were prepared to investigate the effect of the substitution ratio of heavyweight waste glass on the flexural behavior. The size of the RC member for the flexural test is 150 mm × 205 mm × 1400 mm (width × depth × length). All RC members were loaded at two points symmetrically about the center section. Cyclically increasing displacement loading was applied on beams and the total load applied by the hydraulic jack was measured with a 500 kN capacity load cell. A load spreader beam was used to distribute the total load at two loading points such that a constant moment region of length 400 mm was obtained. The details related to the test set up are shown in Figure 1.

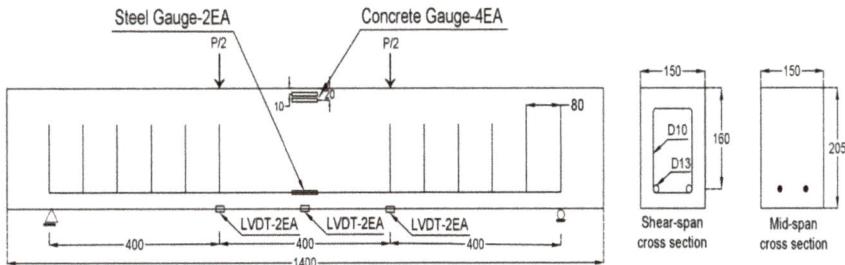

Figure 1. Reinforcement detail and cross-section of the RC member.

Meanwhile, the mechanical properties of the reinforcement used in this study are shown in Table 5. The nominal diameter of the primary reinforcement is 13 mm and the yield strength (f_y) is 430 MPa. The yield strength of the stirrups with a diameter of 10 mm is 480 MPa. The stirrup spacing is 80 mm.

Table 5. Mechanical properties of reinforcement used in RC member.

Diameter	Elastic Modulus (GPa)	Yield Stress (MPa)
D 13	200	430
D 10	200	480

When the central displacement at the center of a member reached 0.5Δy, 1.0Δy, 2.0Δy, 3.0Δy and 4.0Δy, the vertical load was removed and reapplied cyclically, where Δy is the yielding displacement of the member as obtained from the load-displacement curve (Figure 2).

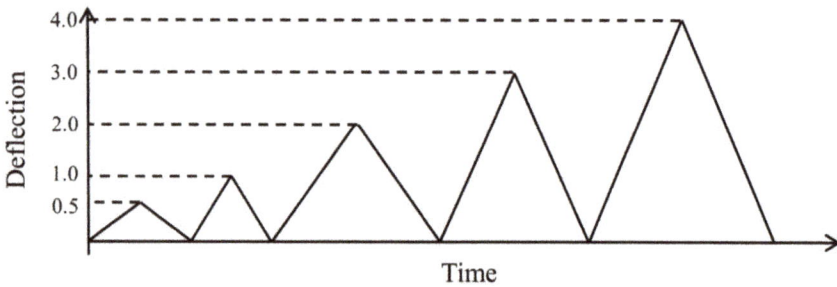

Figure 2. Schemed applied cyclic loading history.

To obtain the vertical displacement under an increasing load, an LVDT was installed vertically on the bottom of the center section. In addition, the steel strain at each stage was measured using a steel strain gauge. To obtain the concrete strain at the compressive zone, the strain gauge of concrete was attached to the 10 mm, and 20 mm lines from the edge of the compressive zone.

3. Results and Discussion

3.1. Effect of Heavyweight Waste Glass on Compression Properties of Concrete

Prior to the flexural test of the RC members, an experiment was conducted to evaluate the compression properties of the concrete specimens. The test results of the compressive strength and elastic modulus of the concrete specimens replaced heavyweight waste glass are presented in Figure 3. Likewise, results of previous studies [11,12,20,21] indicated that the compressive strength increased with decreased water-binder ratio, and the use of mineral admixture, regardless of the heavyweight waste glass substitution ratio. However, when the fine aggregate was replaced by heavyweight waste glass in concrete, the compressive strength decreased, whereas it decreased by approximately 9.8–28% as compared to the 0% substitution case. Namely, when the heavyweight waste glass substitution ratio increased, the compressive strength decreased. This phenomenon is most likely due to the poorer adhesion between the smooth surface of the glass and the cement past [11,22]. Furthermore, the elastic modulus decreased with increased of substitution ratio of heavyweight waste glass. As such, the elastic modulus was determined based on the compressive strength.

Meanwhile, the compressive strength and elastic modulus of the 45FA20 and 45BFS50 case specimens demonstrated better results than the 45OPC case at 91 days. The mineral admixture could be attributed to the pozzolanic reaction at a later ages.

In general, the compressive strength of normal concrete (OPC) increases with age. Particularly, the influence of water content on concrete strength is obvious; however, sealed curing was executed for RC members in this study. Owing to the difference between previous research [20] and the cuing condition of this study, therefore, the increment in the compressive strength of the OPC case was little after 28 days curing period.

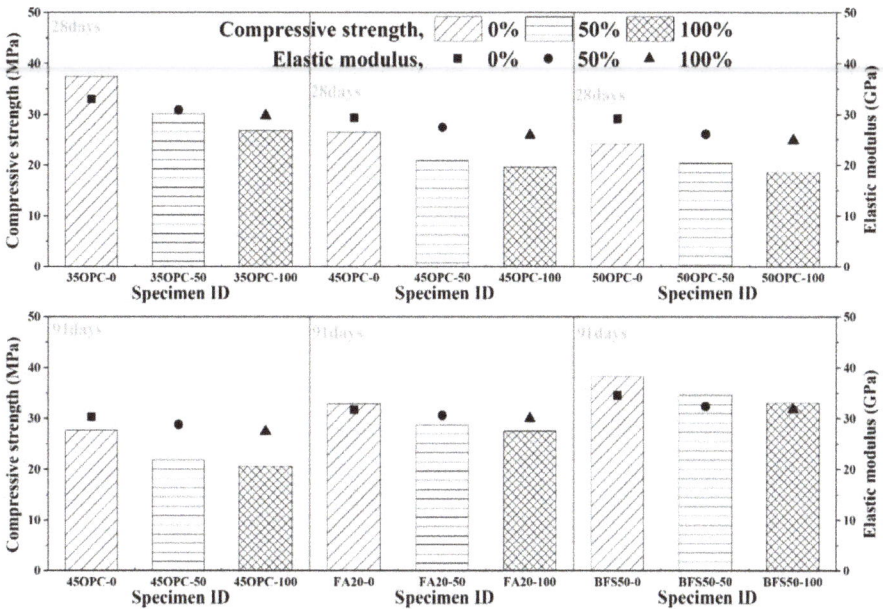

Figure 3. Compression properties of concrete specimen.

3.2. The Effect of Heavyweight Waste Glass on Cracking of Concrete

3.2.1. Crack Pattern

To evaluate the crack pattern, a crack map was drawn during the flexural test at different load levels. The Figure 4 shows a typical crack occurrence pattern in this test. From the results, all RC members began with the appearance of flexural cracks after the crack opening at the bottom of the center of the RC member, which grew vertically, regardless of the W/B ratios and substitution ratios. Furthermore, the initial flexural cracks were developed in the pure bending zone; as the load increased gradually, the shear-flexural crack appeared. The flexural cracks are extended in the vicinity of the pure bending zone and the failure of the RC members occurred owing to concrete crushing in the compression zone.

Figure 4. *Cont.*

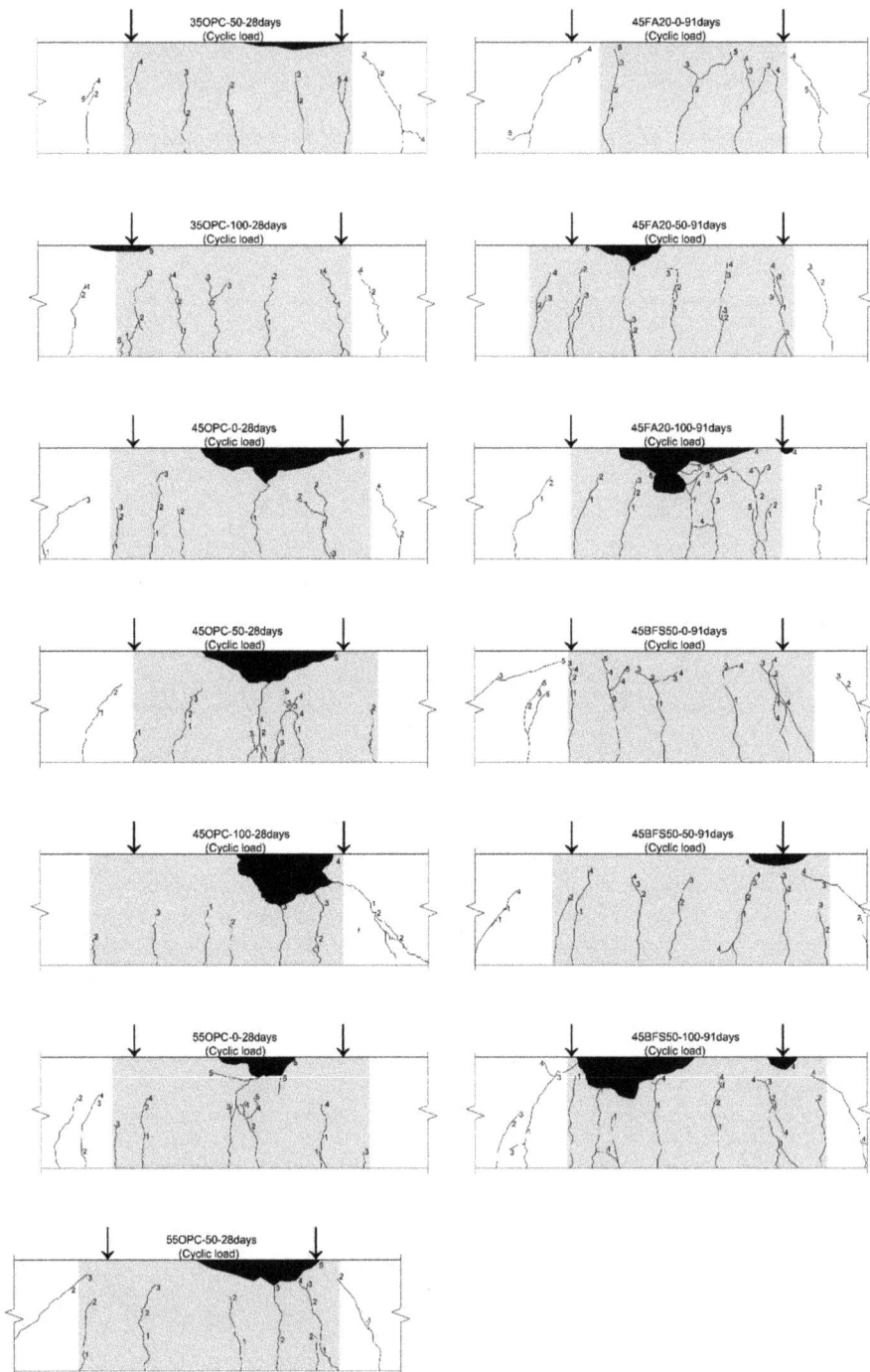

Figure 4. Typical crack occurrence pattern.

3.2.2. Crack Number and Failure Mode

Table 6 shows the number of cracks, mean cracking spacing, and area of compressive crushing determined in the vicinity of the pure bending zone. According to the test results, when the heavyweight waste glass substitution ratio increased, the area of the compressive crushing increased; however, the mean cracking spacing decreased. This is because the compressive strength decreased as the heavyweight waste glass substitution ratio increase. However, 45FA20, 45BFS50 case demonstrated less crushing by the addition of admixture. Meanwhile, W/B 55% case demonstrated the different results. The W/B 55% case exhibits a low compressive strength; therefore, the number of cracks decreased and the mean crack spacing increased with heavyweight waste glass substitution. The area of compression crushing in the RC member contributed to the lower bond strength between the concrete replaced heavyweight waste glass as fine aggregate and rebar compared with that of non-substituted concrete. Therefore, the bond strength influenced crack propagation in the RC member.

Table 6. Results of experimental test for RC member (Crack).

Specimen ID	35OPC			45OPC			55OPC		
H.R	0	50	100	0	50	100	0	50	100
Number of cracks (ea)	5	5	7	5	6	6	7	6	4
Mean. crack spacing (mm)	101.69	101.46	84.27	103.85	92.67	85.31	79.11	98.99	134.15
Area of compression crushing (mm^2)	710.6	1447.2	1120.7	9450.2	8668.7	11,988.7	3167.1	6250.3	9214.2
Specimen ID	45FA20			45BFS50					
H.R	0	50	100	0	50	100			
Number of cracks (ea)	4	7	6	6	7	7			
Mean. crack spacing (mm)	108.57	81.18	71.91	90.54	86.51	79.73			
Area of compression crushing (mm^2)	-	2488.6	8917.2	-	1761.5	10521.2			

3.3. Load-Displacement Relationship

The displacement of the RC member was measured at each load levels by the LVDT installed at the center of the RC member. The load-displacement curves of the RC members are shown in Figure 5. Furthermore, Table 7 shows the initial cracking load, yielding load, peak load, and displacement for each specimen.

The initial cracking load could be specified at the end point of the linear relationship between the load and relative displacement of the RC members. Furthermore, the yielding load is determined from the reading values of the strain gauges placed on the reinforcing rebar at mid-span. The peak load is the maximum load of the load-displacement curve. As shown in Figure 5 and Table 7, the initial crack load was the largest for the 35OPC-0 case, and it was affected by the water-binder ratio and heavyweight waste substitution ratio.

In general, the yield point depends on the tensile capacity of the reinforcement. The yielding load is considered to be similar regardless of the heavyweight waste glass substitution ratio. However, the yield point is changed slightly, and there was no trend. Therefore, the structural test error was included. Meanwhile, the maximum load of the RC member was affected by the heavyweight waste glass substitution ratio. The load capacity of the RC member was reduced gradually on the large deformed stage when all the fine aggregate was replaced by heavyweight waste glass.

Particularly, the 55OPC case exhibited a relatively low strength than other W/B ratios; therefore, even if the heavyweight waste glass is substituted with only 50% of the fine aggregate, the load capacity is reduced. The mineral admixture, however, improved the load capacity of the RC member substituted heavyweight waste glass as all the fine aggregate. The load-displacement curves of the RC members were affected by the heavyweight waste glass substitution ratio. However, the effect of heavyweight waste glass could be reduced using a relatively low water-binder ratio or mineral admixture.

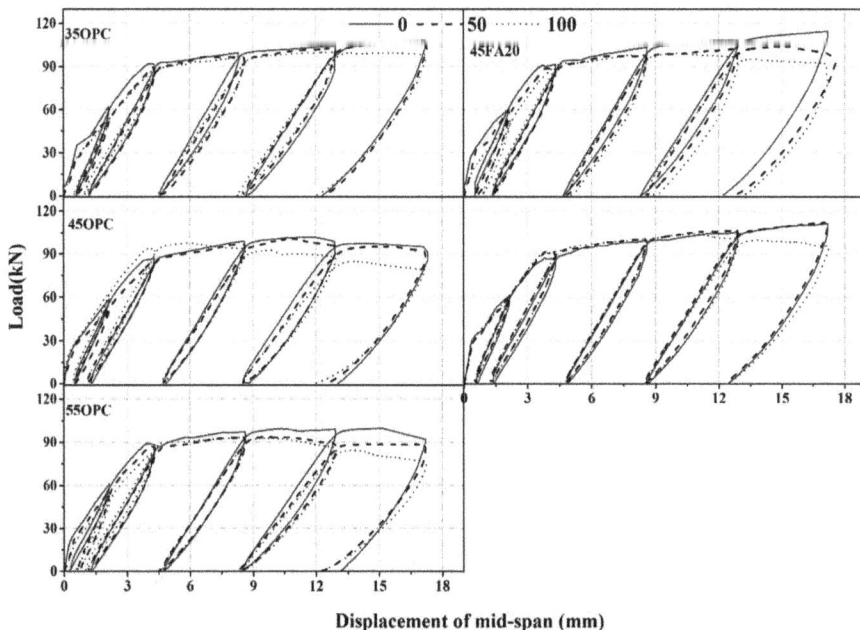

Figure 5. Load-displacement curves for RC beam.

Table 7. Experimental results of ductility index of RC member.

Specimen ID		35OPC			45OPC			55OPC		
H.R		0	50	100	0	50	100	0	50	100
Initial crack	P_{cr} (kN)	34.15	25.36	23.85	22.22	29.88	18.36	22.31	23.57	22.03
	Δ_{cr} (mm)	0.62	0.59	0.49	0.31	0.69	0.18	0.32	0.66	0.48
Yield Point	P_y (kN)	91.87	90.02	89.94	85.95	86.11	93.77	89.23	82.36	87.76
	Δ_y (mm)	4.00	4.30	4.58	3.83	4.28	3.96	3.91	4.13	4.43
Peak point	P_u (kN)	108.24	107.40	99.93	101.97	100.85	97.81	99.82	93.57	93.16
	Δ_u (mm)	17.19	17.19	12.57	11.74	10.98	5.92	15.13	8.46	8.33
Ductility Index		4.3	4.0	2.74	3.07	2.57	1.49	3.87	2.05	1.88
Specimen ID		45FA20			45BFS50					
H.R		0	50	100	0	50	100			
Initial crack	P_{cr} (kN)	27.72	31.26	25.07	33.58	31.15	28.73			
	Δ_{cr} (mm)	0.33	0.55	0.72	0.50	0.47	0.38			
Yield Point	P_y (kN)	91.49	89.79	91.16	87.13	91.95	91.12			
	Δ_y (mm)	4.31	3.95	3.72	3.8	3.88	3.73			
Peak point	P_u (kN)	114.51	103.99	99.21	111.38	112.38	104.25			
	Δ_u (mm)	17.11	15.10	9.95	17.2	17.2	10.9			
Ductility Index		4.46	3.82	2.67	4.53	4.43	2.92			

According to ACI 363 [23], ductility is explained as a ratio of the deflection (or cross-section curvature) at the ultimate load to the deflection (or curvature) at the load producing yield of the reinforcement (Equation (1)).

$$\mu = \Delta u / \Delta y \tag{1}$$

where, μ is the ductility index; Δu is the beam deflection at the ultimate load; Δy is the beam deflection at the yield load of the RC member.

The ductility index is shown in Table 7. From the results, the ductility index of the RC members is affected significantly by the heavyweight waste glass substitution ratio. This implies that the increment of the substitution ratio of heavyweight waste glass in the RC member caused the reduced flexural ductility. As mentioned earlier, these phenomena were affected by the load capacity decrement at the large deformed stage. However, the ductility index of the RC members using heavyweight waste glass can be improved by a low water-binder ratio or mineral admixture.

3.4. Bending Moment-Curvature Relationship

The accurate determination of the bending moment-curvature curve of the RC members is a reliable indicator of the load capacity of the concrete structure. Therefore, the measured values from the concrete strain gauges and embedded rebar strain gauges were compared. Furthermore, the values of individual gauges were measured until their readings became unreliable owing to the cracking on the concrete surface. Figure 6 shows the applied bending moment-curvature curve at the mid-span of the RC members. The results demonstrate that the slopes of the moment-curvature were similar before the initial cracking occurred, and the flexural rigidity did not exhibit any difference. When the heavyweight waste glass is replaced with fine aggregate, however, the moment-curvature curves changed. Namely, the heavyweight waste glass substitution ratio did affect the flexural rigidity of the RC members.

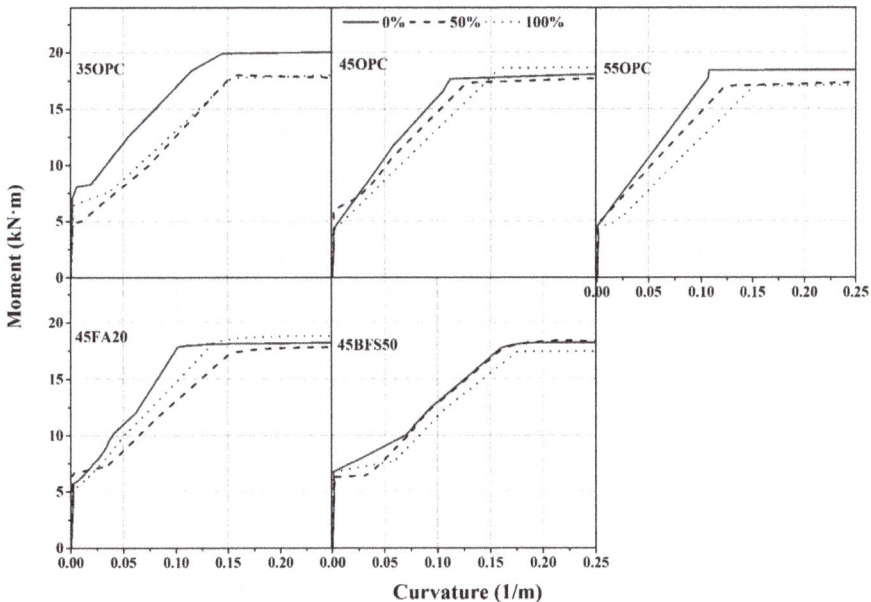

Figure 6. Moment-curvature curve for RC beam.

The bending moment of the initial cracking load corresponds to 3.6–6.8 kN·m in all RC members. The moment-curvatures exhibited a nonlinear relationship after curvature of the first cracking load.

In the nonlinear region, the curvature of the all the RC members increased suddenly. Conclusively, the moment-curvature curves of the RC member were reduced significantly when all the fine aggregate were replaced by heavyweight waste glass.

3.5. Location of the Neutral Axis

The experimental neutral axis depth of the RC member was obtained from the experimentally measured strain gauge values in the concrete and rebar. Figure 7 shows that the relationship between the moment and neutral axis depth determined until the rebar yielded. The results indicated that the neutral axis depth is immediately increased until 2–3 kN·m; subsequently, it converges to a constant value until the maximum moment is achieved. This trend is observed in all RC members, regardless of the water-binder ratio and heavyweight waste glass substitution ratio.

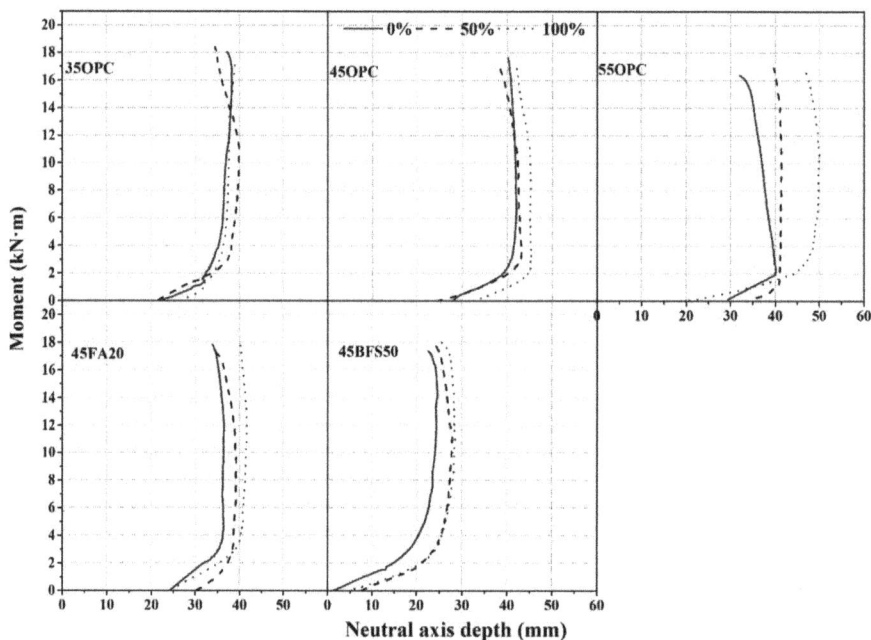

Figure 7. Location of the neutral axis depth.

However, the neutral axis depth is increased with heavyweight waste glass substitution ratio and water-binder ratio. Thus, the neutral axis depth of the rectangular stress block in the RC member is expected to increase with decreasing compressive strength. This trend is also shown in the RC member using the mineral admixture; however, the increase in the neutral axis depth could be alleviated. Therefore, to use the heavyweight waste glass as fine aggregate, the high compressive strength concrete is required.

4. Conclusions

The flexural behaviors of reinforced concrete member substituted heavyweight waste glass as fine aggregate under cyclic loading were examined. The conclusions obtained from this study are as follows:

(1) The heavyweight waste glass substitution ratio affected the crack occurrence patterns; further, when all the fine aggregate is replaced by heavyweight waste glass in the RC members, the possibility of sudden failure increased owing to concrete crushing in the compression zone.

(2) The maximum load and ductility index of the RC members were affected by the heavyweight waste glass substitution ratio.

(3) The flexural rigidity of the RC member was reduced when the fine aggregate was replaced by heavyweight waste glass.

(4) In conclusion, the flexural behavior of RC members was affected by the heavyweight waste glass substitution ratio; however, the flexural performance could be improved using mineral admixture as a binder or a low water-binder ratio.

Author Contributions: All authors contributed to the experiments, analyses, and writing of the paper.

Funding: This research was supported by grant (18CTAP-C129756-02) from the Technology Advancement Research Program (TARP) funded by the Ministry of Land, Infrastructure and Transport of the Korean government.

Conflicts of Interest: The authors declare no conflict of interest.

References

1. Choi, S.Y.; Choi, Y.S.; Yang, E.I. Characteristics of volume change and heavy metal leaching in mortar specimens recycled heavyweight waste glass as fine aggregate. *Constr. Build. Mater.* **2018**, *165*, 424–433. [CrossRef]
2. Mehta, P.K.; Monteiro, P.J.M. *Concrete: Microstructure, Properties, and Materials*; McGraw-Hill: New York, NY, USA, 2014.
3. Chen, H.-J.; Wu, C.-H. Influence of Aggregate Gradation on the Engineering Properties of Lightweight Aggregate Concrete. *Appl. Sci.* **2018**, *8*, 1324. [CrossRef]
4. Kurda, R.; de Brito, J.; Silvestre, J. Combined Economic and Mechanical Performance Optimization of Recycled Aggregate Concrete with High Volume of Fly Ash. *Appl. Sci.* **2018**, *8*, 1189. [CrossRef]
5. Ogrodnik, P.; Szulej, J. The Assessment of Possibility of Using Sanitary Ceramic Waste as Concrete Aggregate—Determination of the Basic Material Characteristics. *Appl. Sci.* **2018**, *8*, 1205. [CrossRef]
6. Yao, Z.; Ling, T.C.; Sarker, P.K.; Su, W.; Liu, J.; Wu, W.; Tang, J. Recycling difficult-to-treat e-waste cathode-ray-tube glass as construction and building materials: A critical review. *Renew. Sustain. Energy Rev.* **2018**, *81*, 595–604. [CrossRef]
7. Disfani, M.M.; Arulrajah, A.; Bo, M.W.; Hankour, R. Recycled crushed glass in road work applications. *Waste Manag.* **2011**, *31*, 2341–2351. [CrossRef] [PubMed]
8. Kim, I.S.; Choi, S.Y.; Yang, E.I. Evaluation of durability of concrete substituted heavyweight waste glass as fine aggregate. *Constr. Build. Mater.* **2018**, *184*, 269–277. [CrossRef]
9. De Castro, S.; De Brito, J. Evaluation of the durability of concrete made with crushed glass aggregates. *J. Clean. Prod.* **2013**, *41*, 7–14. [CrossRef]
10. Ling, T.-C.; Poon, C.-S. Development of a method for recycling of CRT funnel glass. *Environ. Technol.* **2012**, *33*, 2531–2537. [CrossRef] [PubMed]
11. Choi, S.Y.; Choi, Y.S.; Won, M.S.; Yang, E.I. Evaluation on the Applicability of Heavy Weight Waste Glass as Fine Aggregate of Shielding Concrete. *J. Korea Inst. Struct. Maint. Insp.* **2015**, *19*, 101–108.
12. Choi, S.Y.; Choi, Y.S.; Yang, E.I. Effects of heavy weight waste glass recycled as fine aggregate on the mechanical properties of mortar specimens. *Ann. Nucl. Energy* **2017**, *99*, 372–382. [CrossRef]
13. Shao, Y.; Lefort, T.; Moras, S.; Rodriguez, D. Studies on concrete containing ground waste glass. *Cem. Concr. Res.* **2000**, *30*, 91–100. [CrossRef]
14. Shi, C.; Zheng, K. A review on the use of waste glasses in the production of cement and concrete. *Resour. Conserv. Recycl.* **2007**, *52*, 234–247. [CrossRef]
15. Tan, K.H.; Du, H. Use of waste glass as sand in mortar: Part I—Fresh, mechanical and durability properties. *Cem. Concr. Compos.* **2013**, *35*, 118–126. [CrossRef]
16. Wang, H.-Y. A study of the effects of LCD glass sand on the properties of concrete. *Waste Manag.* **2009**, *29*, 335–341. [CrossRef] [PubMed]
17. ASTM. *Standard Test Method for Sieve Analysis of Fine and Coarse Aggregates*; ASTM International: West Conshohocken, PA, USA, 2006.
18. ASTM. *Standard Test Method for Compressive Strength of Cylindrical Concrete Specimens*; ASTM International: West Conshohocken, PA, USA, 2005.

Appl. Sci. **2018**, *8*, 2208

19. ASTM. *Standard Test Method for Static Modulus of Elasticity and Poisson's Ratio of Concrete in Compression*; ASTM International: West Conshohocken, PA, USA, 2010.

20. Kim, Y.M.; Choi, S.Y.; Kim, I.S.; Yang, E.I. A study on the Mechanical Properties of Concrete using Electronic Waste as Fine Aggregate. *J. Korea Inst. Struct. Maint. Insp.* **2018**, *22*, 90–97.

21. Kou, S.C.; Poon, C.S. Properties of self-compacting concrete prepared with recycled glass aggregate. *Cem. Concr. Compos.* **2009**, *31*, 107–113. [CrossRef]

22. Lee, C.T. Production of alumino-borosilicate foamed glass body from waste LCD glass. *J. Ind. Eng. Chem.* **2013**, *19*, 1916–1925. [CrossRef]

23. American Concrete Institute. *Report on High-Strength Concrete Reported*; American Concrete Institute: Farmington Hills, MI, USA, 2010.

applied
sciences

MDPI

Article

An Experimental Study on Dynamic Mechanical Properties of Fiber-Reinforced Concrete under Different Strain Rates

Yuexiu Wu [1], Wanpeng Song [2,3,*], Wusheng Zhao [3,*] and Xianjun Tan [3,*]

[1] Key Laboratory of Rock Mechanics in Hydraulic Structural Engineering, Ministry of Education, Wuhan University, Wuhan 430072, China; yuexiuwu@whu.edu.cn
[2] Engineering Technology Research Institute of CCTEB Group Co., Ltd., Wuhan 430064, China
[3] State Key Laboratory of Geomechanics and Geotechnical Engineering, Institute of Rock and Soil Mechanics, Chinese Academy of Sciences, Wuhan 430071, China
* Correspondence: songwanpeng555@163.com (W.S.); wszhao@whrsm.ac.cn (W.Z.); xjtan@whrsm.ac.cn (X.T.); Tel.: +86-13349913380 (W.S.); +86-15872386874 (W.Z.); +86-15972092588 (X.T.)

Received: 17 September 2018; Accepted: 10 October 2018; Published: 12 October 2018

Abstract: Fiber-reinforced concrete (FRC) has a great advantage in earthquake-resistant structures, as compared with regular concrete. However, there are many difficulties in the construction and maintenance of concrete structures due to the high density and easy corrosion of the steel fiber in commonly used steel FRC. With the development of polymer material science, polyvinyl alcohol (PVA) fiber has been rapidly promoted for use in FRC because of its low density, high strength, and large elongation at break value. Dynamic uniaxial compression and splitting tensile experiments of FRC with PVA fiber were carried out with two matrix strengths (i.e., C30 and C40), which were blended with PVA fibers with a length of 12 mm in different volume contents (0, 0.2, 0.4, and 0.6%), at the age of 28 days, under different strain rates (i.e., 10^{-5}, 10^{-4}, 10^{-3}, and 10^{-2} s^{-1}). The results show that PVA has an obvious enhancing and toughening effect on concrete, which can improve its brittle properties and residual strength. With increasing strain rate, the compressive strength, split tensile strength, and elastic modulus increase to a certain extent, while the toughness index and the peak strain decrease to a certain degree. The post-peak deformation characteristic changes from a brittle failure of sudden caving to a ductile failure with dense cracking. The effect of PVA is different when enhancing the concrete with two different matrix strengths. The lower the matrix strength, the more obvious the enhancement effect of the fiber, showing characteristics of a higher compressive strength and low split tensile strength in FRC with low strength and a smoother post-peak stress–strain curve.

Keywords: seismic load; strain rate; fiber-reinforced concrete; dynamic mechanical property

1. Introduction

Concrete is a porous, brittle material widely used in civil engineering. It has a high compressive strength but poor tensile strength, impact resistance, and toughness, which results in a weak resistance to cyclic, impact, seismic, and explosive loads. Therefore, many scholars have been exploring ways to improve the tensile performance of concrete. One of the most promising methods of modification is to add an appropriate amount of chaotic fibers to plain concrete, which can improve the tensile strength, stiffness, fatigue life, and ductility of the concrete, based on the influence of the fiber on the initial crack initiation and propagation [1–3]. There are a wide variety of fibers that can be used for the reinforcement of concrete—i.e., metallic fibers, organic fibers, and inorganic fibers—according to material composition. These fibers mainly include steel fiber, glass fiber, polypropylene fiber, polyvinyl alcohol (PVA) fiber, basalt fiber, and coir fiber [4,5]. Presently, there are many studies on the enhancing and toughening effect of fiber-reinforced concrete (FRC) under static loads. Song and Hwang [6] studied the characteristics of the compressive strength, split tensile strength, and fracture modulus of concrete with different steel fiber contents. It was found that compressive strength was the highest with a fiber content of 1.5%, and the split strength, toughness index, and fracture modulus increased with increasing fiber content. Yazici et al. [7] studied the characteristics of compressive strength, split tensile strength, flexural strength, and ultrasonic velocity of steel FRC with different aspect ratios and volume contents. Some conclusions have also been drawn, that the addition of steel fiber could significantly improve the split and bending strength, while the improvement in compressive strength was not obvious and ultrasonic velocity showed a downward trend. Vajje and Krishnamurthy [8] focused on the characteristics of natural fiber concrete with different types of fibers, including basalt fiber, jute, sisal, hemp, banana, and pineapple. The results showed that the properties of FRC were related to the material properties of the fiber itself. There are also many studies on natural fiber concrete, which show that the addition of natural fibers results in a certain degree of improvement in the split tensile strength, bending strength, and absorbing energy of the concrete, with the fiber content in concrete relating to the fiber type [9–11]. Shafiq et al. [12] carried out a three-point bending experiment on PVA and basalt FRC beams with different contents (1–3%). The results showed that PVA fibers could significantly improve the post-peak bending behavior of concrete beams compared with basalt fiber. The PVA concrete beams showed deflection hardening characteristics with 3% content, while the basalt fiber concrete beams showed deflection softening characteristics with high content. However, the diameter of the PVA fiber (0.66 mm) was larger than that of the basalt fiber (0.018 mm). The different bending properties of FRC beams should have a certain relationship with fiber diameter.

From the above studies, it is evident there have been many achievements in testing fiber concrete with different kinds of fiber under static loads. However, concrete is a brittle material with a high sensitivity to strain rate [13,14]. Concrete structures will inevitably encounter a variety of dynamic loads during their service period, such as the wind load suffered by ground constructions and bridges, the hydrodynamic pressure encountered by dams and maritime terminals, and the seismic loads encountered by engineering structures in strong earthquake areas. Many studies have shown that the properties of concrete under a dynamic load are quite different from those under a static load. Therefore, it must be irrational to use the concrete strength parameters under static loads to design concrete structures that may be subjected to dynamic loads during service. It is important to study the strength and deformation characteristics of concrete or fiber concrete under dynamic loads. According to the existing research results [4,15], the strain rate of a concrete structure under different loads can be further divided into the following types: creep load (10^{-8}–10^{-6} s^{-1}), static load (10^{-6}–10^{-5} s^{-1}), vehicle load (10^{-5}–10^{-4} s^{-1}), seismic load (10^{-4}–10^{-2} s^{-1}), impact load (10^{-2}–10^{2} s^{-1}), and explosion and high-speed collision load (10^{2}–10^{4} s^{-1}). Abrams et al. [16] first carried out the compression experiments of concrete under static load (with a strain rate of approximately 8×10^{-6} s^{-1}) and dynamic load (with a strain rate of ~2×10^{-4} s^{-1}) and found that there was a strain rate sensitivity for the compressive strength of concrete. In 1917, many scholars carried out a variety of dynamic experimental studies on the mechanical properties of concrete and FRC. Cook et al. [17] used the drop hammer experimental system to study the dynamic mechanical properties of coir FRC and found that the impact index of fiber concrete increased with the increase of fiber length and content. Zhang et al. [18] conducted three-point bending tests on notched beams of steel FRC under a large range of loading rates by using both a servo-hydraulic machine (with a loading rate of ~10^{-3}–1 mm/s) and a drop-weight impact device (with a loading rate of approximately 10^{2}–10^{3} mm/s). The experimental results showed that the rupture energy and the peak load increased with increasing loading rate, and the growth values at a low loading rate were smaller than those at a high loading rate. This was due to the viscous effects of free water at lower rates and the inertia effect and greater fiber pullout energy at high rates. Dong et al. [1] studied the mechanical properties of basalt fiber-reinforced recycled aggregate under different replacement ratios and contents of basalt fiber (e.g., concrete failure modes, compressive strength, tensile strength, elastic modulus, Poisson's ratio, and ultimate strain under static conditions) and some mechanical properties under cyclic loading and unloading. The experimental results showed that the basalt fiber enhanced the mechanical properties of recycled aggregate concrete.

Currently, the most commonly used fibers in concrete are steel fibers, which have significantly improved the tensile properties of concrete. However, the addition of steel fiber not only improves the tensile strength, but also increases the weight of the concrete structure [19]. The development of polymer materials science has led to the fabrication of the newly developed PVA, which is a kind of synthetic fiber with a low price, high strength, and high elastic modulus. It has good hydrophilicity and a high bonding strength with cementitious material. This not only effectively inhibits early cracks in the concrete, but also improves the strength, toughness, and durability of the concrete [20]. At present, the enhancing properties of PVA have been confirmed in the application of Engineered Cementitious Composite (ECC) concrete, which does not contain coarse aggregate [21–23]. However, due to the large content of PVA in concrete (2%) without coarse aggregate, the cost of concrete is so high that the application is still limited to the key parts of structures subjected to large forces in engineering. This is not conducive to large-scale promotion of PVA fiber concrete. Now, the available dynamic characteristics of PVA FRC with coarse aggregate also concentrate on high strain rates, such as impact loads, explosions, and high-speed collision loads. There are still few studies on the dynamic characteristics at the strain rate of seismic loads for fine PVA FRC that are suitable for testing the strength of an engineering structure.

Strong earthquake activity has brought huge losses to the western region of infrastructure construction in China, which is an earthquake-prone country. Due to the poor tensile properties of conventional concrete, the traditional support structures are prone to drawing, bending, and shearing under the action of a seismic load, so it is necessary to develop high tensile performance in an underground structure to reduce damage taken in strong earthquake areas. This is of great significance for the design of concrete structures. In this paper, the dynamic experiments of PVA concrete, with two matrix strengths designed for the engineering of structures, are carried out to study the strengthening and toughening effect of PVA in different contents under quasi-static state and dynamic loads, which will be useful for the application of FRC in earthquake prone areas.

2. The Sample Preparation for the FRC with PVA

2.1. Experimental Materials and Production

2.1.1. PVA Fiber

The experimental fiber is the TQ-II -II type of hardened anti-cracking synthetic fiber. The fibers are bunched monofilament, white, safe, and non-toxic. The detailed parameters and actual picture are shown in Table 1 and Figure 1, respectively. According to the test report by the National Textile and Garment Quality Supervision Inspection Center (Zhejiang) (No. 201509666 document), the measured mechanical indicators of PVA used in the experiment meet industry requirements. The specific test results are shown in Table 2.

Table 1. Physical and mechanical parameters of polyvinyl alcohol (PVA) fiber.

Fiber Shape	Density (g/cm^3)	Fiber Diameter (μm)	Fiber Length (mm)	Tensile Strength (MPa)	Elastic Modulus (GPa)	Elongation at Break (%)	Acid and Alkali Resistance
Bunchy monofilament	1.30	15–25	12	≥1200	≥30	5–20	Strong

Table 2. Test report results of PVA fiber.

Serials No.	Test Items	Standard Requirement	Measured Value	Individual Assessment
1	Tensile strength (MPa)	≥1200	1721.2	Qualified
2	Initial elastic modulus (GPa)	≥30	35.7	Qualified
3	Elongation at break (%)	5–20	7.0	Qualified

Figure 1. Image of PVA.

2.1.2. Mixture Design Proportions and Mixing of FRC

The mix design is in accordance with the Chinese standards outlined in "Specification for mix proportion design of ordinary concrete" (JGJ55-2011) [24] and "Steel fiber-reinforced concrete" (JG/T 472-2015) [25]. After several adaptations and experiments, concrete with two matrix strengths (i.e., C30 and C40) with different fiber contents was designed. The specific design parameters are shown in Table 3. In the table, the types of concrete are named after two matrix strengths and different fiber contents. For example, C40PVA0.4 indicates that the matrix strength was 40 MPa and the volume content of PVA was 0.4%. In order to minimize the effect of the coarse aggregate and fine aggregate on the experimental results, the difference between the two contents was made to be small, and the sand ratio was fixed at 35%. The water/cement ratios for the two matrix strengths concrete were designed to maintain a constant of 0.53 and 0.49, respectively, so as to reduce the effect of water/cement on the experimental results.

Table 3. Mix proportions of fiber-reinforced concrete (FRC) (1 m^3).

Type	Cement (kg)	Water (kg)	Fine Aggregate (kg)	Coarse Aggregate (kg)	W/C	Sand Ratio	PVA Volume Content
C30PVA0	377.6	200	643.1	1194.3	0.53	35%	0%
C30PVA0.2	377.6	200	643.1	1194.3	0.53	35%	0.2%
C30PVA0.4	377.6	200	643.1	1194.3	0.53	35%	0.4%
C30PVA0.6	377.6	200	643.1	1194.3	0.53	35%	0.6%
C40PVA0	438.8	215	611.17	1135.03	0.49	35%	0%
C40PVA0.2	438.8	215	611.17	1135.03	0.49	35%	0.2%
C40PVA0.4	438.8	215	611.17	1135.03	0.49	35%	0.4%
C40PVA0.6	438.8	215	611.17	1135.03	0.49	35%	0.6%

It has been found that the key factor in the success of the experiment is the dispersion of the fiber in concrete during the process of multiple adaptation of the fiber concrete. According to the relevant research, we can see that the smaller-diameter fiber is less likely to be dispersed in the concrete; hence, the amount of fine fibers in concrete should not exceed a certain number. Based on a number of mixing experiments and the mixing experience of fiber concrete outlined in the literature, a fast and efficient laboratory mixing method is put forward. First, an appropriate amount of coarse aggregate and fine aggregate is put into the forced mixer machine to dry mix for 30 s, and the PVA fiber and cement is put into the pot to dry mix for 2 min. Then, the fiber and cement mixture are placed in the forced mixer to dry mix with the coarse aggregate and fine aggregate mixture until the cement and fiber are mixed evenly, and the designated water quantity is added for wet mixing for 3 min. After the above process is complete, the fiber will be distributed evenly in the mixed FRC without the occurrence of the knot phenomena, meeting construction requirements. The specific construction process is illustrated in Figure 2.

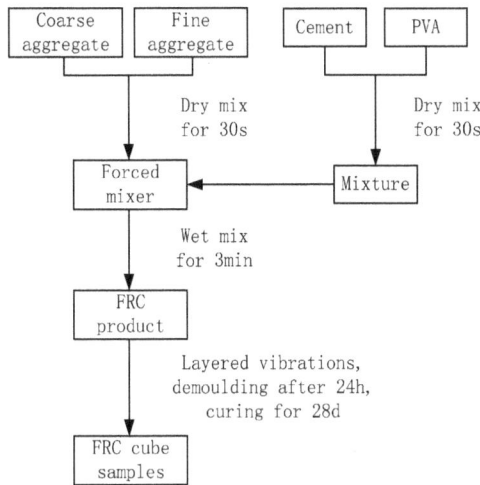

Figure 2. Flow chart for FRC mixing and pouring.

The mixed fiber concrete is placed in a standard plastic sample mold, with dimensions of 150 × 150 × 150 mm, in three layers and vibrated for 2 min. Then, it is covered with cling film to prevent moisture from evaporating. After 24 h, the samples need to be demolded and put into the standard curing room for water conservation at a temperature of 20 ± 2 °C and a humidity greater than or equal to 95%. After curing for 28 days, the samples are ready for the relevant mechanical experiments.

2.2. Experimental Program

The dynamic compression and splitting tensile mechanics experiment of fiber concrete under a medium strain rate were carried out by using the RMT-201 rock and concrete mechanics experiment system developed by the Wuhan Institute of Rock and Soil Science at the Chinese Academy of Sciences. In the experiment, a 150 × 150 × 150 mm plastic mold was used to cast the concrete, and samples of the two kinds with different sizes were prepared by drilling from the mold. The experiment was carried out with four kinds of fiber volume contents (i.e., 0, 0.2, 0.4, and 0.6%) and four different strain rates (i.e., 10^{-5}, 10^{-4}, 10^{-3}, and 10^{-2} s^{-1}). The dynamic compression experiments were conducted with a cylindrical sample with the size of Φ50 × 100 mm, 3 samples per group, and a total of 48 samples. The sample for the dynamic splitting experiment was a cylinder sample with a size of Φ50 × 30 mm. There were 3 samples per group, and a total of 48 samples. The total number of dynamic experiments was 192, with two different sizes for two matrix strengths of concrete.

3. Experimental Results and Discussion

3.1. Compression Strength

The experimental results of the compressive strength for FRC at different strain rates are shown in Figures 3 and 4. In general, the compressive strength of FRC increases with the increase of strain rate, showing an obvious rate sensitivity, which agrees with the existing research [13,14,16–18]. At the quasi-static strain rate (i.e., a strain rate of 10^{-5} s^{-1}), the strength of the two kinds of plain concrete meet the design requirements. The addition of PVA can lead to an increase in compressive strength with the two matrix strengths. Under the quasi-static state, the maximum growth values for the FRC of C30 and C40 are 15.1% and 8.7%, respectively, relative to the two types of plain concrete. With the increase of strain rate, there is a significant difference in the increase of the compressive strength with different fiber volume contents. The two types of concrete with a fiber content of 0.2% show the maximum growth at different strain rates, which is related to the most uniform distribution of PVA fibers in concrete, similar to previous findings [6].

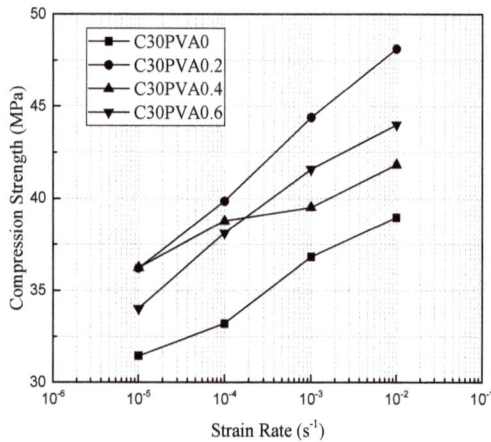

Figure 3. Uniaxial compressive strength of FRC (C30).

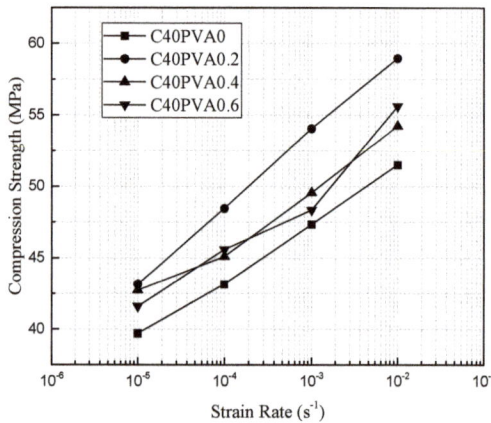

Figure 4. The uniaxial compressive strength of FRC (C40).

From the experimental results, it can be seen that the matrix strength of concrete is another important factor that affects the compressive strength of FRC, in addition to fiber volume content and loading rate. The higher the matrix strength of the concrete, the smaller the effect of PVA on the increase in compressive strength, which is consistent with experimental results from literature [18,26]. Taking the FRC with a volume content of 0.2% as an example, the growth values of uniaxial compressive strength for a matrix strength of C30 are 15.1, 20.9, 20.6, and 23.5%, respectively for each strain rate tested compared with plain concrete at the same strain rates. For C40, the values are 8.7, 12.3, 14.2, and 8.1% for each strain rate, respectively, which are significantly lower than those of C30. The reason for this phenomenon is related to the bond strength and the matrix strength between the fiber and the concrete [1]. Under the same conditions, the reinforcing effect of the fiber is constant. The higher the matrix strength, the lower the proportion of the fiber reinforcement in the high strength concrete. This, in turn, will weaken the fiber-reinforcing effect.

In order to describe the dynamic strength characteristics of FRC under different loading rates, a series of empirical formulas are proposed, including a logarithmic function [27], exponential function [28], and Fib model code [29], which have many practical applications. In the uniaxial compression experiment, as the reinforcing effect of the concrete with 0.2% content is the best, the empirical formula of dynamic compressive strength with different matrix strengths is established by fitting the relevant experimental data for the FRC with 0.2% PVA content. The fitting dynamic impact factor formulas and the fitting curves are shown below, as in Figure 5. The fitting results clearly confirm the above conclusions that the higher the matrix strength, the weaker the fiber-reinforcing effect.

$$DIF_c = \frac{f_c}{f_{cs}} = 1.1591 + 0.0258\lg\left(\frac{\dot{\varepsilon}}{\dot{\varepsilon}_s}\right) \text{ for C30PVA0.2} \qquad (1)$$

$$DIF_c = \frac{f_c}{f_{cs}} = 1.0951 + 0.0193\lg\left(\frac{\dot{\varepsilon}}{\dot{\varepsilon}_s}\right) \text{ for C40PVA0.2} \qquad (2)$$

where DIF_c, f_c, f_{cs}, $\dot{\varepsilon}$, and $\dot{\varepsilon}_s$ are the dynamic impact factor of compression strength, dynamic compression strength, quasi-static compressive strength, dynamic strain rate, and quasi-static compression strength, respectively.

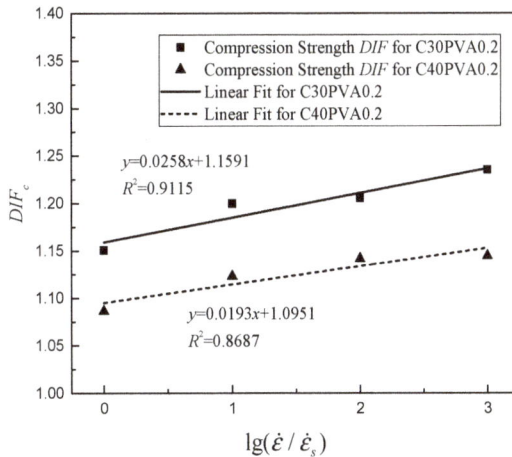

Figure 5. Fitting curves of compressive strength DIF_c for C30PVA0.2 and C40PVA0.2.

3.2. Split Tensile Strength

The experimental results of the split tensile strength of fiber concrete at different strain rates are shown in Figures 6 and 7. Similar to the law of the compressive strength, the addition of PVA can also increase the tensile strength of concrete. In the quasi-static state (i.e., a strain rate of 10^{-5} s^{-1}), the addition of PVA significantly improved the tensile strength of FRC compared with plain concrete, as also seen in previous findings [28]. When the matrix strength of concrete is C30, the growth values are 19.4, 10.9, and 20.3% for fiber volume contents of 0.2, 0.4, and 0.6%, respectively. When it is C40, the growth values are 17, 10.9, and 7%, respectively. The tensile strength of FRC also has a rate sensitivity compared with plain concrete. As strain rate increased, the splitting tensile strength with different contents increased to some extent, while the growth rate increased first and then decreased later, which did not follow the law of the compressive strength. In the low range of strain rates (e.g., 10^{-5} and 10^{-4} s^{-1}), the bridging effect of the fiber plays a major role in concrete. At this time, the increase of the tensile strength of fiber concrete depends mainly on the bonding force between the fiber and the cementitious material. As the strain rate increases (e.g., to 10^{-3} or 10^{-2} s^{-1}), the increase of the loading rate exceeds the expansion rate of the cracks inside the concrete. Hence, the concrete aggregate is directly cut off, and the bridging effect of the fiber is relatively weakened, resulting in the tensile strength of FRC showing a downward trend, as compared with that of plain concrete. In general, the bridging effect of the fiber plays a major role at low strain rates. With the increase of strain rate, the bridging effect of the fiber is weakened, leading to the change of the concrete fracture form from the destruction of cementitious material under a low strain rate to the direct cut of coarse aggregate a under high strain rate. This is confirmed in the existing literature [30,31] and shown in Figure 8.

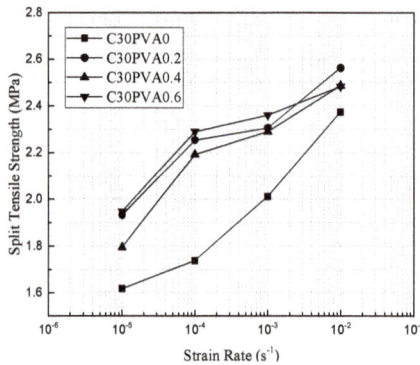

Figure 6. Split tensile strength of FRC (C30).

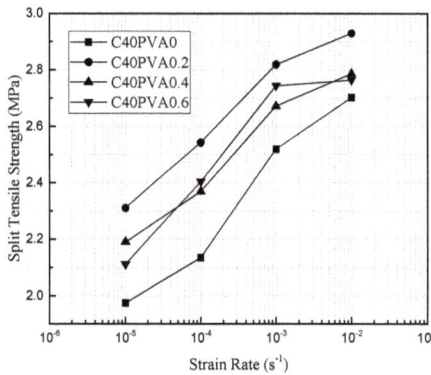

Figure 7. Split tensile strength of FRC (C40).

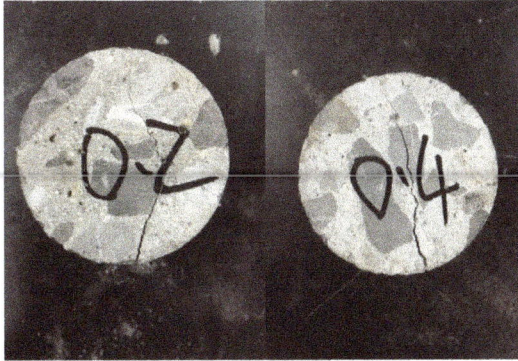

Figure 8. Two kinds of split tensile failure modes of FRC.

According to the fitting method of the dynamic impact factor in the concrete compression experiment, the formula for the dynamic tensile strength of fiber concrete with 0.2% content is established by using polynomial fitting. The fitting formula and curves are shown below, as in Figure 9. From the fitting results, the conclusion that a higher matrix strength is directly related to weaker fiber enhancement is equally applicable to splitting tensile strength. However, unlike the fitting curves of compressive strength, the curves of splitting tensile strength tend to increase first and then decrease. The dynamic impact factor of split tensile strength of fiber concrete reaches the maximum at a strain rate of 10^{-4} s^{-1} for both matrix strengths.

$$DIF_t = \frac{f_t}{f_{ts}} = 1.211 + 0.0783 \lg\left(\frac{\dot{\varepsilon}}{\dot{\varepsilon}_s}\right) - 0.0426\left[\lg\left(\frac{\dot{\varepsilon}}{\dot{\varepsilon}_s}\right)\right]^2 \text{ for C30PVA0.2} \qquad (3)$$

$$DIF_t = \frac{f_t}{f_{ts}} = 1.1766 + 0.0087 \lg\left(\frac{\dot{\varepsilon}}{\dot{\varepsilon}_s}\right) - 0.0139\left[\lg\left(\frac{\dot{\varepsilon}}{\dot{\varepsilon}_s}\right)\right]^2 \text{ for C40PVA0.2} \qquad (4)$$

where DIF_t, f_t, and f_{ts} are the dynamic impact factor of split tensile strength, dynamic split tensile strength, and quasi-static split tensile strength, respectively. $\dot{\varepsilon}$ and $\dot{\varepsilon}_s$ have the same meaning as previously described above.

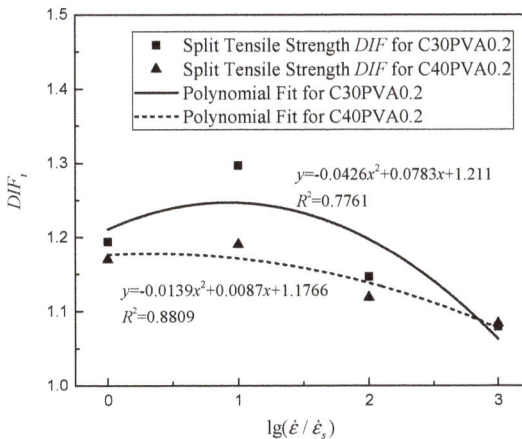

Figure 9. Fitting curves of split tensile strength DIF_t for C30PVA0.2 and C40PVA0.2.

3.3. Elastic Modulus and Peak Strain

The elastic modulus of FRC at different strain rates is shown in Figures 10 and 11. The results show that the elastic modulus of FRC increases with the increase in the strain rate, similar to the compressive strength and split tensile strength [1,7,18,26]. The increasing range of the elastic modulus is different in FRC with different matrix strengths. In all cases, the elastic modulus with 0.2% fiber content is the highest for both the matrix strengths, similar to previous findings [6] The lower the matrix strength, the more obvious the effect of the fiber is on the elastic modulus of the FRC [18]. The increase of the elastic modulus indicates that the ability of concrete to bear the elastic deformation of the load is reduced, which is beneficial to the non-destructive instability of the structure without causing excessive elastic deformation under load.

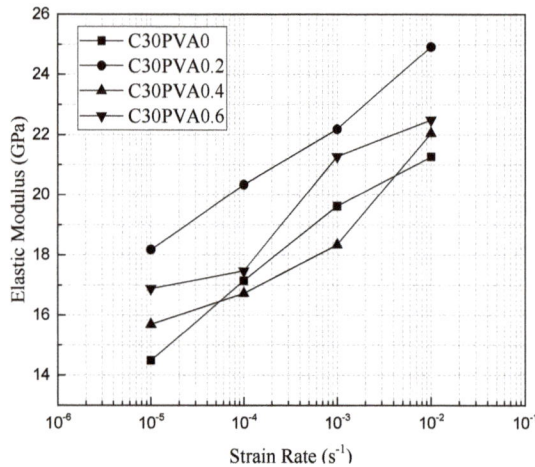

Figure 10. Elastic modulus of FRC (C30).

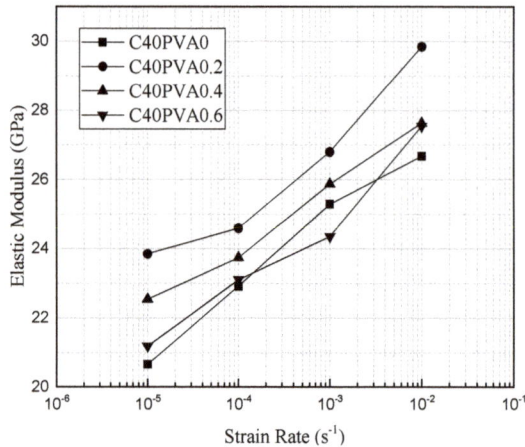

Figure 11. Elastic modulus of FRC (C40).

The peak strain of fiber concrete at different strain rates is shown in Figures 12 and 13. The peak strain here refers to the strain corresponding to the peak stress. Contrary to the trends of the strength and elastic modulus, the peak strain of concrete decreases with an increase of strain rate. The peak strain of the fiber concrete can be enhanced under each strain rate compared with plain concrete, indicating that the addition of the fiber can improve the ability of the concrete to bear the deformation [15]. In fiber concrete, the peak strain with 0.2% fiber content is smaller than that with the other two fiber content levels, which is directly related to the maximum elastic modulus of concrete with 0.2% content.

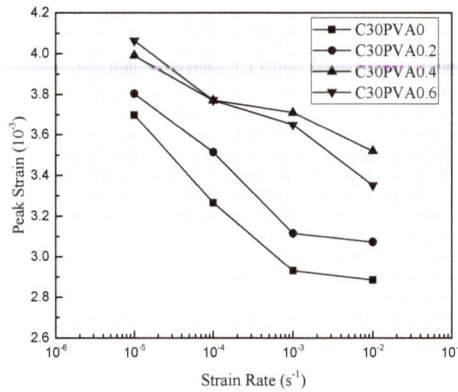

Figure 12. Peak strain of FRC (C30).

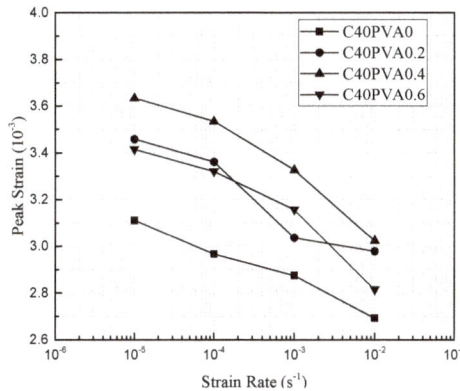

Figure 13. Peak strain of FRC (C40).

3.4. Deformation Characteristics

The typical stress–strain curves of concrete with two matrix strengths at different loading strain rates are shown in Figure 14. For the two kinds of plain concrete, the brittle characteristics of concrete became more obvious with increasing strain rate, which has been confirmed in the literature [27]. When the strain rate reaches 10^{-2} s^{-1}, the failure modes have characteristics of brittle fracture, with the stress–strain curve showing a cliff-like landing. The addition of PVA can significantly improve the post-peak mechanical properties of concrete, showing a significant reduction in the drop rate per unit of time after undergoing peak stress. The stress and strain curves are smoother, indicating that the properties of the concrete change from brittle to ductile, which is of great significance in improving the seismic performance of concrete [12]. From the analysis of the typical stress and strain curves, it is found that the addition of PVA has little effect on the curve at the upward section, while showing a certain effect at the downward section. With the increase in the content of PVA, the descending rate gradually slows down at the downward section of the curve, indicating that PVA has a significant effect in improving toughness of fiber concrete [20,21]. For all types of concrete, the slopes of the stress–strain curve at the downward section increase gradually with the increase of strain rate, showing that the bridging effect of the fiber on the concrete decreases. At this point, the loading rate plays a greater role in the failure behavior of the concrete.

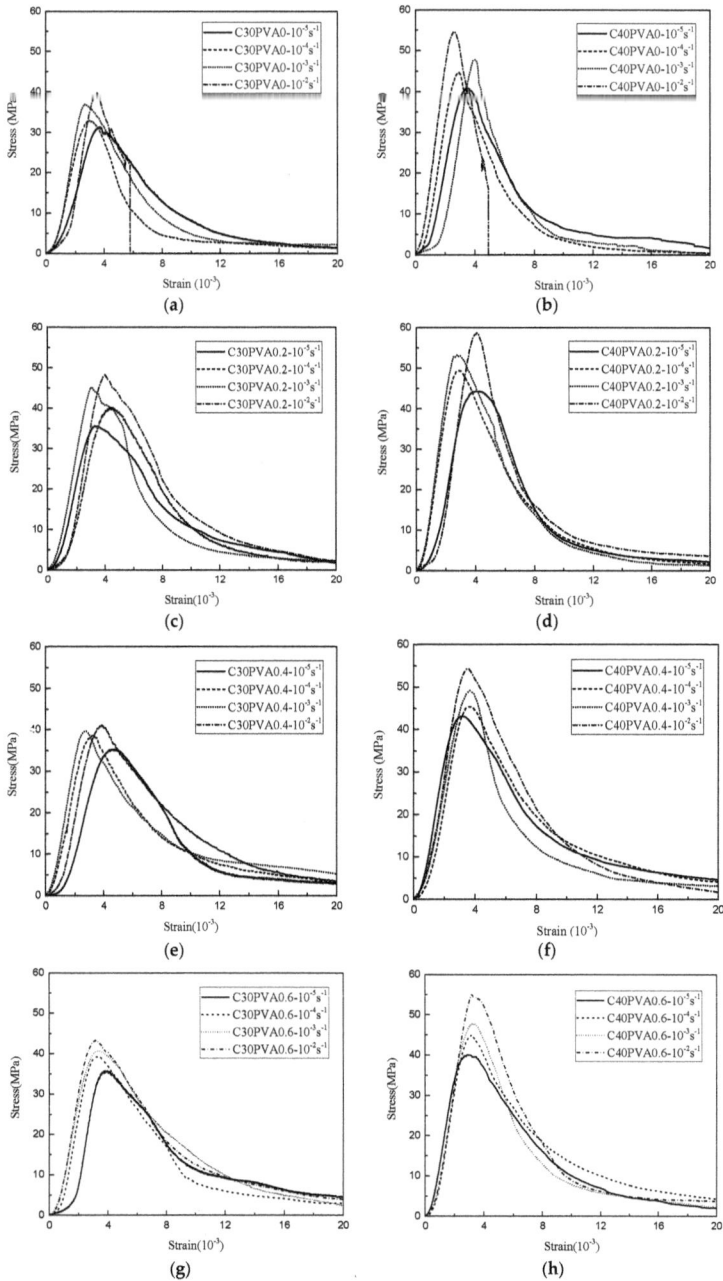

Figure 14. Typical stress–strain curve under different strain rates for (**a**) C30PVA0, (**b**) C40PVA0, (**c**) C30PVA0.2, (**d**) C40PVA0.2, (**e**) C30PVA0.4, (**f**) C40PVA0.4, (**g**) C30PVA0.6, and (**h**) C40PVA0.6.

Another aspect of the change in the post-peak mechanical properties is the residual strength and the failure mode. As shown in Figures 15 and 16, the two matrix strengths of plain concrete appear to undergo the phenomena of fall-block and caving after destruction. When the concrete block is made into a bulk, the residual strength is nearly zero. However, when the fiber concrete is crushed, the evenly distributed fibers begin to bear the load. Due to the bridging effect of the fiber, the concrete structure can maintain a relatively complete form with a number of small cracks on the surface, and the residual strength remains at approximately 3–5 MPa without brittle damage, unlike the collapse of plain concrete.

Figure 15. Typical failure modes of plain concrete under a dynamic load.

Figure 16. Typical failure modes of FRC under a dynamic load.

Therefore, the deformation characteristics of fiber concrete are different from those of plain concrete. The addition of fiber can improve the brittle properties of the concrete, including the fall-block and collapse of the concrete, the obvious decrease of the drop rate per unit time after undergoing peak stress, and the residual strength [27]. The stress–strain curves of concrete with different fiber contents are similar, and the deformation behavior of concrete at a high strain rate is more brittle than that at a low strain rate. The reinforcing effect of fiber on the mechanical properties of concrete is different for the two matrix strengths. The concrete material with low matrix strength is more ductile than the one with a higher matrix strength. The same trend is evident with the reinforcing effect of fiber improving the uniaxial compressive strength [18,26].

3.5. Toughness Index

In the evaluation of the post-peak mechanical properties of concrete, the use of the ductility index-peak strain to reflect the toughness of the material has a certain one-sidedness. Due to the large dispersion of concrete materials, there may be some errors in the peak strain of the experiment results;

hence, the use of a non-dimensional relative index to describe the toughness of the material is more reasonable. According to the method of defining the toughness of steel FRC in the literature [32,33], the formula for the toughness index can be defined for PVA-reinforced concrete as follows:

$$\eta = \frac{W_f}{W_e} \tag{5}$$

where W_f is the area of OCD, which is defined the area surrounded by the limit strain of $20 \times E^{-3}$ in this paper; W_e is the elasticity energy consumed by concrete materials, which is defined by the area of OAB at the strain relative to 0.85^*f_c; and f_c is the peak stress. A detailed diagram is shown in Figure 17.

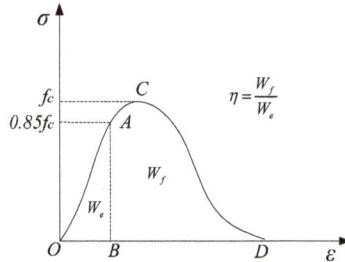

Figure 17. Diagram of the toughness index of concrete.

The above formula does not only account for both the elastic energy absorbed during the elastic phase and the plastic energy absorbed during the plastic phase of the fiber concrete, but also eliminates the energy calculation error caused by the dispersion of the concrete, which is reasonable for describing the toughness of the concrete [32]. The calculated results of the toughness indices according to the formula are shown in Table 4 and Figures 18 and 19 below.

Table 4. Calculated results of toughness index for FRC at different strain rates.

Type	Strain Rate (s^{-1})	W_f (J/m³)	W_e (J/m³)	Toughness Index
C30PVA0	10^{-5}	201.21	29.02	6.93
C30PVA0	10^{-4}	157.87	25.06	6.30
C30PVA0	10^{-3}	160.90	29.24	5.50
C30PVA0	10^{-2}	112.61	28.41	3.96
C30PVA0.2	10^{-5}	237.59	28.59	8.31
C30PVA0.2	10^{-4}	203.66	28.95	7.03
C30PVA0.2	10^{-3}	252.82	40.15	6.30
C30PVA0.2	10^{-2}	244.93	50.11	4.89
C30PVA0.4	10^{-5}	286.49	29.64	9.66
C30PVA0.4	10^{-4}	288.44	34.27	8.42
C30PVA0.4	10^{-3}	259.54	34.39	7.60
C30PVA0.4	10^{-2}	233.84	36.57	6.39
C30PVA0.6	10^{-5}	266.11	25.28	10.53
C30PVA0.6	10^{-4}	317.60	35.56	8.93
C30PVA0.6	10^{-3}	263.17	31.09	8.47
C30PVA0.6	10^{-2}	259.54	34.79	7.46
C40PVA0	10^{-5}	198.86	30.50	6.52
C40PVA0	10^{-4}	200.75	35.28	5.69
C40PVA0	10^{-3}	190.13	38.66	4.92
C40PVA0	10^{-2}	163.00	44.82	3.64
C40PVA0.2	10^{-5}	269.46	35.58	7.57
C40PVA0.2	10^{-4}	297.19	41.69	7.13
C40PVA0.2	10^{-3}	327.58	47.43	6.91
C40PVA0.2	10^{-2}	298.17	52.14	5.72

Table 4. *Cont.*

Type	Strain Rate (s^{-1})	W_f (J/m^3)	W_e (J/m^3)	Toughness Index
C40PVA0.4	10^{-5}	332.30	34.11	9.74
C40PVA0.4	10^{-4}	320.76	37.03	8.66
C40PVA0.4	10^{-3}	312.84	47.58	6.58
C40PVA0.4	10^{-2}	270.63	51.37	5.27
C40PVA0.6	10^{-5}	362.71	35.42	10.24
C40PVA0.6	10^{-4}	351.34	37.54	9.36
C40PVA0.6	10^{-3}	347.59	41.10	8.46
C40PVA0.6	10^{-2}	338.79	47.57	7.12

Note: W_f and W_e in the table are averages of three experimental results.

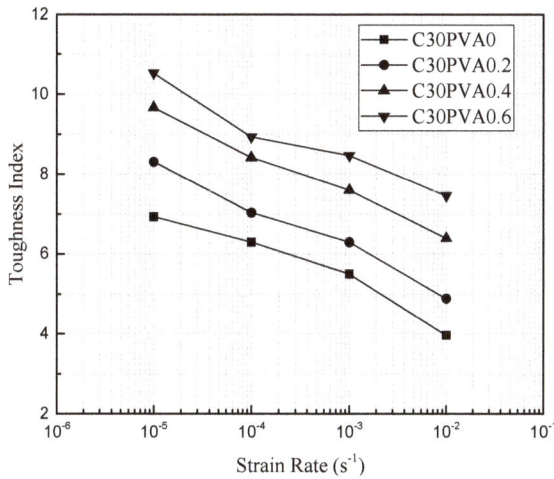

Figure 18. Toughness index of FRC (C30).

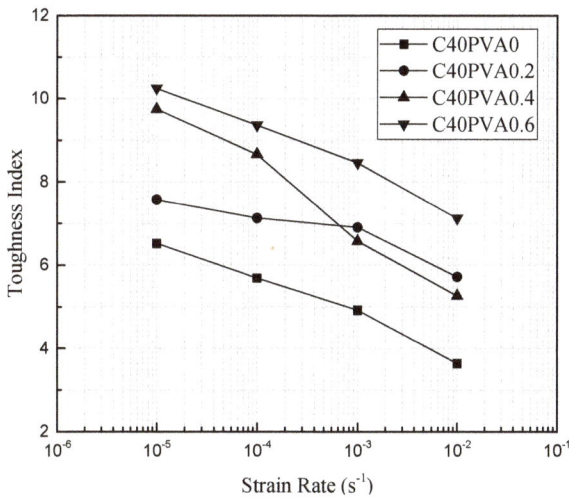

Figure 19. Toughness index of FRC (C40).

The results show that the toughness index of FRC with two matrix strengths decreases with the increase of the strain rate, and concrete material is more brittle at a high strain rate [27]. It can be seen from Table 4 that the elastic strain energy absorbed by the FRC increases with the increase of the strain rate. The increase of the strain rate leads to the increase of the strength and elastic modulus of the concrete. There is a corresponding increase in the area surrounded by the elastic phase of the concrete stress–strain curve, indicating that the elastic strain energy of the concrete increased.

The addition of PVA can significantly enhance the toughness of concrete. In the quasi-static state (i.e., a strain rate of 10^{-5} s^{-1}), the toughness indices of FRC (C30) with 0.2, 0.4, and 0.6% content are increased by 19.9, 39.4, and 51.8%, respectively, compared to those of the plain concrete. The values are 16.2, 49.4, and 57.1%, respectively, for C40. The enhancement effect of the fiber on the concrete toughness is more obvious with the increase of fiber content. However, at a high strain rate (i.e., 10^{-2} s^{-1}), the toughness index of FRC (C30) is increased by 23.3, 61.3, and 88.2%, respectively, and the values increased 57.3, 44.9, and 95.6%, respectively, for C40. The improvement of the strain rate will lead to a decrease in the toughness index to a certain extent. However, the toughness index will be less reduced compared to the decrease in plain concrete, which shows that the toughness of the concrete with PVA decreases more smoothly with increases of the strain rate compared to plain concrete. These results illustrate that the toughness of FRC is better than that of plain concrete under the strain rate range of seismic load [20]. This is of great significance to the application of FRC in seismic design. Considering the effect of the fiber on the compressive strength, splitting strength, elastic modulus, and peak strain of concrete, as well as the cost factors and dispersion technology of the fiber, the concrete with a 0.2% PVA content is recommended for actual applications, as it can meet the engineering requirements.

4. Conclusions

In order to investigate the dynamic mechanical properties of FRC with PVA on strain rates corresponding to seismic loads (i.e., 10^{-5}, 10^{-4}, 10^{-3}, and 10^{-2} s^{-1}), the dynamic compression and splitting tensile mechanics experiment for two kinds of matrix strengths (i.e., C30 and C40) with four kinds of fiber volume contents (i.e., 0, 0.2, 0.4, and 0.6%) were carried out by using the RMT-201 rock and concrete mechanics experiment system. The physical and mechanical properties were obtained, and the following conclusions can be drawn.

(1) PVA has some enhancement and improvement effects on the concrete, mainly regarding the improvement of the compressive strength, splitting tensile strength, the toughness index, and the post-peak mechanical properties of the stress and strain curves at the descending stage. The addition of PVA can also significantly improve the failure behavior of the concrete, which changes from the fall-block and caving of plain concrete to a relatively complete form of FRC with a residual strength of 3–5 MPa. The enhancing effect of PVA on the concrete differs for two matrix strengths. The lower the matrix strength, the more obvious the reinforcing effect of the fiber is on the concrete.

(2) PVA FRC is a rate-sensitive material similar to plain concrete. The uniaxial compressive strength, splitting tensile strength, and elastic modulus of concrete increase with increasing strain rate, while the peak strain of concrete decreases, indicating that the FRC under a high strain rate is more brittle than that under a low strain rate.

(3) The PVA FRC with a 0.2% volume content has greater advantages than the other two kinds of fiber concrete in improving concrete's mechanical properties. Considering cost factors and construction convenience, concrete with a 0.2% PVA content is recommended in engineering applications.

Concrete workability should be guaranteed. Moreover, the durability issue should be deeply analyzed. In this work, to highlight the influence of strain rates on dynamic mechanical properties, these two parts were ignored. However, at a wider level, further research is also required.

Author Contributions: W.Y.: experimental tests and writing-original draft preparation; S.W.: experimental tests and discussion of the results; Z.W.: elaboration data and discussion of the results; T.X.: discussion of the results and writing-review & editing.

Acknowledgments: The financial support from National Basic Research Program of China (973 Program) (No: 2015CB057906), the natural science foundation of China (No: 51679172, 41130742, 51574180, 51579238), Hubei Provincial Natural Science Foundation of China (2018CFA012), and Youth Innovation Promotion Association CAS.

Conflicts of Interest: The authors declare no conflicts of interest.

References

1. Dong, J.F.; Wang, Q.Y.; Guan, Z.W. Material properties of basalt fibre reinforced concrete made with recycled earthquake waste. *Constr. Build. Mater.* **2017**, *130*, 241–251. [CrossRef]
2. Richardson, A.; Coventry, K. Dovetailed and hybrid synthetic fibre concrete–impact, toughness and strength performance. *Constr. Build. Mater.* **2015**, *78*, 439–449. [CrossRef]
3. Su, H.; Xu, J. Dynamic compressive behavior of ceramic fiber reinforced concrete under impact load. *Constr. Build. Mater.* **2013**, *45*, 306–313. [CrossRef]
4. Yan, Z.; Zhu, H.; Ju, J.W. Behavior of reinforced concrete and steel fiber reinforced concrete shield TBM tunnel linings exposed to high temperatures. *Constr. Build. Mater.* **2013**, *38*, 610–618. [CrossRef]
5. Asokan, P.; Osmani, M.; Price, A.D.F. Assessing the recycling potential of glass fibre reinforced plastic waste in concrete and cement composites. *J. Clean. Prod.* **2009**, *17*, 821–829. [CrossRef]
6. Song, P.S.; Hwang, S. Mechanical properties of high-strength steel fiber-reinforced concrete. *Constr. Build. Mater.* **2004**, *18*, 669–673. [CrossRef]
7. Yazici, Ş.; İnan, G.; Tabak, V. Effect of aspect ratio and volume fraction of steel fiber on the mechanical properties of SFRC. *Constr. Build. Mater.* **2007**, *21*, 1250–1253. [CrossRef]
8. Vajje, S.; Krishnamurthy, N.R. Study on addition of the natural fibers into concrete. *Int. J. Sci. Technol. Res.* **2013**, *2*, 213–218.
9. Sivaraja, M.; Velmani, N.; Pillai, M.S. Study on durability of natural fibre concrete composites using mechanical strength and microstructural properties. *Bull. Mater. Sci.* **2010**, *33*, 719–729. [CrossRef]
10. Agopyan, V.; Savastano, H., Jr.; John, V.M.; Cincotto, M.A. Developments on vegetable fibre–cement based materials in São Paulo, Brazil: An overview. *Cem. Concr. Compos.* **2005**, *27*, 527–536. [CrossRef]
11. Ali, M.; Liu, A.; Sou, H.; Chouw, N. Mechanical and dynamic properties of coconut fibre reinforced concrete. *Constr. Build. Mater.* **2012**, *30*, 814–825. [CrossRef]
12. Shafiq, N.; Ayub, T.; Khan, S.U. Investigating the performance of PVA and basalt fibre reinforced beams subjected to flexural action. *Compos. Struct.* **2016**, *153*, 30–41. [CrossRef]
13. Grote, D.L.; Park, S.W.; Zhou, M. Dynamic behavior of concrete at high strain rates and pressures: I. experimental characterization. *Int. J. Impact Eng.* **2001**, *25*, 869–886. [CrossRef]
14. Malvar, L.J.; Ross, C.A. Review of strain rate effects for concrete in tension. *ACI Mater. J.* **1998**, *95*, 735–739.
15. Soufeiani, L.; Raman, S.N.; Jumaat, M.Z.B.; Alengaram, U.J.; Ghadyani, G.; Mendis, P. Influences of the volume fraction and shape of steel fibers on fiber-reinforced concrete subjected to dynamic loading—A review. *Eng. Struct.* **2016**, *124*, 405–417. [CrossRef]
16. Abrams, D.A. Effect of rate of application of load on the compressive strength of concrete. *Proc. ASTM* **1917**, *17*, 364–377.
17. Cook, D.J.; Pama, R.P.; Weerasingle, H. Coir fibre reinforced cement as a low cost roofing material. *Build. Environ.* **1978**, *13*, 193–198. [CrossRef]
18. Zhang, X.X.; Abd Elazim, A.M.; Ruiz, G.; Yu, R.C. Fracture behaviour of steel fibre-reinforced concrete at a wide range of loading rates. *Int. J. Impact Eng.* **2014**, *71*, 89–96. [CrossRef]
19. Jiang, C.; Fan, K.; Wu, F.; Chen, D. Experimental study on the mechanical properties and microstructure of chopped basalt fibre reinforced concrete. *Mater. Des.* **2014**, *58*, 187–193. [CrossRef]
20. Deng, Z. *High Performance Synthetic Fiber Concrete*; Science Press: Beijing, China, 2003.
21. Holschemacher, K.; Höer, S. Influence of PVA fibers on load carrying capacity of concrete with coarse aggregates. In Proceedings of the 7th International RILEM Symposium on Fibre Reinforced Concrete: Design and Applications, Chennai, India, 17–19 September 2008; pp. 219–229.
22. Tosun-Felekoğlu, K.; Felekoğlu, B.; Ranade, R.; Lee, B.Y.; Li, V.C. The role of flaw size and fiber distribution on tensile ductility of PVA-ECC. *Compos. Part B Eng.* **2014**, *56*, 536–545. [CrossRef]
23. Suryanto, B.; Maekawa, K.; Nagai, K. Predicting the Creep Strain of PVA-ECC at High Stress Levels based on the Evolution of Plasticity and Damage. *J. Adv. Concr. Technol.* **2013**, *11*, 35–48. [CrossRef]

24. Banthia, N.; Mindess, S.; Trottier, J.F. Impact resistance of steel fiber reinforced concrete. *ACI Mater. J.* **1996**, *93*, 472–479.

25. China Academy of Building Research (CABR). *JGJ 55-2011: Specification for Mix Proportion Design of Ordinary Concrete*; China Architecture Building Press: Beijing, China, 2011.

26. National Standards of the People's Republic of China. *JG/T472-2015: Steel Fiber Reinforced Concrete*; Standards Press of China: Beijing, China, 2015.

27. Zhang, Y. *Experimental Study of the Mechanical Properties of Concrete under Different Strain Rates*; Beijing University of Technology: Beijing, China, 2012.

28. Mihashi, H.; Wittmann, F.H. Stochastic approach to study the influence of rate of loading on strength of concrete. *HERON* **1980**, *25*, 1–54.

29. Fib Model Code. *Fib Model Code for Concrete Structures 2010*; Ernst Sohn: Berlin, Germany, 2010.

30. Wu, S.; Chen, X.; Zhou, J. Tensile strength of concrete under static and intermediate strain rates: Correlated results from different testing methods. *Nucl. Eng. Des.* **2012**, *250*, 173–183. [CrossRef]

31. Shah, S.P. Experimental methods for determining fracture process zone and fracture parameters. *Eng. Fract. Mech.* **1990**, *35*, 3–14. [CrossRef]

32. Zhang, H.; Gao, Y.; Li, F. Experimental study on dynamic properties and constitutive model of polypropylene fiber concrete under high strain rate. *J. Central South Univ. (Sci. Technol.)* **2013**, *44*, 3464–3473.

33. Yi, C.; Fan, Y.; Zhu, H.; Wang, J.Q. Research on fatigue damage of concrete under uniaxial compression on the basis of toughness. *Eng. Mech.* **2010**, *27*, 113–119.

applied
sciences

MDPI

Article

Effect of Recycled Aggregate Quality on the Bond Behavior and Shear Strength of RC Members

Thanapol Yanweerasak [1], Theang Meng Kea [2], Hiroki Ishibashi [3] and Mitsuyoshi Akiyama [3,*]

[1] Faculty of Engineering at Sriracha, Kasetsart University Sriracha Campus, 199, Tungsukla, Sriracha, Chonburi 20230, Thailand; thanapol.yan@gmail.com
[2] KTL Veng Sreng Construction Co., Ltd, 25, St. 271, Phsar Doeum Thkov, Chamkarmon, Phnom Penh 12307, Cambodia; theangmeng@yahoo.com
[3] Department of Civil and Environmental Engineering, Waseda University, 3-4-1, Okubo, Shinjuku-Ku, Tokyo 169-8555, Japan; hirokiishibashi@toki.waseda.jp
* Correspondence: akiyama617@waseda.jp; Tel.: +81-352-862-694

Received: 28 September 2018; Accepted: 22 October 2018; Published: 25 October 2018

Abstract: During the aggregate crushing process, natural aggregate and clinging mortar from existing concrete will inevitably produce small cracks and weak bonds between the aggregate and the existing cement mortar. The weaknesses of the existing cement mortar, adhered to a natural aggregate, negatively affect the properties of a recycled aggregate concrete, which prevents its application in reinforced concrete (RC) structures. Recycled aggregate can be classified into several categories, according to its physical and mechanical properties. The properties of concrete incorporated with the recycled aggregate of various qualities can be controlled, and the variability in its strength can also be reduced. This study aims to promote the application of recycled aggregate by investigating the effects of recycled aggregate quality (i.e., water absorption and the number of fine particles) classified by the Japanese Industrial Standards (JIS) on material properties, mechanical properties, and shear behavior of RC beams with recycled aggregate.

Keywords: recycled aggregate quality; bond strength; shear behavior; aggregate interlock mechanism; size effect

---

1. Introduction

Concrete materials have become the second most-consumed type of material in the world [1]. Due to the considerable annual demand for concrete and aggregate, the supply of these materials in the near future is suspect. The global demand for construction aggregates is estimated to increase by 5.2%, annually, to 48.3 billion tons, until 2020. A large percentage of the global aggregate supply (approximately 67%, in 2015) is consumed in Asia-Pacific countries. The sustainable development of these materials relies on the judicious use of natural resources. The use of recycled aggregate from crushed concrete waste, for the production of new concrete, is one alternative offering an effective green solution by prolonging the lifespan of natural aggregate. Although the advantages of recycled aggregate are widely recognized, its applications remain limited. Several researchers have investigated the effects of recycled aggregate on the properties and behavior of concrete. For example, several authors investigated the effect of the parent concrete properties on the properties of recycled aggregate concrete [2,3]. Serifou et al. [4] investigated concrete made with fine and coarse aggregates recycled from fresh concrete waste. Abdullahi et al. [5] investigated the effect of aggregate type on the compressive strength of concrete.

The weaknesses and mortar content of recycled aggregate depend on the type, size, and strength of the natural aggregate, as well as the parent concrete strength, crusher types, and crushing processes. A parent concrete of a higher strength creates a stronger mortar and a better bond between the mortar and the natural aggregate, leading to improvements in the properties of the recycled aggregate concrete. However, the higher mortar content could have greater negative effects on the recycled aggregate concrete properties. It is extremely difficult to obtain information on the recycled aggregate, parent concrete, and aggregate production from recycling factories because concrete waste is typically obtained from a variety of sources. Therefore, another approach should be developed to more widely understand and control the effects of recycled aggregate quality on concrete properties. Such an approach would expand the application of recycled aggregate concrete in structural concrete members. Zhao et al. [6] proposed the study of stress–strain behavior of the fiber-reinforced plastic (FRP)-confined recycled aggregate concrete. The effects of replacement by recycled coarse aggregate, on the axial and the lateral strain of a concrete cylinder, were examined. The experimental results show that the specimens that use 20% replacement, by a coarse, recycled aggregate, behave similarly to that of normal concrete but specimens that use a 100% replacement show a lower strength and significantly different stress–strain response. Xu et al. [7] investigated the mechanical behaviors of recycled aggregate, under a tri-axial compression of the FRP-confined column; the influence of a recycled aggregate content and its source were also examined.

In the field of civil engineering, concrete is always reinforced with steel rebar, and the bond strength and interacting behaviors between the concrete and steel rebar are among the most important requirements of reinforced concrete (RC) design and construction. The bond behavior of concrete is controlled by the concrete type, rebar arrangement, loading conditions, and construction details [8]. The bonding strength between natural and recycled aggregate concrete and rebar are similar, although the compressive strength exhibits a downward trend with increases in the recycled coarse aggregate replacement ratio [9]. Creazza and Russo [10] presented a model for predicting crack-width on the basis of bond stress, with different reinforcement amounts and concrete strength. It was confirmed that the effect of reinforcement amounts on the crack-width was significant. Thus, the compressive strength alone can no longer be used as the main parameter to estimate the bond strength between the recycled aggregate concrete and the steel rebar. The recycled aggregate has a lower density than natural aggregate, due to the lower density and porous nature of the existing adhered cement mortar.

Fathifazl et al. [11] and Kim et al. [12] reported that the shear strength of RC beams with recycled aggregate concrete was lower than that of RC beams with natural aggregate concrete, with the same testing procedure and mix proportions, even though there were no significant differences between the failure modes and the crack patterns. Sato et al. [13] conducted loading tests on RC members using a recycled aggregate to evaluate whether recycled aggregate can be applied to concrete structures. The experimental results showed that the effects of recycled aggregate on the ultimate flexural and ductility capacities were rarely observed because the beams failed in pure flexure or flexure-shear, after the yielding of the tensile reinforcing bars. Therefore, the ultimate flexural strength could be estimated by using conventional structural analysis.

There is limited published experimental data on the structural behavior of RC components that use recycled aggregate. When conducting structural experiments on RC components using recycled aggregate, it is important to consider that recycled aggregate is produced from various RC members of building structures. The application of a recycled aggregate to structural concrete depends on the resulting properties. In this study, to clearly understand the effects of the properties of the recycled aggregate on the structural performance, the quality of the recycled aggregate, classified as high, medium, or low-quality, based on Japanese Industrial Standards (JIS), depending on the water absorption and the number of fine particles, is taken as the main test parameter. The bond behavior between the deformed steel rebar and concrete, and the shear behavior of the RC beams, without shear reinforcement, were experimentally investigated.

2. Experimental Program

2.1. Specimen for Bond Testing

To more thoroughly understand the effects of the recycled aggregate quality on the bond behavior of the interaction between the recycled aggregate concrete and the steel rebar, different water-to-cement ratios (0.30, 0.45, 0.60, and 0.75) and different diameters of the deformed steel rebar (13 and 19 mm) were considered. Due to the large variation within the specimens, the specimens were cast with 13-mm or 19-mm-diameter deformed steel rebar for the pull-out tests, to understand the failure behavior of the recycled aggregate concrete bond strength at the ultimate load and in the post-peak region. The specimens for the bond tests were developed, such that bond stresses for different steel diameters could be compared by maintaining the embedment-to-steel diameter ratio, following the procedure presented by Yamao et al. [14] and Hong et al. [15]. They reported that the bond stress of specimens with short embedment depths exhibits a higher strength than those with long embedment depths, for a given steel diameter. The concrete specimens, made with natural and recycled aggregate concrete, for the pull-out tests, were cubic in shape, with a side length of 150 mm. The concrete was poured into the mold, as shown in Figure 1a, to keep the rebar stable and at the correct position, in the RC beam.

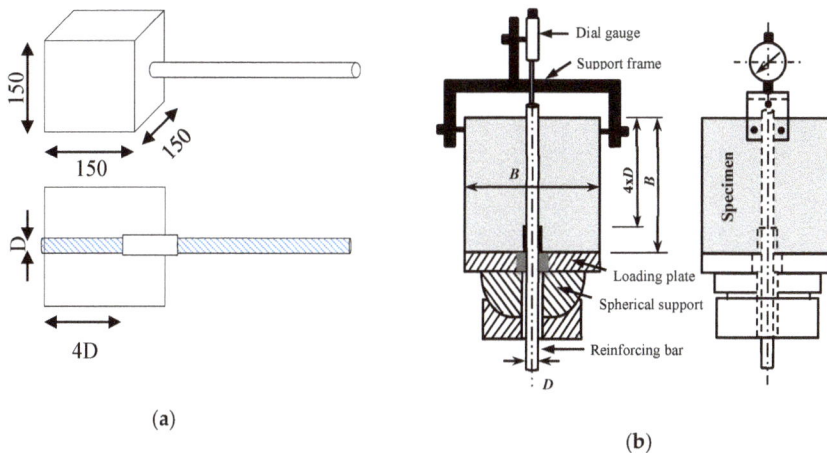

(a)

(b)

Figure 1. Specimen installation for the pull-out test using the Japan Society of Civil Engineers (JSCE). (**a**) Dimension of specimen and (**b**) specimen installation.

The 13- and 19-mm deformed bars were embedded in the concrete, with a length that was four times the steel diameter, and the remaining portion of the bars was covered by plastic piping to prevent the steel surface from contacting the concrete. This short embedment length allowed a uniform bond-stress distribution to be assumed. Sixty-four specimens were prepared for the pull-out tests of the steel rebar embedded in the natural and recycled aggregate concrete. For each natural (N) aggregate concrete and concrete with high-quality (H), medium-quality (M), and low-quality (L) recycled coarse aggregate replacement, three specimens were cast with a 13-mm-diameter deformed steel rebar and one specimen was cast with a 19-mm-diameter deformed steel rebar to consider the effects of the steel bar diameter on the bond behavior. The tensile load provided by the hydraulic universal testing machine was recorded automatically. The loading rate for the deformed rebar with diameters of 13 and 19 mm was varied from 2 to 6.6 MPa/min. Following the experimental procedure specified by JSCE-G 503-1998 "Bond strength testing of reinforcement and concrete by pull-put" [16], the vertical slip of the steel rebar (i.e., the relative displacement of the steel rebar and the concrete specimen) was measured using a dial gauge, installed at the top of the concrete specimen, as shown in Figure 1b.

2.2. Specimens for Shear Testing

The shear behavior of the RC beams using a recycled aggregate concrete was investigated using two groups of RC rectangular beams without shear reinforcement. The details of the specimen are shown in Figure 2. The RC beams were subjected to four-point bending tests. All of the beam specimens were designed such that they would fail in diagonal shear failure and not shear compression failure. The different-sized beam specimens were designed to consider the fact that the shear resistance of RC beams typically decreases with an increase in the effective depth of the RC beam [17]. For larger beams, the aggregate particle size is relatively smaller than the beam cross-section. Thus, the aggregate interlocking mechanism is not effective in improving the shear resistance in larger beams.

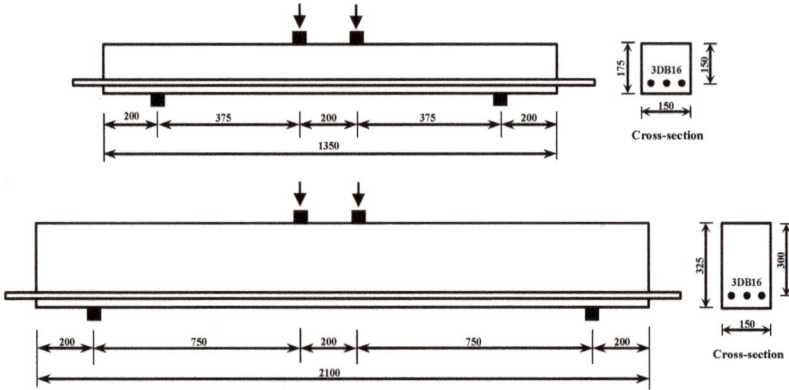

Figure 2. Specimen dimensions and loading points for the shear testing (unit: mm).

Each group of specimens for the shear strength test had four beam specimens with different coarse aggregate replacements: Natural aggregate and high-, medium-, and low-quality recycled aggregate. The details of the specimens are provided in Table 1. The suffixes B60 and S60 represent large and small specimens, respectively, with water-to-cement ratios of 0.60. The concrete cover depth of all specimens was 25 mm, and the cross-sectional area of the longitudinal rebar was 596 mm^2, enough to prevent flexural failure. The beam length was extended by 200 mm at each end to prevent tensile bond failure of the steel rebar. The shear span-to-effective depth ratio (a/d) was kept constant at 2.5 for both the large and small specimens.

Table 1. Details of the specimen for the shear tests. N: natural aggregate concrete; H: high-quality; M: medium-quality; L: low-quality (L).

Specimen	b (mm)	d (mm)	h (mm)	a (mm)	Pure Bending (mm)	Length (mm)	a/d	ρ (%)
NS60 HS60 MS60 LS60	150	150	175	375	200	2100	2.5	2.65
NB60 HB60 MB60 LB60		300	325	750		1350		1.32

Two strain gauges were attached to the steel rebar before casting the concrete to confirm that the yielding of steel rebar would not occur. Two other strain gauges were attached to the concrete, subjected to a compression zone, at the mid-span, to estimate the change in the neutral axis and the concrete crushing. Three dial gauges were used to measure the beam deflections; the dial gauges were located at the (1) mid-span of the beam, (2) left mid-shear span, and (3) right mid-shear span. During the experiment, the load corresponding to the first crack expected to occur at the pure bending region was measured. Then, the first diagonal crack in the shear span region, after the appearance of the flexural crack, was also recorded.

2.3. Material Properties

This study used a natural coarse aggregate and three qualities of the recycled coarse aggregate, with particle sizes ranging from 5 to 20 mm. The relationship between the cumulative percentage passing and the particle sizes for the different aggregate qualities is shown in Figure 3. Natural sand was used as the fine aggregate for all concrete mixes. The detailed physical properties of the coarse aggregates are provided in Table 2. Although the densities of the natural and high-quality recycled coarse aggregate are highly similar, the water absorption of the high-quality recycled aggregate is slightly higher than that of the natural aggregate. The cross-section of the coarse aggregate is shown in Figure 4.

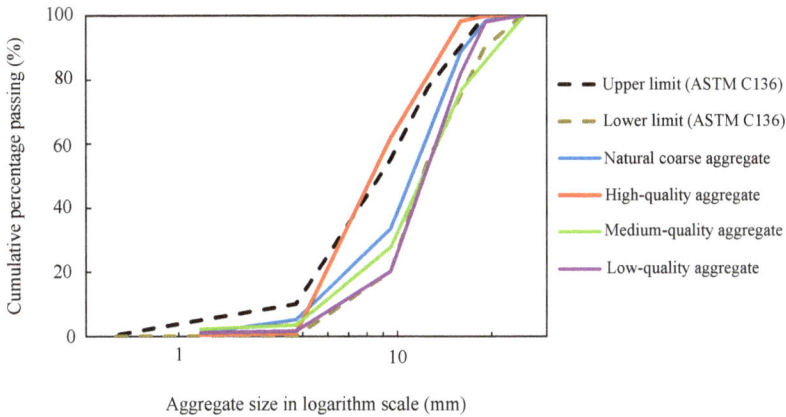

Figure 3. Size distributions of the coarse natural and recycled aggregates.

(a) (b) (c) (d)

Figure 4. Cross-sections of the natural and recycled coarse aggregates. (**a**) Natural coarse aggregate; (**b**) high-quality recycled coarse aggregate; (**c**) medium-quality recycled coarse aggregate; (**d**) low-quality recycled coarse aggregate.

Table 2. Physical properties of the coarse aggregate. SSD: Saturated-surface dry; OD: Oven-dry.

Aggregate	Notation	SSD Specific Gravity	OD Specific Gravity	Water Absorption %	Moisture Content %
Natural fine aggregate	NF	2.55	2.47	2.14	3.47
Natural coarse aggregate	N	2.63	2.60	1.02	1.00
Recycled coarse aggregate	H	2.63	2.60	1.16	0.70
	M	2.55	2.47	3.06	2.80
	L	2.41	2.27	5.92	4.19

The texture of the natural aggregate is highly similar to that of coarse basalt aggregate, whereas the high-quality recycled aggregate is uniform and similar to marble stone. The medium- and low-quality recycled aggregates are highly similar to the natural and high-quality recycled aggregates, respectively. The main difference between them is the amount of old mortar that is still adhered to the particles. The strength of the recycled aggregate concrete is controlled by (1) the aggregate strength, (2) the bond between the aggregate and adhered old mortar, (3) the bond between the aggregate and the new mortar, (4) the bond between the new and the old mortar, and (5) the new mortar strength. The low-quality recycled aggregate could be easily identified as having the largest amount of adhered old mortar. The recycled aggregates used in this experiment were classified by the JIS, according to their properties, as shown in Table 3.

Table 3. Classification of the recycled aggregate based on the Japanese Industrial Standards (JIS).

Properties	Natural Aggregate	Recycled Aggregate		
		High-Quality	Medium-Quality	Low-Quality
	JIS A5308	JIS A5021	JIS A5022	JIS A5023
	Coarse aggregate			
Oven-dry density (g/cm3)	≥2.5	≥2.5	≥2.3	—
Water absorption (%)	≤3.0	≤3.0	≤5.0	≤7.0
Solid content (%)	—	≥55	≥55	—
Amount of fine particle (%)	≤1.0	≤1.0	≤1.5	≤2.0
Chloride amount (%)		≤0.04	≤0.04	≤0.04

3. Experimental Results

3.1. Properties of Fresh Concrete

For the bond strength tests, the slump test of the natural and recycled aggregate concrete with different water-to-cement ratios is shown in Figure 5. There were no noticeable differences in the slump among the different recycled aggregate qualities, for different water-to-cement ratios, even though there was an approximate 1% difference in their air content. However, the concrete with a water-to-cement ratio of 0.30 had a notably low slump. A water reduction agent must be used to improve the workability of both the natural and the recycled aggregate concrete. Therefore, the loss of workability related to the higher water absorption of recycled aggregate could be neglected when recycled aggregates were presoaked. To illustrate the shear strength test, the slump and air content test results are shown in Figure 6. The slump decreased steadily, based on the quality of the recycled aggregate. The slump of the concrete with a low-quality recycled aggregate was 25% lower than for the concrete with a natural aggregate for the large-beam specimens. In contrast, the air content of the concrete with a low-quality recycled aggregate was 17% higher than that of the natural concrete. There was no a significant difference in the air content or the slump for the small beams, except that LS60 had the lowest slump of 18 cm. The slumps of the different types of fresh concrete were consistently higher than the estimated value of 12 cm, even though the air content was lower than the target value of 6%. Thus, the concrete was not sufficiently solid to maintain the air content in the concrete.

3.2. Properties of Hardened Concrete

3.2.1. Compressive Strength

The compressive strength of the hardened concrete with different water-to-cement ratios is shown in Figure 7. The compressive strength of concrete did not depend on the quality of the recycled aggregate for the lower-strength concrete, and the differences became more dramatic as the concrete strength increased. For the lower-strength concrete, the ultimate strength might not be governed by the strength of the aggregate or the cement mortar. The average compressive strength increased from 30% to 50% with a 15% decrease in the water-to-cement ratio, regardless of the recycled aggregate quality.

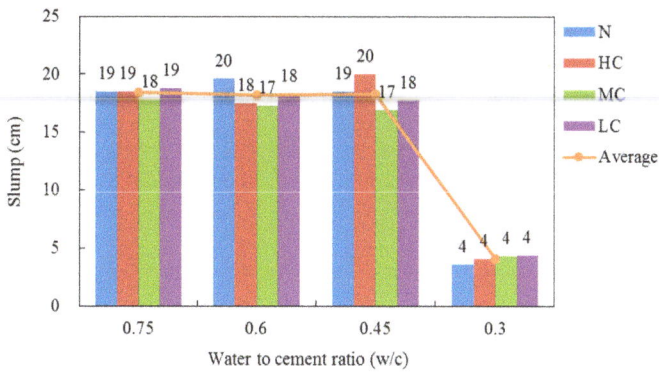

Figure 5. Slump of fresh concrete with different water-to-cement ratios. N: natural aggregate concrete; HC: high-quality recycled aggregate concrete; MC: medium-quality recycled aggregate concrete; LC: low-quality recycled aggregate concrete.

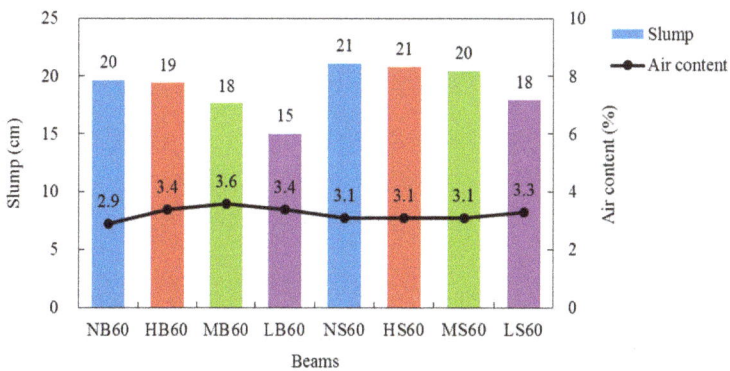

Figure 6. Slump and air content of fresh concrete.

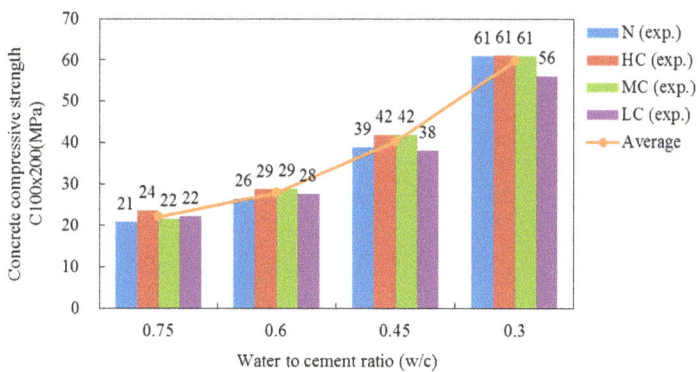

Figure 7. Compressive strengths under different water-to-cement ratios.

3.2.2. Mortar Compressive Strength

The compressive strength of the mortar with different water-to-cement ratios is shown in Figure 8. The compressive strength of the mortar exhibited the same trend as the concrete compressive strength. However, the mortar compressive strength was higher than the concrete compressive strength, regardless of the aggregate type. Viso et al. [18] found that the size effects of the cylinder compressive strength of the cylinder specimens were negligible. Thus, the compressive strength of all concrete specimens was governed by the aggregate strength, the bond strength between the aggregate and mortar, or the bond strength between the old and new mortar. An example is the case of the natural aggregate. Although the mortar compressive strength of the natural aggregate was higher, the compressive strength of the natural aggregate concrete was highly similar to the strength of the low-quality recycled aggregate concrete.

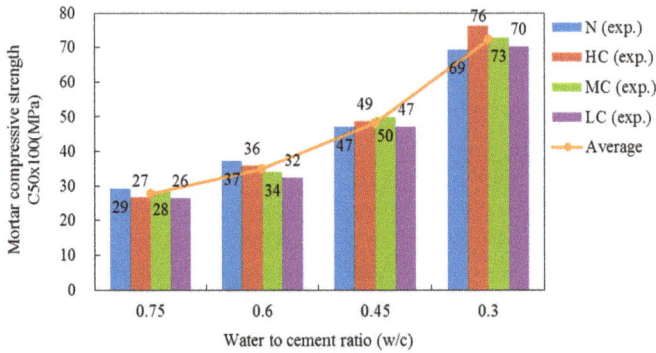

Figure 8. Compressive strength of mortar with different water-to-cement ratios.

3.2.3. Split Tensile Strength

The effects of the recycled aggregate quality on the concrete tensile strength with different water-to-cement ratios are shown in Figure 9. Overall, the split tensile strength for each concrete specimen increased with decreases in the water-to-cement ratio. The compressive strength of the mortar and concrete specimens exhibited the same trend. However, the split tensile strength of the natural aggregate concrete was highly similar to that of the low-quality recycled aggregate concrete. The exception was the concrete with a water-to-cement ratio of 0.60, where the split tensile strength of the natural aggregate concrete was the highest in the group but with the lowest compressive strength. The difference in the split tensile strength could be better understood by considering the compressive strength of the mortar.

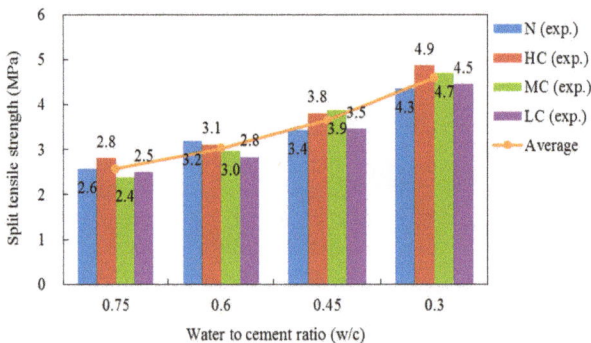

Figure 9. Split tensile strength of concrete with different water-to-cement ratios.

3.2.4. Elastic Modulus

- Static Modulus

The experimental results of the static elastic modulus for the concrete are shown in Figure 10. Although the compressive strengths of the natural and the low-quality recycled aggregate concrete were highly similar, the static elastic modulus of the natural aggregate concrete was higher than that of the low-quality recycled aggregate concrete and even higher than those of the high- and medium-quality recycled aggregate concrete, except for the case with a water-to-cement ratio of 0.60. This difference in trend could be due to a lower stiffness of the recycled aggregate resulting from the relatively weak adhesion of the adhered old mortar.

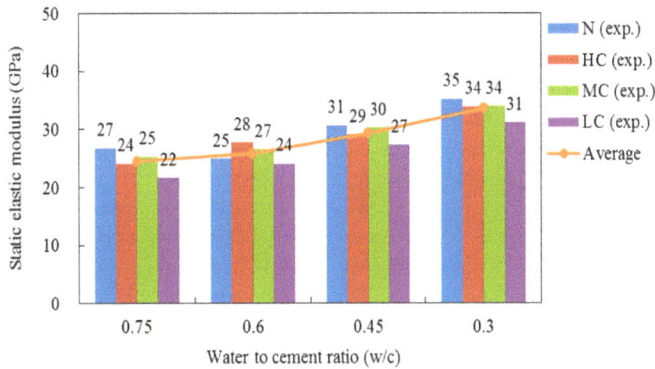

Figure 10. Static modulus at 28 days with different water-to-cement ratios.

- Dynamic Modulus

The dynamic elastic moduli of the concrete measured with the resonance vibration method are shown in Figure 11. The slope of the curve was largely flat when the water-to-cement ratio changed from 0.60 to 0.45. Compared to the natural aggregate concrete, the dynamic modulus of the concrete with a low-quality recycled fine aggregate was 24.3%, 20.5%, 17.5%, and 15.6% lower for the different water-to-cement ratios, indicating that the difference between the dynamic modulus of the natural and low-quality recycled aggregate concrete decreased with a decrease in the water-to-cement ratio or with an increase in the compressive strength. The dynamic modulus of the low-quality recycled coarse aggregate concrete was approximately 8 GPa lower than that of the natural aggregate concrete.

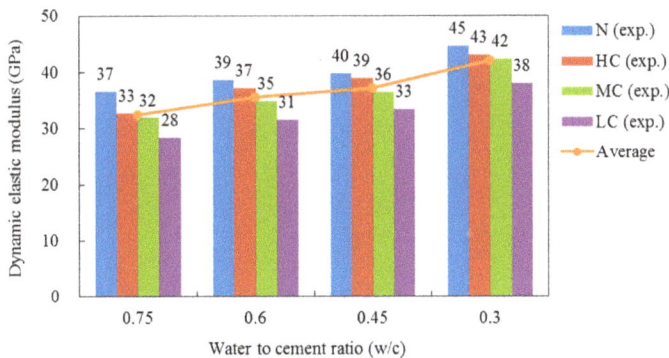

Figure 11. Dynamic modulus of concrete at 28 days with different water-to-cement ratios.

3.2.5. Drying Shrinkage

The results of the drying shrinkage of the natural and recycled aggregate concrete with different water-to-cement ratios are illustrated in Figure 12. The variation in temperature and humidity between specimens was reduced because the four different-quality aggregate concrete and the natural aggregate concrete were cast and cured at the same time. It could be concluded that the ultimate shrinkage of both the natural and recycled aggregate concrete decreased as the water-to-cement ratio decreased. The same trend occurred for the mechanical and elastic behavior of the concretes made of the recycled-concrete coarse aggregates. The drying shrinkage of the low-quality recycled coarse aggregate concrete was the highest, whereas the shrinkage of the concrete with the medium- and high-quality recycled aggregate were highly similar in terms of their water-to-cement ratios—0.75 and 0.60. The increase in drying shrinkage could be explained by the adhered old cement mortar contributing to an increase in the volume of the paste (old + new), as proposed by Tavakoli et al. [19] and Kou et al. [20]. For the case of a water-to-cement ratio of 0.60, the drying shrinkages of the low-, medium-, and high-quality recycled aggregate were 902, 749, and 724 μm/m, which were 39.2%, 15.6%, and 11.7% higher, respectively, compared to that of the natural aggregate concrete (648 μm/m) over the same period. The drying shrinkage for specimen HC60 was slightly higher than that of concrete made by 50% high-quality recycled coarse aggregate replacement and a water-to-cement ratio of 0.64, after 56 weeks, as determined by Morohashi et al. [21]. For the case of a water-to-cement ratio of 0.45, the quality of the recycled aggregate had nearly no effect on the drying shrinkage of concrete. However, for the higher-strength concrete, the drying shrinkage of the low-quality recycled aggregate concrete was lower than that of the high- and medium-quality recycled aggregate concrete but higher than that of the natural aggregate concrete.

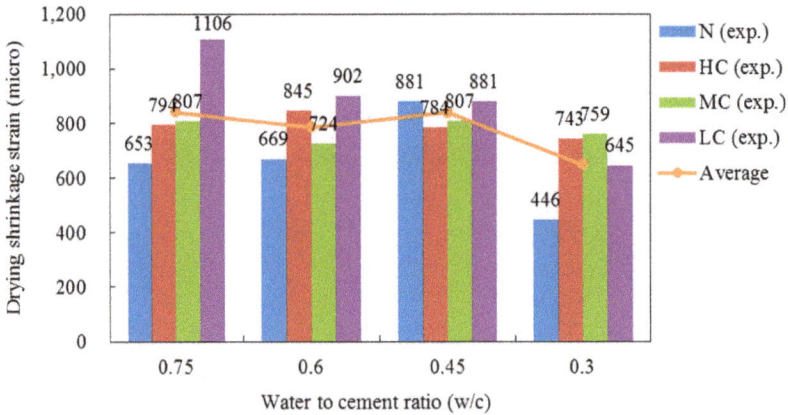

Figure 12. Ultimate drying shrinkage strain of concrete with different water-to-cement ratios.

3.3. Bond Behavior

According to the experimental results of the pull-out test, the bond strength could be expressed as

$$\tau_{max} = \frac{P_{max}}{\pi d l_a} \tag{1}$$

where τ_{max} was the peak bond stress between the concrete and steel rebar (MPa), P_{max} was the peak load indicated by the universal testing machine (N), d was the diameter of the steel rebar (mm), and l_a was the embedment length of the steel rebar (mm) and was equal to four times the steel diameter.

3.3.1. Effect of the Material Properties on the Bond Strength

The bond strength between the concrete and deformed steel rebar for different water-to-cement ratios and for 13 and 19 mm diameters are illustrated in Figure 13. The bond strength between the concrete and deformed bars tend to increase with decreases in the water-to-cement ratio. Unlike the compressive and split tensile strength of the concrete, when the water-to-cement ratio decreased from 0.6 to 0.45, the bond strength between the concrete and the deformed bars appeared to be constant. This trend could be explained by the increase in drying shrinkage having a tendency to decrease with the water-to-cement ratio and resulted in a gentler slope of the dynamic elastic modulus. The bond strengths between the concrete and the 13- and 19-mm-diameter deformed bars had nearly the same ultimate value and trends, indicating that the steel diameter had no effect on the bond strength between the concrete and steel rebar, when the embedded length was in proportion to the steel diameter (four times the steel diameter). In addition, the difference in the bond strengths of concrete with different types of coarse aggregate replacement was consistent with the differences in other material properties, such as the concrete compressive strength, split tensile strength, and static elastic modulus. Compared to the recycled aggregate replacement ratio, the recycled aggregate quality had a greater effect on the bond strength between the steel rebar and the concrete with the recycled coarse aggregate replacement, even though the bond strength between the deformed bars and the recycled aggregate concrete were comparable to that of the natural aggregate concrete, due to the mechanism of the rib of the deformed bars. For example, the bond strength of the concrete with the low-quality recycled coarse aggregate was always the lowest (up to a 24.5% reduction in bond strength for the worst case). However, the attenuation in strength could be ignored for the high- and medium-quality recycled aggregate concretes. In some cases, the bond strength of the concrete with the high- and medium-quality recycled coarse aggregate replacement was higher than the bond strength of the natural aggregate concrete (8.34% and 5.27% higher for the high-quality and the five medium-quality recycled coarse aggregate concrete, respectively, with a water-to-cement ratio of 0.30).

3.3.2. Bond Stress-Slip Relation

There was a clear difference in the bond strength between the deformed steel rebar and concrete made with natural and recycled aggregates of different qualities.

With the 13-mm-diameter steel rebar, the majority of the concrete specimens did not fail at the ultimate load, and the post-peak behavior of the bond stress-slip curve could be observed. The average bond stress–strain relationships between the steel rebar and concrete, made with different water-to-cement ratios are plotted in Figure 14a–d. The bond behavior between the steel rebar and the concrete with natural and recycled aggregate and with different water-to-cement ratios are highly similar, except for the concrete with a water-to-cement ratio of 0.30. Although the ultimate bond strengths of the HC30 and the MC30 specimens were higher than that of the N30, the bond stresses of the HC30 and MC30 specimens decreased rapidly in the post-peak region and became lower than that of N30. In contrast, the ultimate bond strength of LC30 was the lowest but became the largest in the post-peak region. When the water-to-cement ratio decreased, the stiffness in the elastic region tended to increase, resulting in a higher ultimate bond strength. The adhesive strength between the concrete and steel rebar also increased with an increase in the water-to-cement ratio, but the magnitude was small, compared to the anchorage mechanism of the rib of the deformed bars.

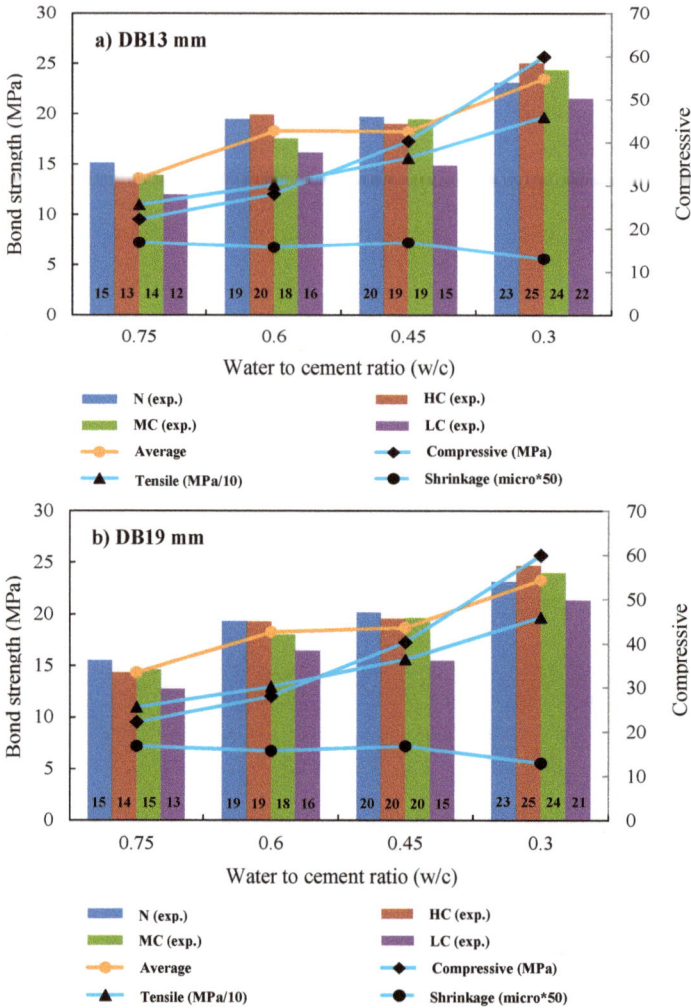

Figure 13. Bond strength between the steel rebar and concrete with different water-to-cement ratios and bar diameters. (**a**) DB13 mm and (**b**) DB19 mm.

With the 19-mm-diameter steel rebar, all of the specimens broke at the maximum tensile load. Therefore, the post-peak behavior of the bond between the concrete and the steel rebar could not be obtained for this diameter. The experimental results are shown in Figure 15a–d. The slope of the bond stress-slip curve was steeper for the lower water-to-cement ratios. The slip value of specimens with the low-quality recycled aggregate was larger, and a clear difference could be observed for the case when the water-to-cement ratio was 0.45. However, although the bond strength of the specimens with a water-to-cement ratio of 0.45 was nearly the same as that of the specimens with a water-to-cement ratio of 0.60, there was no significant improvement in strength.

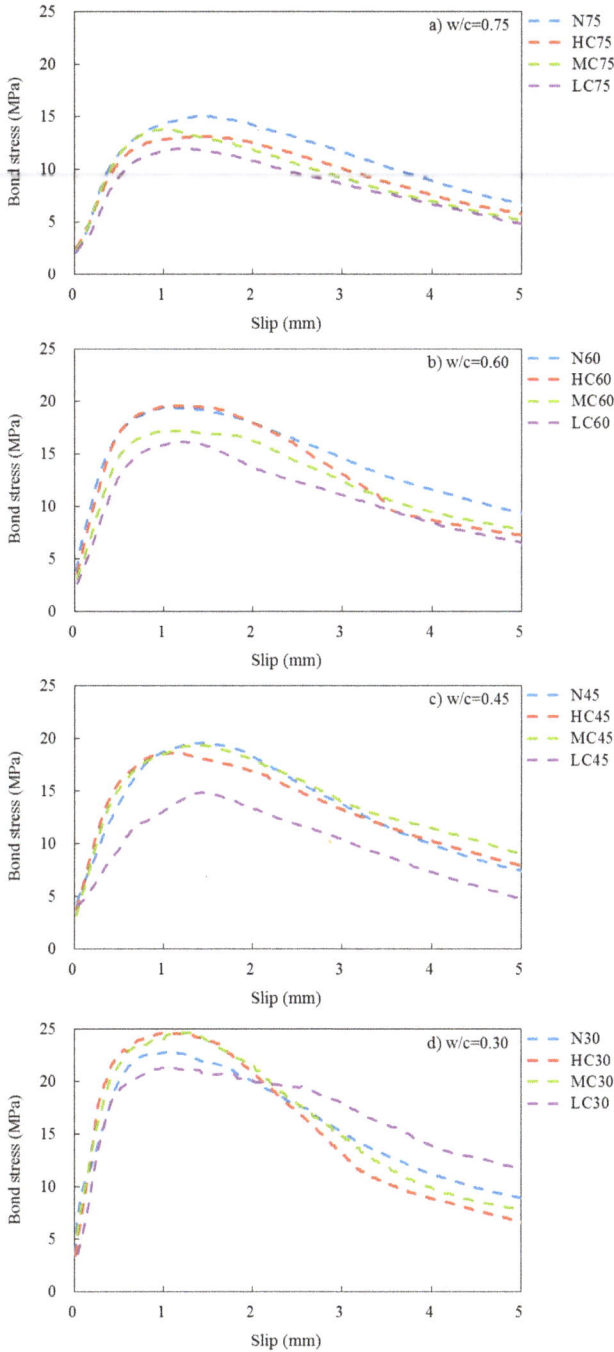

Figure 14. Bond stress-slip relationship of the concrete with a 13-mm-diameter steel rebar with different water-to-cement ratios. (**a**) For w/c = 0.75; (**b**) for w/c = 0.60; (**c**) for w/c = 0.45; (**d**) for w/c = 0.30.

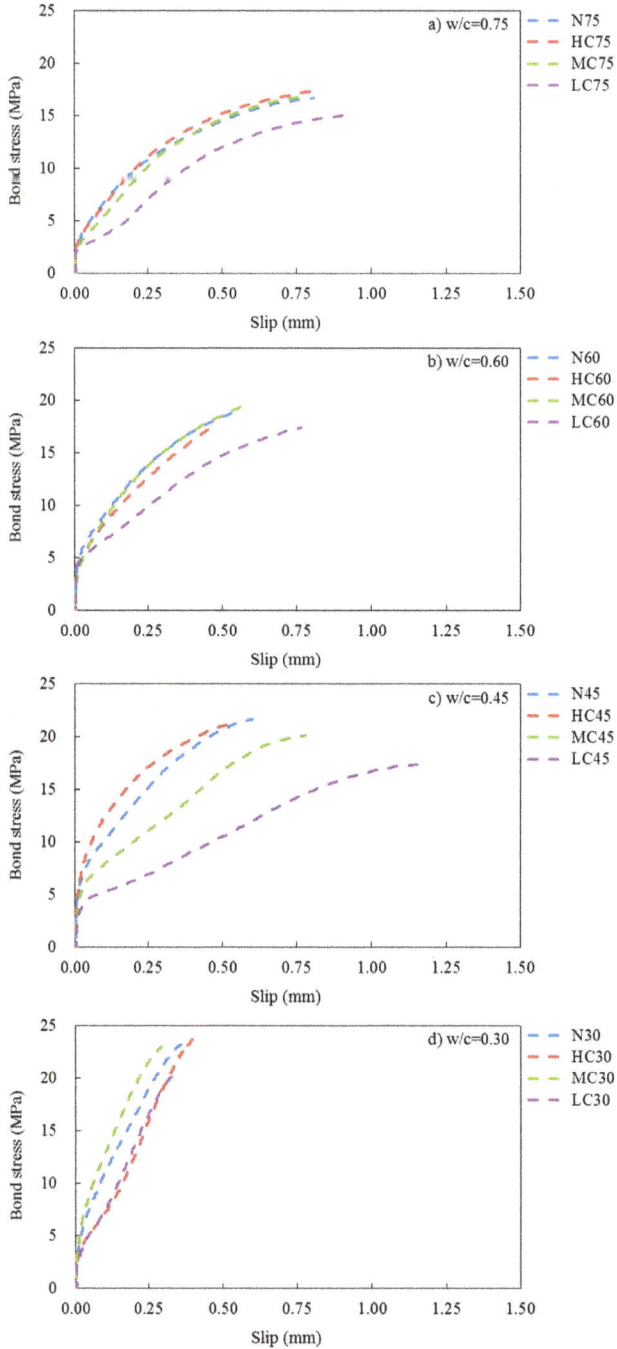

Figure 15. Bond stress-slip relationship of the concrete with a 19-mm-diameter steel rebar with different water-to-cement ratios. (**a**) For w/c = 0.75; (**b**) for w/c = 0.60; (**c**) for w/c = 0.45; (**d**) for w/c = 0.30.

3.4. Shear Behavior

3.4.1. Shear Failure Mechanism

All of the beam specimens hade similar failure modes, exhibiting an initial flexural crack mainly in the pure bending region and subsequent flexural cracks, away from the middle span. As the load increased, the flexural cracks extended into the diagonal cracks within the shear span and typically at the mid-height of the beam. After the formation of large diagonal cracks, the load dropped slightly and the deflection increased considerably. Then, brittle failure occurred, which was consistent with the literature. Fonteboa et al. [22] reported that the shear failure of the RC beams, without shear reinforcement was extremely brittle, and the final failure occurred after the formation of a large diagonal crack. Table 4 compares the experimental load of the large and small RC beams at the occurrence of the first crack, the first diagonal crack, the first peak, and the final peak. The failure modes of the large and small RC beams are shown in Figures 16 and 17, respectively. The loads at which the first large diagonal crack occurred for the RC beams with different recycled aggregate qualities are shown in Figure 16. Load-deflection relationships of the large and small RC beams were illustrated in Figures 18 and 19, respectively. For the small RC beam, the loads at which the first large diagonal crack occurred for NS60, HS60, MS60, and LS60 are 79.1, 71.7, 72.6, and 60 kN, respectively, corresponding to the load-deflection relationship of the small RC beams, as illustrated in Figure 19. The shear behavior of the RC beams with natural and recycled aggregate concrete was highly similar. There are no significant differences between the failure modes, cracking patterns, or shear performance of the RC beams, using the natural or recycled aggregate concrete. The final beam failure due to shear occurred after the formation of a large diagonal crack, except for the beam NB60. Immediately before reaching the maximum shear load, several beam specimens exhibited several vertical cracks that form at the beam top and/or in the shear compression zone.

Table 4. Comparison of the experimental load of the reinforced concrete (RC) beam shear testing.

Notation	Experimental Load (kN)			
	First Crack	**First Diagonal Crack**	**First Peak**	**Final Peak**
NB60	70	90	91	93
HB60	50	68	101	102
MB60	51	64	87	110
LB60	30	50	84	99
NS60	43	54	79	103
HS60	38	51	72	88
MS60	29	45	-	73
LS60	22	53	60	75

(a)

Figure 16. *Cont.*

(b)

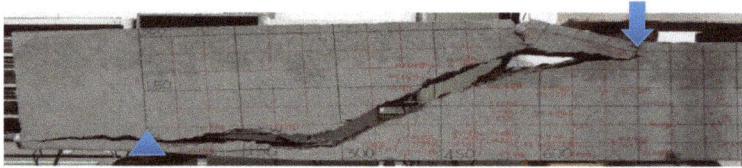

(c)

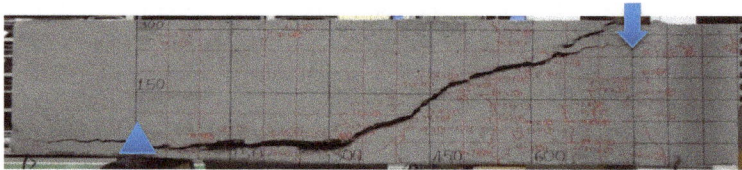

(d)

Figure 16. Failure mode of the large reinforced concrete (RC) beams. (**a**) Beam NB60; (**b**) beam HB60; (**c**) beam MB60; and (**d**) beam LB60.

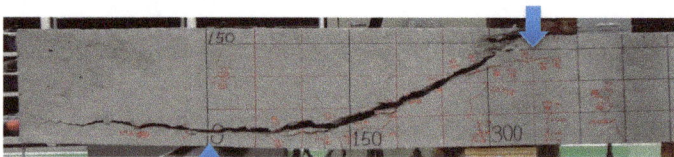

(a)

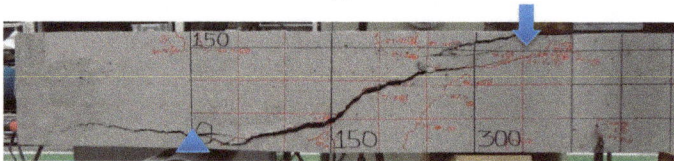

(b)

(c)

Figure 17. *Cont.*

(d)

Figure 17. Failure mode of the small RC beams. (**a**) Beam NS60; (**b**) beam HS60; (**c**) beam MS60; and (**d**) beam LS60.

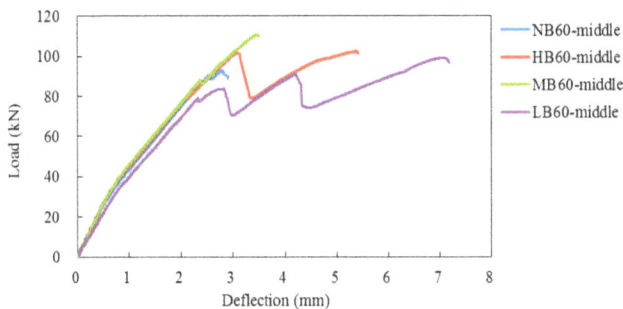

Figure 18. Load-deflection relationship for the large RC beams.

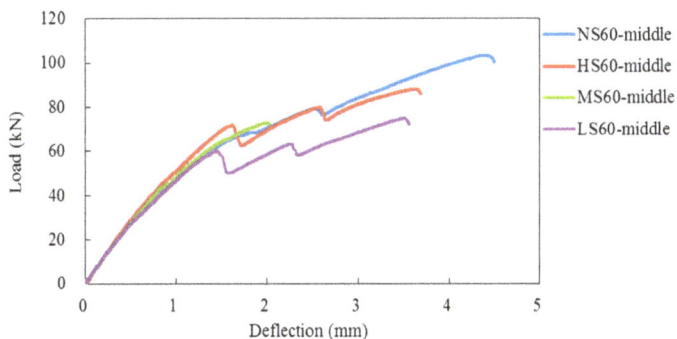

Figure 19. Load-deflection relationship for the small RC beams.

3.4.2. Effect of the Material Properties on the Shear Behavior

The ultimate loads of the shear strength of the RC beams from the experiment are shown in Figure 20 along with their material properties, including (1) the concrete compressive strength from the cylinder specimen (100 mm in diameter and 200 mm in height), (2) the mortar compressive strength from the cylinder specimen (50 mm in diameter and 100 mm in height), and (3) the static and (4) dynamic elastic moduli of the concrete. The total maximum shear strength of the large beams, NB60, HB60, MB60, and LB60, was 93, 102, 110, and 99 kN, respectively. The total maximum shear strengths of all of the large beams, using the recycled aggregate, were higher than that of the NB60 natural aggregate concrete beam, without any attenuation in the observed shear strength. In addition, the shear strength of MB60 was the highest among the group and 7.8% higher than that of HB60. This high shear strength could be due to the 4% and 4.9% higher compressive strengths of the concrete and mortar, respectively, compared to those of HB60. The concrete compressive strength of HB60 was 32 MPa, 6.2% higher than that of NB60 at 30.2 MPa, even though the compressive strength of the NB60 mortar was slightly higher. The natural aggregate had a relatively smooth surface and round shape, which weakened the bond between the cement matrix and the aggregate. For the small beams, the beam width was kept constant at 150 mm, and the effective depth decreased. The influence of the recycled aggregate could be clearly observed in the small beams, even though the concrete compressive strengths of HS60, MS60, and LS60 were 12.4%, 12.6%, and 5.7% higher than that of NS60, respectively.

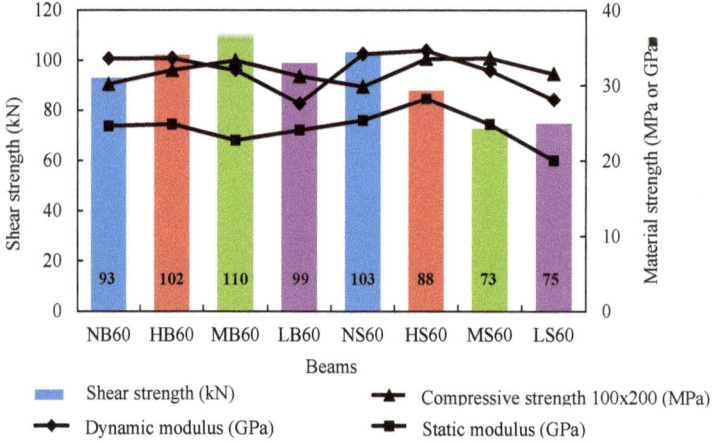

Figure 20. Shear strength of the RC beams and material properties of the concrete.

3.4.3. Size Effects on Shear Behavior

The shear strength of the RC beams increased with increases in the beam width and/or effective depth. However, the increase in the shear strength was not proportional to the increase in the beam size, even though the shear span-to-effective depth ratio was constant, as shown in Table 5, which presents the experimental data of the shear strength test of the RC beam, without shear reinforcement, using a natural aggregate (Taylor [23]). The beam size doubled by maintaining a shear span-to-effective depth ratio (*a*/*d*) of 3.0 and a reinforcement ratio (*ρ*) of 1.35%. Taylor reported that the average shear stress of the half-sized beams increased by a factor of 1.14 or 1.21, compared to that of the original beams. In contrast, the current experimental results indicated that the average shear stress of the small beams was 1.7 times higher than that of the large beams, regardless of the recycled aggregate quality. Dowel action contribution with the different *ρ* might have caused a difference of the shear strength between the small and large beams. The ultimate shear strength increased as the beam size increased. However, the shear stress calculated by Equation (2) was almost identical. In addition, the v_u of the small specimens was typically higher than that of the large specimens. In the JIS, the ultimate shear strength of the RC beam due to diagonal shear failure was calculated using Equation (3).

$$v_u = \frac{V_u}{b_w d} \tag{2}$$

$$V_u = 0.2 f_c'^{1/3} (100 p_w)^{1/3} \left(\frac{10^3}{d}\right)^{1/4} \left(0.75 + \frac{1.4d}{a}\right) b_w d \tag{3}$$

where v_u was the ultimate shear stress (MPa), V_u was the ultimate shear strength due to diagonal shear failure (N), f_c' was the compressive strength of concrete (MPa), $p_w = A_s / b_w d$ was the longitudinal tensile reinforcement ratio, d and b_w were the effective depth and width of the rectangular cross-section beam (mm), respectively, and a was the shear span (mm).

Table 5. Experimental results of the shear strength of the RC beam using natural aggregate concrete as given in Taylor [23].

No	Section (mm)			f'_c (MPa)	$V_{u,exp}$ (kN)	$V_{u,com}$ (kN)	$V_{u,exp}/V_{u,com}$	v_u	
	b	d	h					(MPa)	Average
A1	400	930	1000	29.2	358.4	313.71	1.14	0.84	0.82
A2				25.5	328.4	300.10	1.09	0.81	
B1	200	465	500	27.2	104.2	91.09	1.14	0.98	0.99
B2				24.9	87.3	88.44	0.99	0.95	
B3				32.1	85.3	96.26	0.89	1.04	
C1	100	232	250	25.6	22.5	26.50	0.85	1.14	1.13
C2				25.6	24.0	26.50	0.91	1.14	
C3				27.5	27.5	27.14	1.01	1.17	
C4				20.8	22.5	24.72	0.91	1.07	
C5				22.4	27.0	25.34	1.07	1.09	
C6				28.8	27.5	27.56	1.00	1.19	

Table 6 shows the experimental results of the shear strength of the RC beams, using a natural and recycled coarse aggregate concrete, without the shear reinforcement. The effective depth of the large beams was two times larger than that of the small beams. However, the large and small beams had the same cross-sectional area of reinforcement, resulting in half the reinforcement ratios for large beams, at 1.32%, versus 2.65% for the small beams. The shear strengths obtained experimentally were typically smaller than the estimated values, which ranged from 0.85 to 1.27.

Table 6. Experimental results of the shear strength of the RC beams using a natural and recycled aggregate concrete, without web reinforcement.

Notation	Section (mm)			a/d	ρ (%)	f'_c (MPa)	$V_{u,exp}$ (kN)	$V_{u,com}$ (kN)	$V_{u,exp}/V_{u,com}$	v_u	
	b	d	h							(MPa)	Average
NB60	150	300	325	1.32		30.17	46.39	54.45	0.85	1.03	1.12
HB60						32.03	51.05	55.55	0.92	1.13	
MB60						33.33	55.10	56.29	0.98	1.22	
LB60				2.5		31.21	49.45	55.07	0.90	1.10	
NS60		150	175	2.65		29.86	51.55	40.65	1.27	2.29	1.88
HS60						33.56	43.90	42.27	1.04	1.95	
MS60						33.62	36.30	42.29	0.86	1.61	
LS60						31.55	37.35	41.41	0.90	1.66	

For the large beams, the ultimate shear strength was consistent with the compressive strength of concrete, regardless of the effect of the recycled aggregate quality. However, the current estimation equation appeared to overestimate the shear strength of the large beam for all types of aggregate. Although the specimens had enough anchorage length and the local rebar slip at the anchorage region was not observed during the loading test, the splitting cracks as shown in Figure 16, might have affected the shear strength.

For the small beams, the ultimate shear strength of the natural aggregate increased from 46.39 to 51.55 kN, whereas the shear strength of all specimens made of recycled aggregate decreased to 43.90, 36.30, and 37.35 kN, for the high-, medium-, and low-quality recycled aggregate, respectively, irrespective of the compressive strength. The effects of the recycled aggregate quality could be clearly observed among the small beams. The aggregate interlocking mechanism had a greater influence on the shear strength of the smaller beam sizes. Similarly, the shear strength of the RC beams with different specimen sizes, shear span-to-effective depth ratios, and recycled aggregate types, found in the literature, are shown in Table 7. The recycled aggregate used in each experiment was classified, based on its density and water absorption, using the JIS.

Table 7. Shear strength of the RC beam using a recycled aggregate concrete, from the literature.

References	Section (mm)		a/d	ρ (%)	f'_c (MPa)	$V_{u,exp}$ (kN)	$V_{u,com}$ (kN)	$V_{u,exp}/V_{u,com}$	Class
	b	d							
Ji et al. [24]	170	270	2.0	1.10	31.6	66.65	60.79	1.09	N
	170	270	2.0	1.10	39.66	59.98	65.01	0.92	H
Kim et al. [12]	200	300	2.5	1.90	31.8	75.5	83.36	0.91	N
	200	450	2.5	1.90	31.8	106.9	112.92	0.95	N
	200	600	2.5	1.90	31.8	125.9	140.19	0.90	N
	300	450	2.5	1.90	31.8	156.7	171.41	0.91	N
	400	600	2.5	1.90	31.8	256.4	280.38	0.91	N
	200	300	2.5	1.90	34.9	72.9	85.98	0.85	H
	200	450	2.5	1.90	34.9	96.4	116.47	0.83	H
	200	600	2.5	1.90	34.9	125.1	144.61	0.87	H
	300	450	2.5	1.90	34.9	159.8	176.81	0.90	H
	400	600	2.5	1.90	34.9	256.6	289.21	0.89	H
Ikegawa et al. [25]	200	260	2.31	1.10	32.5	71.9	64.9	1.11	N
	200	260	2.31	1.10	34.8	57.15	66.40	0.86	L
	200	260	2.31	1.10	44.1	65.4	71.85	0.91	L
	200	260	2.31	1.10	42.4	81.4	70.92	1.15	L
Fathifazl et al. [11]	200	309	2.59	1.62	34.4	110.1	81.68	1.35	N
	200	300	1.50	1.00	43.5	195.7	95.99	2.04	M
	200	300	2.00	1.50	43.5	174.1	94.65	1.84	M
	200	305	3.93	2.46	43.5	106.3	86.22	1.23	M
	200	300	1.50	1.00	36.9	187.2	90.87	2.06	L
	200	300	2.00	1.50	36.9	165.6	89.60	1.85	L
	200	309	2.59	1.62	36.9	104.0	83.65	1.24	L
	200	305	3.93	2.46	36.9	83.8	81.62	1.03	L

3.5. Serviceability

Structural design, particularly for buildings, must consider the serviceability of the structures by satisfying the allowable deflection criteria, even though the ultimate limit state was generally verified in advance. The diversity of serviceability conditions could not all be handled by simply limiting the flexibility of the structures. Each individual condition must be addressed by the designer, for specific performance requirements.

For beams NB60, HB60, MB60, and LB60, the mid-span deflections corresponding to the ultimate load were 2.80, 5.37, 3.49, 7.08 mm, respectively. The corresponding mid-span deflection was not consistent with the shear strength. However, the shear strength of the RC beams made of recycled aggregate concrete was higher than that of the natural aggregate concrete beam, but the deflections of the high-, medium-, and low-quality recycled aggregate concrete beams were 1.92, 1.25, and 2.53 times higher, respectively. Therefore, careful consideration should be given when designing concrete structures using a low-quality recycled aggregate.

A better relationship between the shear strength and corresponding mid-span deflection could be achieved for the small beams. A higher shear strength was associated with a larger mid-span deflection. Beam MS60 had a 29.6% lower shear strength and a 54.9% lower deflection, than the beam NS60, indicating the improved performance of the medium-quality recycled aggregate, despite its lower shear strength. In contrast, beams HS60 and LS60 had 17.6% and 20.4% lower deflections and 14.8% and 27.5% lower shear strengths, respectively, than the beam NS60, which was nearly the same rate as that for the reduction in strength.

4. Discussions

Based on the results of the investigation of the effects of recycled aggregate quality on bond and shear behaviors of RC beams, using a recycled aggregate concrete, the following points can be discussed.

Concrete made with lower quality recycled aggregate always has a lower slump due to its higher water absorption capacity and excessive moisture loss from the aggregate, during a one-day drying period, before the concrete is mixed, resulting in a compressive strength comparable with natural concrete, without any attenuation in strength.

The bond strength of the concrete with the low-quality recycled coarse aggregate is always the lowest, but the bond strength deterioration could be ignored for the high- and medium-quality recycled aggregate concrete. In some cases, the bond strength of the concrete with the high- and medium-quality recycled coarse aggregate replacement is higher than that of the natural aggregate concrete. In addition, the bond between the concrete and different diameters of deformed rebar (13 and 19 mm) have highly similar bond behaviors and ultimate bond stresses, when the embedment length is maintained at the same ratio of four times the steel diameter.

Shear strength of recycled concrete beams decreases as the effective depth decreases. In contrast, the ultimate shear strength of natural concrete beams increases, even for smaller beam size. The estimated shear strength of all beams are larger than the experimental results, except for beam NS60 and HS60, having $V_{u,exp}/V_{u,com}$ of 1.27 and 1.04, respectively. The effect of high-quality of recycled aggregate on shear strength should not be considered. In this study, the shear strength of the large beams—HB60, MB60, and LB60—were found to be 10%, 17.5%, and 6.6% higher than the strength of beam NB60, respectively. This could be explained by a higher compressive strength of concrete with a recycled aggregate. However, the shear strength of small beams, HS60, MS60, and LS60 was 14.8%, 30.5%, and 27.5% lower than the strength of beam NS60, respectively. It could be concluded that the recycled aggregate qualities have a large effect on the shear behavior of small size RC beams, due to an aggregate interlock mechanism and this mechanism was negligible for large-size RC beams. However, further research is needed to investigate the effect of aggregate interlock for recycled concrete on the shear strength of the RC beam. In addition, Etxeberria et al. [26] found that the shear strength of RC beams with 25, 50, and 100% recycled coarse replacements was 3.7% higher, 15.4% lower, and 16.0% lower, respectively, than that of the RC beam made of natural aggregate concrete. Therefore, the recycled aggregate quality (medium- and low-quality) had an approximately two-fold larger effect than the 50% recycled aggregate replacement. In contrast, the effect of the high-quality recycled aggregate was almost equal to that of the 50% recycled aggregate replacement.

5. Conclusions

The following conclusions could be drawn based on the results of the current investigation, considering the effects of recycled aggregate quality on the material properties, bond behavior, and the shear strength of RC members, using a recycled aggregate concrete.

(1) Regarding the properties of fresh concrete with different recycled aggregate replacements and qualities, especially concrete with a recycled fine aggregate, the target workability could not be achieved easily, due to the various physical and mechanical properties of recycled aggregates. For instance, due to a higher fineness modulus of high- and medium-quality recycled fine aggregates, compared with the low-quality recycled fine aggregate, concrete requires approximately, a 3% greater mix of water, than concrete made by a low-quality recycled fine aggregate. As a result, a slump of fresh concrete made with a lower quality recycled fine aggregate has a lower slump, even though all coarse and fine aggregates are presoaked before the concrete mix. In contrast, concrete made with a lower quality recycled coarse aggregate has a comparatively higher slump. It might be due to the excessive water on the aggregate surface that could not be dried out completely by pieces of cloth. To overcome this, all coarse aggregates are dried out automatically one day before the concrete mix. As a result, the workability is almost the same, regardless of the effect of recycled aggregate qualities. However, all slump values are approximately 18 cm higher than the target value of 12 cm.

(2) The compressive strength of concrete with different recycled aggregate replacements and qualities is normally lower than the strength of natural concrete, followed by the flexural strength,

split tensile strength, static and dynamic elastic modulus, and the hardened concrete density. Even though the compressive strength of a high-quality recycled coarse aggregate concrete was 14% higher than that of natural concrete, only a little strength improvement could be observed for other material properties. In general, the strength of a high-quality recycled aggregate concrete is the highest followed by a medium- and low-quality recycled aggregate concrete. The dynamic elastic modulus is always lower for the lower quality recycled aggregate, irrespective of recycled aggregate replacements. Drying shrinkage of recycled concrete is generally higher than that of natural concrete.

(3) Bond strength of both natural and recycled concrete increased with a decrease in water to cement ratio but there was no strength development, when water to cement ratio decreased from 0.60 to 0.45. It could be explained by a sharp drop in density increment, which has a tendency to increase, and almost the same drying shrinkage and dynamic elastic modulus of the concrete for a water to cement ratio of 0.60 and 0.45, even though the compressive strength, split tensile strength, static elastic modulus, and the compressive strength of mortar increased smoothly with a decrease in water to cement ratio. High- and medium-quality recycled coarse aggregate have a very little effect on the bond strength of concrete and are sometimes higher. However, the bond strength of low-quality recycled coarse aggregate was always the lowest one. The variation among specimens for bond stress was very large, compared with the material properties test.

(4) Shear behavior of the RC beam with a natural and recycled concrete is very similar. The final beam failure due to shear occurs after the formation of a large diagonal crack. Just before, the maximum load, some beam specimens have several vertical cracks that form at the beam top or at the shear compression zone. The effects of recycled aggregate qualities could be revealed for the case of a small size beam. It might be due to a comparable effect of aggregate interlock mechanism and this effect is too small for the large-sized RC beams.

Further research is needed to conduct the numerical simulation for understanding the bond behavior and shear strength of RC specimens, with the recycled aggregates. It is also necessary to conduct an additional experiment for estimating the long-term structural performance of RC members, using the recycled aggregates.

Author Contributions: Conceptualization, M.A.; Methodology, T.Y. and T.M.K.; Validation, M.A.; Investigation, H.I., Writing-Original Draft Preparation, T.Y. and T.M.K.; Writing-Review & Editing, H.I. and M.A.; and Supervision, M.A.

Funding: This work was supported by JSPS KAKENHI Grant Numbers JP 16KK0152 and 16H04403. The opinions and conclusions presented in this paper are only those of the authors.

Conflicts of Interest: The authors declare no conflict of interest.

References

1. Klee, H. The Cement Sustainability Initiative. World Business Council for Sustainable Development 2009, Switzerland. Available online: http://ficem.org/pres/csi_co2_accounting_and_reporting_protocol.pdf (accessed on 8 April 2013).
2. Akbarnezhad, A.; Ong, K.C.G.; Tam, C.T.; Zhang, M.H. Effects of the parent concrete properties and crushing procedure on the properties of coarse recycled concrete aggregates. *J. Mater. Civ. Eng.* **2013**, *25*, 1795–1802. [CrossRef]
3. Padmini, A.K.; Ramamurthy, K.; Mathews, M.S. Influence of parent concrete on the properties of recycled aggregate concrete. *Constr. Build. Mater.* **2009**, *23*, 829–836. [CrossRef]
4. Sérifou, M.; Sbartaï, Z.M.; Yotte, S.; Boffoué, M.O.; Emeruwa, E.; Bos, F. A study of concrete made with fine and coarse aggregates recycled from fresh concrete waste. *J. Constr. Eng.* **2013**, *2013*, 317182. [CrossRef]
5. Abdullahi, M. Effect of aggregate type on compressive strength of concrete. *Int. Civ. Struct. Eng.* **2012**, *2*. [CrossRef]
6. Zhao, J.L.; Yu, T.; Teng, J.G. Stress-strain behavior of FRP-confined recycled aggregate concrete. *J. Compos. Constr.* **2015**, *19*. [CrossRef]

7.   Xu, J.J.; Chen, Z.P.; Xiao, Y.; Demartino, C.; Wang, J.H. Recycled aggregate concrete in FRP-confined columns: A review of experimental results. *Compos. Struct.* **2017**, *174*, 277–291. [CrossRef]

8.   Creazza, G.; Russo, S. A new model for predicting crack width with different percentages of reinforcement and concrete strength classes. *Mater. Struct.* **1999**, *32*, 520–524. [CrossRef]

9.   Xiao, J.; Falkner, H. Bond behavior between recycled aggregate concrete and steel rebar. *Constr. Build. Mater.* **2007**, *21*, 395–401. [CrossRef]

10.   Kim, S.W.; Yun, H.D. Influence of recycled coarse aggregates on the bond behavior of deformed bars in concrete. *Eng. Struct.* **2013**, *48*, 133–143. [CrossRef]

11.   Fathifazl, G.; Razaqpur, A.G.; Isgor, O.B.; Abbas, A.; Fournier, B.; Foo, S. Shear strength of reinforced recycled concrete beams without stirrups. *Mag. Concr. Res.* **2009**, *61*, 477–490. [CrossRef]

12.   Kim, S.W.; Jeong, C.Y.; Lee, J.S.; Kim, K.H. Size effect in shear failure of reinforced concrete beams with recycled aggregate. *J. Asian Archit. Build. Eng.* **2013**, *12*, 323–330. [CrossRef]

13.   Sato, R.; Maruyama, I.; Sogabe, T.; Sogo, M. Flexural behavior of reinforced recycled concrete beams. *J. Adv. Concr. Technol.* **2007**, *5*, 43–61. [CrossRef]

14.   Yamao, H.; Chou, L.; Niwa, J. Experimental study on bond stress slip relationship. *Jpn. Soc. Civ. Eng.* **1984**, *343*, 219–228. [CrossRef]

15.   Hong, S.; Park, S.K. Uniaxial bond stress-slip relationship of reinforcing bars in concrete. *Adv. Mater. Sci. Eng.* **2012**, *12*. [CrossRef]

16.   Japan Society of Civil Engineers (JSCE). Standard Specification for Design and Construction of Concrete Structures (Standards), August 1991. Available online: http://www.jsce.or.jp/committee/concrete/e/JGC15_Standard%20Specifications_Design_1.0.pdf (accessed on 7 August 2013).

17.   Syroka-Korol, E.; Tejchman, J. Experimental investigations of size effect in reinforced concrete beams failing by shear. *Eng. Struct.* **2014**, *58*, 63–78. [CrossRef]

18.   Viso, J.R.D.; Carmona, J.R.; Ruiz, G. Shape and size effects on the compressive strength of high-strength concrete. *Cem. Concr. Res.* **2008**, *38*, 386–395. [CrossRef]

19.   Tavakoli, M.; Soroushian, P. Drying shrinkage behavior of recycled aggregate concrete. *Concr. Int.* **1996**, *18*, 58–61.

20.   Kou, S.C.; Poon, C.S.; Chan, D. Influence of Fly ash as cement replacement on the properties of recycled aggregate concrete. *J. Mater. Civ. Eng.* **2007**, *19*, 709–717. [CrossRef]

21.   Morohashi, N.; Sakurada, T.; Yanagibashi, K. Bond splitting strength of high-quality recycled coarse aggregate concrete beams. *J. Asian Archit. Build. Eng.* **2007**, *6*, 331–337. [CrossRef]

22.   Fonteboa, B.G.; Abella, F.M. Shear strength of recycled concrete beams. *Constr. Build. Mater.* **2007**, *21*, 887–893. [CrossRef]

23.   Taylor, H.P.J. Shear strength of large beams. *J. Struct. Div. ASCE* **1972**, *98*, 2473–2490.

24.   Ji, S.K.; Lee, W.S.; Yun, H.D. Shear Strength of Reinforced Concrete Beams with Recycled Aggregates. Tailor Made Concrete Structures-Walraven & Stoelhorst (eds). 2008. Available online: http://www.abece.com.br/web/restrito/restrito/Pdf/CH171.pdf (accessed on 20 September 2013).

25.   Ikegawa, T.; Saito, H.; Ohuchi, H.; Kitoh, H.; Tsunokake, H. Flexural and Shear Failure Tests of Reinforced Concrete Beams with Low Grade Recycled Aggregate. 2009. Available online: http://core.ac.uk/download/pdf/35261895.pdf (accessed on 5 January 2014).

26.   Etxeberria, M.; Marí, A.R.; Vázquez, E. Recycled aggregate concrete as structural material. *Mater. Struct.* **2007**, *40*, 529–541. [CrossRef]

applied
sciences

MDPI

Article

Can We Truly Predict the Compressive Strength of Concrete without Knowing the Properties of Aggregates?

Jorge de Brito [1,*], Rawaz Kurda [1] and Pedro Raposeiro da Silva [2]

[1] Instituto Superior Técnico, Universidade de Lisboa, Av. Rovisco Pais, 1049-001 Lisbon, Portugal; rawaz.saleem@gmail.com

[2] Instituto Superior de Engenharia de Lisboa, R. Conselheiro Emídio Navarro, 1950-062 Lisbon, Portugal; silvapm@dec.isel.ipl.pt

* Correspondence: jb@civil.ist.utl.pt

Received: 29 May 2018; Accepted: 29 June 2018; Published: 5 July 2018

Abstract: This paper is focused on the influence of the geological nature and quality of the aggregates on the compressive strength of concrete and explains why it is important not to ignore the characteristics of aggregates in the estimation of the strength of concrete, even for virgin aggregates. For this purpose, three original (Abrams, American Concrete Institute Manual of concrete practice and Slater) and two modified (Bolomey and Feret) models were used to calculate the strength of concrete by considering results of various publications. The results show that the models do not properly predict the strength of concrete when the characteristics of aggregates are neglected. The scatter between the calculated and experimental compressive strength of concrete, even when made with natural aggregates (NAs) only, was significant. For the same mix composition (with similar cement paste quality), there was a significant difference between the results when NAs of various geological nature (e.g., limestone, basalt, granite, sandstone) were used in concrete. The same was true when different qualities (namely in terms of density, water absorption and Los Angles abrasion) of aggregates were used. The scatters significantly decreased when the mixes were classified based on the geological nature of the aggregates. The same occurred when the mixes were classified based on their quality. For both modified models, the calculated strength of mixes made with basalt was higher than that of the mixes containing other types of the aggregates, followed by mixes containing limestone, quartz and granite. In terms of the quality of the aggregates, the calculated strength of concrete increased (was overestimated) as the quality of the aggregates decreased. The influence of the aggregates on the compressive strength of concrete became much more discernible when recycled aggregates were used mainly due to their more heterogeneous characteristics.

Keywords: compressive strength; models; geological nature of aggregates; quality of aggregates; concrete; recycled aggregates

1. Introduction

1.1. General Facts on Compressive Strength Estimation

Concrete is one of the most consumed materials in the construction industry. Apart from durability, consumers normally demand a target strength. The target strength can be achieved in different ways, e.g., by increasing the cement content, incorporating a specific amount of supplementary cementitious materials (SCM), and decreasing the water content or introducing a superplasticizer (S_P). For that purpose, researchers have proposed several formulas to calculate the strength of concrete [1,2].

Researchers normally focus on the quality of the cement paste to calculate the strength of concrete. However, the strength of concrete not only depends on the strength of the hydrated product and the porosity of the cement paste [3] but also on the properties of the aggregates [4]. Further details regarding the mentioned factors are provided in the following paragraphs.

The strength of the hydrated product in the cement paste depends on the chemical and physical properties of cement [5,6]. The porosity of concrete depends on the air content that is mainly affected by the maximum size of the aggregates, and the particle size distribution in the used materials, mixing procedure, workability, placing, and compaction [4].

The properties of the aggregates directly or indirectly affect the strength of concrete. The influence of the aggregates' properties can be clearly seen in high strength concrete, because failure occurs through the aggregates. For that purpose, a previous study [7] attempted to classify aggregates into four classes (A, B, C, and D) based on their physical properties, mainly by considering their density and water absorption (WA), and mechanical properties (Los Angeles (LA) abrasion) from the results of 116 studies (Table 1). The authors believe the strength of concrete produced with class A (high quality of aggregates) should be higher than that of those with B class, followed by classes C and D.

Table 1. Physical boundaries for each proposed class of aggregates [7].

Aggregate Class	A			B			C			D
	I	II	III	I	II	III	I	II	III	
Minimum oven-dried density (kg/m^3)	2600	2500	2400	2300	2200	2100	2000	1900	1800	No limit
Maximum water absorption (%)	1.5	2.5	3.5	5	6.5	8.5	10.5	13	15	
Maximum LA abrasion mass loss (%)	40			45			50			

In our study, the strength of concrete made with ordinary Portland cement (OPC) and natural aggregates (NA) was calculated using different formulas (Section 1.2) and compared with the actual (experimental) strength. Then, the strength of the NA concrete was classified based on the quality of the aggregates. The first objective of this procedure was to show the importance of the quality of the aggregates rather than their procurement source (e.g., recycled aggregates (RA) versus NA) and confirm the relationship between the strength of concrete and the properties of the aggregates, regardless of the type of virgin aggregates used.

Regarding the indirect effects of the aggregates, studies have concluded that the quality of the hydration products affects the bond between the aggregates and cement paste (Interfacial Transition Zone—ITZ) [8]. Therefore, some of the recent investigations (Section 1.2) have considered the cement content and strength class of cement to calculate the strength of concrete. However, other studies [9] that updated the Bolomey formula [10] realized that, for the same quality of cement paste, the compressive strength of concrete made with rolled natural aggregates (NA) differs from that of concrete made with crushed NA, when both come from the same natural sources. This is because the shape of aggregates also affects the ITZ [11].

In fact, for concrete made with crushed aggregates, the effective water/cement ratio (main contributor to the quality of the cement paste and of the ITZ) should be increased in order to obtain the same workability as concrete made with rolled aggregates [12]. However, all the formulas proposed until then to calculate the strength of concrete failed to consider one of the most effective factors that affects the bond between aggregates and cement paste (ITZ), which is the texture of aggregates. The texture of the aggregates depends on the geological nature of the aggregates (e.g., limestone, basalt, sandstone, granite). For example, basalt can be considered to be a high-quality aggregate due to its high density and high mechanical strength, but its bond with the cement paste may be not strong enough due to its very smooth texture which is relatively similar to that of glass [11]. Thus, the calculated strength may be overestimated. In terms of chemical reactivity, the study of Kong and Du [11] showed that the amount of OH^- that is absorbed by the aggregates and that releases Si^{4+} at the same time depends on the geological nature of the aggregates, and this process significantly affects the pore structure of the hydration product and the ITZ as a result.

Therefore, this study attempted to show the effect of the geological nature of the aggregates on the ultimate strength of concrete and how the scatter between the estimated and the actual strength of concrete decreases if the results take the geological nature of the virgin aggregates into account.

Generally, the geological nature of aggregates affects the ultimate strength of concrete in different ways. When the strength of aggregates is lower than that of the cement paste, the quality of the aggregates controls the ultimate strength, because failure occurs though the aggregates. However, when the strength of the aggregates is higher than that of the cement paste, failure may not only occur in the cement paste itself but also in the ITZ between the aggregates and the cement paste [13]. In fact, the quality of the aggregates needs to be considered a main factor in the estimation of the compressive strength for low and high strength concrete mixes. It is well known that failure normally occurs in the ITZ between the aggregates and the cement paste because the effective water to cement ratio (w/c) of the cement paste around the aggregate particles is higher than that of the other parts of the cement paste [14]. This phenomenon relates to the capillary absorption between cement paste and aggregates which is affected by the WA and density of the aggregates.

Since the effect of SP, binder content and its characteristics, the w/c and aggregate content and the parameters described in [15,16] on concrete have already been studied, this study focuses on the effects of the quality and geological nature of NA on the compressive strength of concrete when all the mentioned parameters are constant, mainly by considering two modified models shown in the next sub-section. Thus, in terms of the aggregates, it is advisable to focus on the two mentioned parameters rather than only on the source of the aggregates, e.g., NA versus RA. To that purpose, this study considered the results of 84 publications for concrete made with virgin aggregates and 41 other publications (Section 2) for concrete made with RA.

1.2. Background of the Compressive Strength Estimation Models

Feret [1] is considered one of the pioneers in devising a formula to calculate the strength of concrete based on the volume of aggregates to the cement ratio and the void index. He was followed by Abrams [2] who calculated the strength of concrete by taking into account the w/c ratio. A study of Hicks [17] simplified Slater's formula [18] and reported that the strength of concrete linearly changes with the w/c ratio. Similarly to Slater's formula, ACI 2000-I [19] calculated the strength of concrete by considering the effect of the w/c ratio.

Powers and Brownyard [3] concluded that the porosity of the cement paste affects the strength of concrete. Thus, they developed the Feret's formula by considering another factor, namely, the "gel (volume of hydration products) to space (capillary porosity)" ratio. Then, Karni [20] added another factor (degree of hydration) to the Powers and Brownyard's formula. Regarding the porosity of the cement paste, Popovics [21] introduced the influence of air content into Bolomey's formula. Thirteen years afterwards, the same researcher [22] worked on Abrams's formula and added the effects of cement's hardening rate and the air content. Since the compressive strength of concrete with the same w/c may vary with its cement content, Popovics [23] then added another factor (cement content) to Abrams's formula. However, the number of concrete samples considered to calibrate these factors and the new model were not significant compared to the previous model. Furthermore, de Larrard [13] updated Feret's formula by considering the maximum paste thickness, the maximum size of the aggregates, the aggregates' packing density, the content of pozzolanic binders and a few types of aggregates. However, de Larrard did not consider the properties of cement, and the number of concrete samples used to calculate the factors was not significant. A recent study [15] proposed an innovative way to calculate the strength of concrete mixes using the M5P model tree algorithm by taking into account the results of nine publications. However, the cement class (e.g., CEM I 35.5 or 42.5) and quality of the aggregates were not considered as main factors in the model.

According to the parameters considered in each of the models suggested by the above researchers, two common facts can be seen. First, some of the formulas calculate the strength of concrete by only taking into account the w/c ratio. However, it is well known that the strength of two concrete mixes with the same w/c may not be equal when their cement contents are different [23]. Therefore, the volume of aggregates to cement ratio and void index (water and air), which depend on the maximum size of the aggregates and the water content, should be considered. Secondly, most formulas fail to consider the strength class of cement. For example, the strength of concrete mixes made with the same cement content is not similar when different cement classes (e.g., CEM I 32.5 or 42.5) are used [24].

The majority of current formulas follow those of Feret [1] and Abrams [2]. In the past decades, a few researchers have attempted to update the mentioned formulas, but their results are still not reliable, since they only considered a few case studies (concrete samples) to calibrate the correction factors. However, Sika-comp recently released a new software (SIKA-mix design) to design concrete mixes by using Faury's [25] method and calculate the strength of concrete by updating Feret's [1] formula. Thus, the new Feret-based formula depends on the ratio of the volume of aggregates to cement, the void index, the strength class of cement (cement class), and the maximum size of aggregates. In addition, a study [9] proposed a modification of Bolomey's [10] formula by taking into account the w/c ratio, the strength class of cement and the aggregates' production method (e.g., crushed or natural). Thus, the two updated formulas (Feret and Bolomey) were used as the main models to analyse the results.

2. Methodology

Three original models (Abrams, Slater and ACI) (Table 2) and two modified models (Bolomey and Feret) (Tables 2 and 3) were used to estimate the compressive strength at 28 days of concrete made with ordinary Portland cement (OPC) and NA. The mentioned formulas were applied to 206 concrete mixes (206 concrete mixes × minimum 3 samples = 618 concrete samples) made with 100% NA sourced from 84 studies (Table 4). After that, the results were classified based on the aggregates' classes (Table 1) proposed by Silva et al. [7]. Then, the geological nature of the aggregates (e.g., limestone, basalt, granite, sandstone) was considered to analyse the results.

Additionally, for each analysis (by considering the aggregates' quality or geological nature), different studies (Table 4) were considered based on the supplementary information given in the publication. For example, the studies that did not provide the geological nature of the aggregates (Table 4—5th row) were still considered in the analysis of the effect of the quality of the aggregates on the compressive strength of concrete when their WAs and densities were provided. In the first stage of this analysis, all the concrete mixes used were made with NA and OPC (without any SCM). In the second stage, some examples (Table 4) containing 100% coarse RA were studied.

Table 2. Formulas to calculate the compressive strength of concrete.

Model	Formula [a],[b]	Notes/Limitations
Abram	$fcm = \frac{A}{B^{w/c}}$	The constants A and B are 96 MPa and 7, respectively. The w/c ratio should be between 0.3 and 1.2
Slater	$fcm = 0.007 \cdot \left(2700 \cdot \frac{C}{w} - 760\right)$	This formula can be used only for mixes without S_P
American Concrete Institute Manual of Concrete Practice (ACI2000-I)	$fcm = 117.07 \cdot e^{-2.572 \cdot w/c}$	The w/c ratio and cement content of the concrete mixes are limited to 0.41–0.82 and 300–360 kg/m^3, respectively. The equation was adapted from the results of the table given in the specification with $R^2 = 0.996$
Bolomey	$fcm = A1 \cdot \left(\frac{1}{w/c} - 0.5\right)$ if $w/c > 0.4$ $fcm = A2 \cdot \left(\frac{1}{w/c} + 0.5\right)$ if $w/c \leq 0.4$	The constants (A1 and A2) depend on the way aggregates are produced and on the strength class of cement (Table 3)
Feret	$fcm = k \cdot \left(\frac{v}{v+I}\right)^2$	The constant, k, depends on the strength class of cement. It is 265–290 and 315–350 for cement class CEM I 32.5 and 42.5, respectively; v is the absolute volume of cement paste and I is the volume index (air and water contents)

[a] c and w are the cement and water contents, respectively. In other words, w/c is the water to cement ratio; [b] fcm is the average compressive strength of the concrete in cylinders (150 mm diameter × 300 mm length) at 28 days (MPa).

Table 3. Constants in Bolomey's formula.

Aggregate	w/c	Constants [a]	Strength Class of Cement (MPa)					
			25	35	40	45	52.5	55
Natural (rolled)	>0.40	A1	13.73	17.65	19.61	20.59	22.0675	22.56
Natural (rolled)	≤0.40	A2	9.32	11.7	12.75	14.22	14.5875	14.71
Natural (crushed)	>0.40	A1	15.2	19.61	21.57	23.54	25.01	25.5
Natural (crushed)	≤0.40	A2	10.3	13.24	14.22	15.69	16.7925	17.16

[a] The constant between the presented strength class was determined by interpolation and the average value of the rolled and crushed aggregates was considered if information on the way the aggregates were produced was not provided.

Table 4. Statistical data from the selected studies.

References [a]	Geological Nature	Aggregates						Cement	Mix Composition			Strength
		Physical Properties	Oven-Dried Particle Density (kg/m³)	WA-24h (%)	Properties	Max. Aggregate Size (mm)	Los Angeles (LA) Abrasion (%)	Cement Type/Strength	Cement Content (kg/m³)	w/c	Water Content (L/m³)	$f_{cm, cyl\ 150 \times 300}$ (MPa)
[26–35]	Basalt	AI	2716–2915	0.68–1.60	Crushed	14–20	17–24	CEM I 42.5–52.5	300–500	0.24–0.70	108–210	20–62
[26,27,31,33,34, 36–80]	Limestone	AI–BI	2277–2739	0.20–5.00	Crushed/ rounded	10–32	10–42	CEM I 32.5–52.5	210–677	0.28–0.86	126–266	10–69
[26,81–84]	Quartz	AI	2624–2780	0.17–1.5	Crushed/ rounded	16–32	24–36	CEM I 42.5	300–500	0.36–0.61	148–214	22–59
[85–87]	Sandstone	AI	2625–2660	0.50–0.94	Crushed	19–20	-	CEM I 42.5	250–463	0.40–0.60	153–185	23–56
[34,50,88–109]	Natural gravel (N.G.) [b]	AI–AII	2511–2719	0.05–2.5	Crushed/ rounded	15–25	20–30	CEM I 42.5	214–635	0.32–0.84	148–259	15–52

[a] The following studies [34–36,42,46–48,50,52–54,57,66,67,70,71,76–78,81,85,87,90,93–95,100,102–105,107,110–117] were considered to study the effects of quality of aggregates on the compressive strength of concrete made with 100% coarse recycled aggregates (RA); [b] N.G.—the geological nature of the natural aggregates (NA) is not given.

It is well known that the shapes and sizes of the concrete samples (e.g., cube or cylinder) affect the results. Therefore, this study took into account these two factors, and the experimental results (non-standard cylinder sizes) were converted to be equivalent to those of a cylinder with a 150 mm diameter × 300 mm length ($f_{cm, \text{cube } 100 \text{ mm}} \rightarrow f_{cm, \text{cube } 150 \text{ mm}} \rightarrow f_{cm, \text{cylinder } 150 \times 300} \leftarrow f_{cm, \text{cylinder } 100 \times 200}$) in order to compare them with the calculated strength. Thus, according to the state-of-the-art method described in FIB Bulletin 42 [118], the "$f_{cm, \text{cube } 150 \text{ mm}}$ to $f_{cm, \text{cube } 100 \text{ mm}}$" ratio is equal to 0.97. Similar results can be seen in other studies [119,120]. After that, $f_{cm, \text{cube } 150 \text{ mm}}$ was converted to $f_{cm, \text{cylinder } 150 \times 300}$, according to EN 1992-1-1:2004 (E). Regarding $f_{cm, \text{cylinder } 100 \times 200}$, the majority of the researchers agree that, up to 33 MPa, the difference between $f_{cm, \text{cylinder } 150 \times 300}$ and $f_{cm, \text{cylinder } 100 \times 200}$ is insignificant. The average difference between them is 2%, according to previous studies [121–123]. For higher strength values (over 33 MPa), the average "$f_{cm, \text{cylinder } 150 \times 300}$ to $f_{cm, \text{cylinder } 100 \times 200}$" ratio is 0.90 (Table 5).

As mentioned above, in this study, the strength of concrete made with OPC and NA was calculated using different formulas (Table 2) and compared with the actual (experimental) strength. Then, the strength of the NA concrete was classified based on the quality of the aggregates. The first objective of this procedure was to show the importance of the quality of the aggregates rather than their procurement source (e.g., recycled aggregates (RA) versus NA) and to confirm the relationship between the strength of concrete and the properties of the aggregates, regardless of the type of virgin aggregates used.

Table 5. The "$f_{cm, \text{cylinder } 150 \times 300}$ to $f_{cm, \text{cylinder } 100 \times 200}$" ratio in different studies.

Studies	$f_{cm, \text{cylinder } 150 \times 300}$ to $f_{cm, \text{cylinder } 100 \times 200}$ Ratio	Strength [MPa]
Malhotra [124]	0.84	~46
Forstie and Schnormeier [125]	0.87	~48
Carrasquillo et al. [126]	0.93	48–80
Lessaed and Aitcin [127]	0.95	35–122
Average over 33 MPa	0.90	

3. Results

Cement has been used as a construction material for many centuries. However, modern cement has been produced only in the last century. Due to the substantial changes made in cement production during the last decades, the quality of cement has significantly improved [128,129]. Therefore, the results of this study focus on the investigations published in the last two decades. Figure 1 shows the relationship between compressive strength and the w/c of concrete mixes made with only NA and OPC. The results show that, for a 95% confidence interval, there is a large amount of scatter between w/c and the compressive strength of concrete. This result was expected due to the quality of cement paste and aggregates. The effect of the quality of the cement paste on concrete strength has been extensively studied for decades, and several formulas have been suggested to relate these two properties. However, the effect of the quality of the aggregates on concrete strength had not previously been sufficiently studied. For this reason, this study focused on the effects of the physical and geological nature of the aggregates in the following sections.

3.1. Effect of the Geological Nature of Natural Aggregates on the Compressive Strength of Concrete

As mentioned in the literature review, the strength of concrete may be affected by different factors, including the quality of the cement paste. Therefore, the original models (Abrams, Slater, and ACI) calculated the strength of concrete based on the w/c (Figure 2). The results showed that the estimated strength of concrete based on the quality of the cement paste is not accurate, and the relationship between the actual and calculated strength in all of the original models was poor (R^2 was -0.31, -0.14 and 0.15 in the Abrams, Slater, and ACI models, respectively) when only w/c was considered to be an influencing factor. This is because the compressive strength of concretes with the same w/c may vary with cement content [23] and cement class [24], which respectively influence the aggregate: cement volume ratio and void index and the quality of the hydration product. The other reason is that the original formulas ignore the quality (e.g., geological nature) of the aggregates. Thus, the relationship between the calculated and actual strength of concrete mixes improved when the original formulas also considered the geological nature of the aggregates (R^2 was $-0.23, 0.03, -0.58, -0.05, 0.52$ and 0.46 with Abram's model, and $0.27, 0.20, -0.57, 0.20, 0.85$ and 0 with Slater's model, and $0.25, 0.40, -0.03, 0.23, 0.81, 0.84$ with the ACI model when the aggregates were classified as basalt, granite, limestone, NA (geological nature not given), quartz, and sandstone, respectively. The negative R^2 is due to the fact that the linear trendline was set to intercept the origin of the axes [130,131]). This behaviour is further discussed in the next paragraphs. Furthermore, for the 95% confidence intervals (red dashed lines in each graph), there is a big scatter between w/c and compressive strength of concrete, and the lower and upper k ($f_{cm, \text{calculated}}/f_{cm, \text{experimental}}$) values in all original models were 0.67 ± 0.02 and 1.67 ± 0.01, respectively.

As mentioned previously, the calculated strength based only on the w/c independently of the cement content and its properties, is not reliable. Therefore, this study essentially focused on the modified models (Bolomey and Feret) rather than the original models (Abrams, Slater, and ACI).

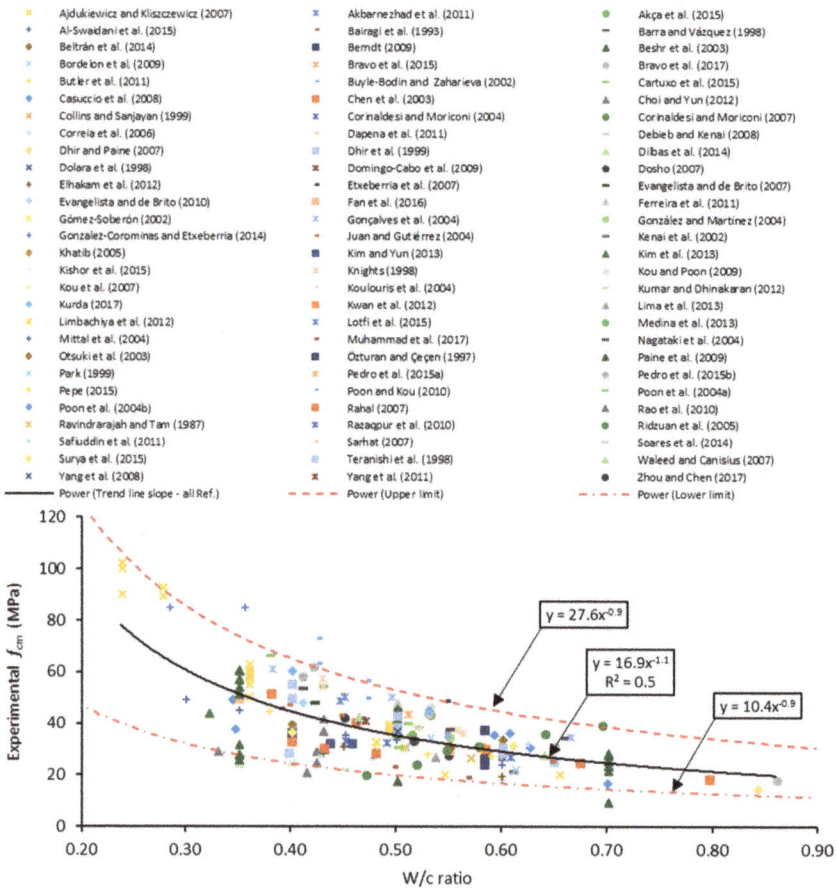

Figure 1. Relationship between the compressive strength "$f_{cm, \text{cylinder 150 mm x 300 mm length}}$" and w/c ratio of concrete.

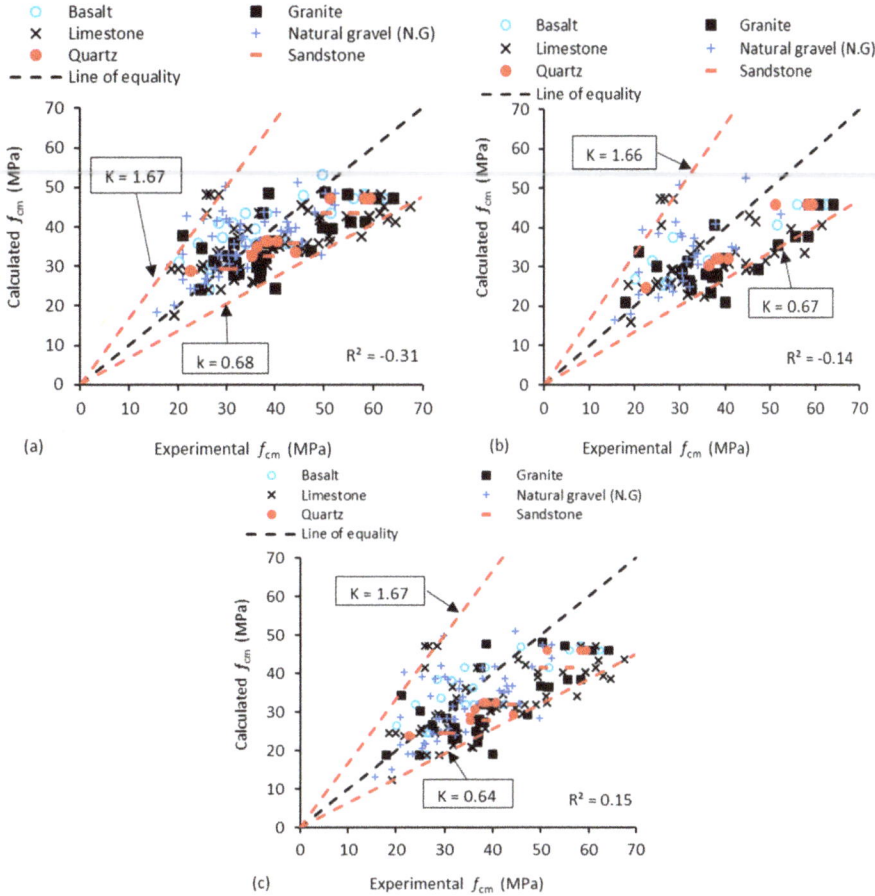

Figure 2. Relationship between the actual and calculated compressive strength of concrete mixes made with NA sourced from different geological natures. The strength values were calculated according to the (**a**) Abrams, (**b**) Slater and (**c**) ACI formulas. $k = (f_{cm, calculated} / f_{cm, experimental})$.

Figure 3a,b show the relationship between the experimental and calculated compressive strength obtained according to the Bolomey and Feret models, respectively. Relative to the original models (R^2 was up 0.15) that only considered the w/c as a main factor, the relationship between the calculated and actual compressive strength significantly improved ($R^2 \approx 0.50$), as seen in Figure 3a. This is due to the additional factors considered in Bolomey's model, namely the strength class of cement and the way that aggregates are produced (Table 2). However, this model neglected the effect of the cement content (volume of aggregates to cement ratio) which significantly affects the results due to its influence on the strength of concrete (for the same w/c, the strength may vary with the cement content [23]). As shown in Figure 3b, by using Feret's model, the relationship between the calculated and experimental strength of concrete significantly improved ($R^2 \approx 0.60$) relative to the original models. This is because of the factors considered in the model, namely, the cement strength and content. However, this model neglected the aggregates' production methods.

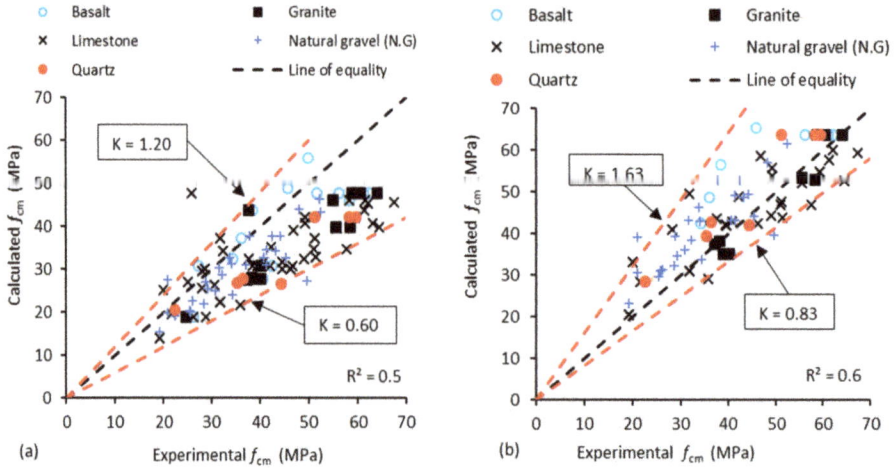

Figure 3. Relationship between the actual and calculated compressive strength of concrete mixes made with NA sourced from different geological natures. The strength values were calculated according to the (**a**) Bolomey and (**b**) Feret formulas. $k = (f_{cm, calculated}/f_{cm, experimental})$.

Even though the relationship between the actual and calculated strength improved by using the modified models (Bolomey and Feret), there was still a big scatter between the mixes (Figure 3). For example, for a 95% confidence interval, the k value ($f_{cm, calculated}/f_{cm, experimental}$) varied between 1.20–0.60 and 1.63–0.83 when the strength was calculated based on the Bolomey and Feret models, respectively. This may be related to the fact that both models neglected the properties of the aggregates as a factor to calculate the strength. To validate this assumption, the mixes were classified according to the geological nature of their aggregates, and the k value was found for each of them according to a 95% confidence interval (Figure 4). As a consequence, the relationship between calculated and actual strengths considerably improved in most cases. For example, the R^2 of concrete mixes made with basalt, granite, limestone, and quartz was about 0.60, 0.80, 0.40, and 0.80 when Bolomey's model was used to calculate the strength, and 0.40, 0.90, 0.60, 0.90 for Feret's model, respectively. Additionally, for both the Bolomey (Figure 4a) and Feret models (Figure 4b), the difference between upper (k1) and lower (k2) boundaries (k value = $f_{cm, calculated}/f_{cm, experimental}$) for a 95% confidence interval of the concrete mixes significantly decreased when the classification was based on the geological nature of the aggregates, except for concrete with limestone aggregates. This is because limestone has a large scatter of characteristics (Table 4), and it is classified in various generic categories, e.g., carboniferous, dolomitic, and calcareous, or it may be a composite (e.g., limestone–quartzitic, limestone–siliceous).

In order to simplify the results, the average of the k1 and k2 values (as seen in Figure 4) was considered to make Figure 5. According to both models, all other parameters being the same, the calculated/actual strength ratio of concrete mixes made with basalt was higher than that of mixes containing other types of the aggregates, followed by mixes containing limestone, quartz, and granite. This ranking, which was repeated for both models, is further explained in the following paragraphs. Furthermore, apart from the above factors, the shape ratio and shape regularity of the aggregates may also affect the strength of concrete. For example, less water content (one of the main contributors to the quality of the cement paste and ITZ) is required to obtain the target slump with rolled aggregates compared to that with angular aggregates [12]. However, the bond between angular aggregates and cement paste is stronger than that in the mixes made with rolled aggregates. However, the bond also depends on the surface texture of the aggregates.

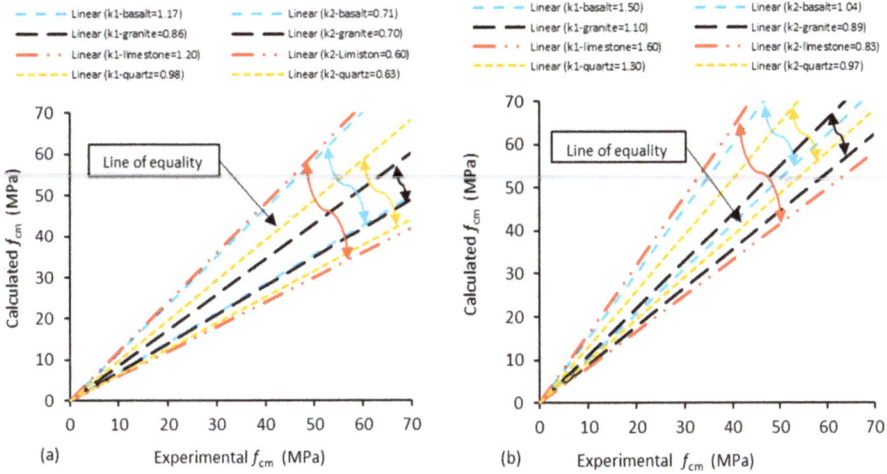

Figure 4. Relationship between the actual and calculated compressive strength of concrete mixes made with NA sourced from different geological natures. k1 and k2 are the upper and lower boundaries for a 95% confidence interval, respectively, according to the (**a**) Bolomey and (**b**) Feret models.

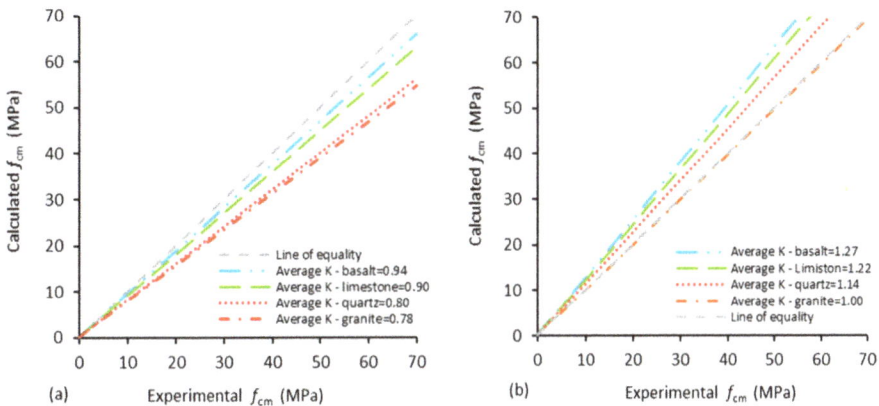

Figure 5. Relationship between the actual and calculated compressive strength of concrete mixes made with NA sourced from different geological natures. The average *k* value is the average of k1 and k2 from Figure 4, according to the (**a**) Bolomey and (**b**) Feret models. The standard deviation for the k values was ≈0.20 in both models.

Ingham et al. [132] studied the shape, form, and texture of various geological natures of aggregates using about 270 cases. Based on their explanations and the classification of Silva et al. [7] (Section 1), Figure 6 was drawn to show all of the factors that may affect the influence of the aggregate on the quality of concrete. The shape ratio and shape regularity of the aggregates also affects the quality of concrete, but they can be controlled by sieving the aggregates. However, the surface texture of the aggregates and their quality and chemical composition may not be easily controlled. Therefore, it is important to identify the geological nature of the aggregates, because it determines the texture [132] and chemical composition [11], and in most cases, the physical characteristics, shape ratio, and regularity.

Notwithstanding a significant standard deviation that may be affected by the selected database, in both models, the calculated/actual strength ratio of concrete containing basalt was generally higher than that of the other aggregates (Figure 5). This is because, generally, basalt has an aphanitic texture or, in some cases, can be considered to be glassy [133]. As a consequence, the bond between basalt aggregates and cement paste will be weaker and failure may occur at ITZ. In other words, the models that calculate strength based on the quality of cement paste (without considering the geological nature of the aggregates) may fail, and relative to mixes containing other aggregates, the strength is overestimated. Furthermore, the chemical composition of the aggregates (rich with silicon or calcium), which influences the way they absorb other elements from the cement paste (and vice-versa), may also affect the interfacial hydrates around the aggregates. For example, limestone is rich in calcium, and therefore, there is more ettringite and calcium–hydroxide in the ITZ, while for basalt, which is rich in silicon, further calcium silicate hydrates are generated in that zone [11].

Unlike basalt aggregates, granites generally have a phaneritic texture [133]. This explains the trends shown in Figure 5, where the calculated/actual strength ratio of concrete containing granite is lower than that of the mixes made with other aggregates. The rougher surface of the granite increases the bond between cement paste and aggregates. Although both basalt and granite have similar chemical compositions (rich in silicon content) because of their surface texture, the ultimate strength of concrete with each aggregate and the same cement paste is different. Furthermore, the surface texture of the aggregates, as well as their chemical composition, shape ratio and regularity, and quality (e.g., WA and density) also affect the ultimate strength of concrete, as discussed in the following sections.

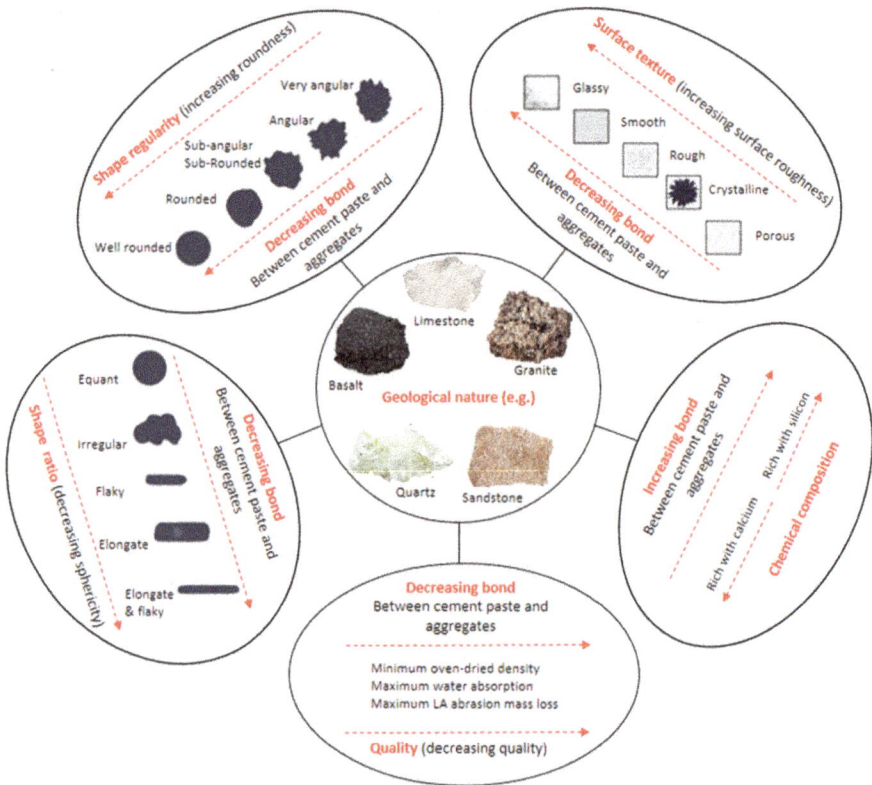

Figure 6. Factors that affect the influence of aggregates on concrete quality.

3.2. Effect of the Quality of Natural Aggregates on the Compressive Strength of Concrete

Concrete mixes were classified based on the aggregates' quality, namely their WA, density, and LA abrasion (Table 1), according to the study by Silva et al. [7]. As for the geological nature (Section 3.1), the results show that the relationship between the calculated and actual strength of concrete (without taking into account the aggregates' classes, the R^2 values for the linear trends were 0.50 and 0.60 for the Bolomey and Feret formulas, respectively) improved when the mixes were classified based on their quality (classes A, B, C, and D). Thus, the R^2 values for the linear trends were 0.72 and 0.62 for the Bolomey model and 0.60 and 0.74 for the Feret model when the mix's aggregates were from classes AI and AII (specified in Table 1), respectively (Figure 7). This shows that the quality of the aggregates may significantly affect the ultimate strength of concrete. Therefore, it is important to identify the quality of the aggregates simultaneously with the other factors shown in Figure 6. According to both models, the scatter between the results (the k values of the all mixes were 0.6–1.30 and 0.80–1.50 when the strength was calculated according to the Bolomey and Feret models, respectively) decreases when the mixes are classified in terms of the quality of their aggregates (Figure 7).

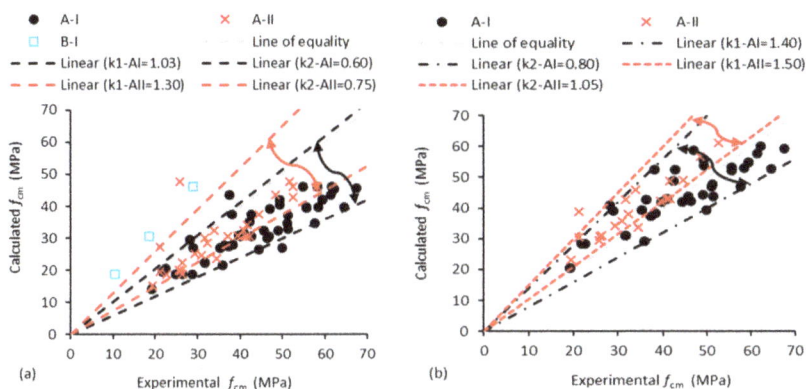

Figure 7. Relationship between the actual and calculated compressive strength of the concrete mixes made with NA, classified in terms of their quality. The physical quality of A-I aggregates is better than that of A-II aggregates, followed by B-I aggregates. k1 and k2 are the upper and lower boundaries for the 95% confidence intervals, respectively, according to the (**a**) Bolomey and (**b**) Feret models.

In order to simplify the results, the average of the k1 and k2 values (as seen in Figure 7) was considered to make Figure 8. According to both models, all other parameters being the same, the calculated/actual strength ratio of concrete increases as the quality of the aggregates (high WA and LA abrasion, and low density "Table 1") decreases. In low-strength concrete, failure normally happens at the ITZ. Apart from the factors discussed in Section 3.1, namely, the shape ratio and regularity, and the surface texture and chemical composition of the aggregates, this can be related to the fact that the w/c of the cement paste around the aggregates is higher than in other parts of the cement paste [14]. The quality of the aggregates, namely their WA and density, which both affect the capillary absorption between the aggregates and the cement paste, is a determinant of the ITZ bond strength. For high strength concrete, failure may not happen in the cement paste (the mechanical properties of the aggregates control concrete's ultimate strength). Therefore, again, the quality of the aggregates, namely their LA abrasion, controls the ultimate strength of concrete. Additionally, relative to the Feret model, the calculated strength using Bolomey's model is either close to the line of equality or below it. This may be related to the aggregates' production method (rolled or crushed) considered in the formula, which helps to prevent overestimation of the compressive strength of the aggregates.

This result also proves the importance of considering the properties of aggregates to calculate the strength of concrete.

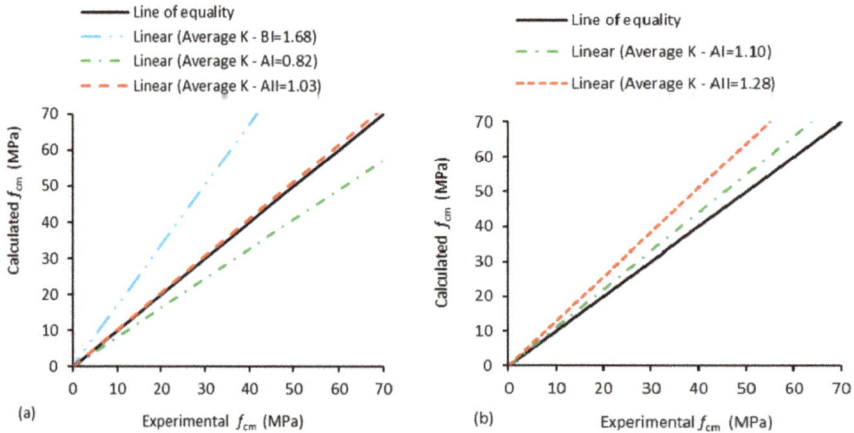

Figure 8. Relationship between the actual and calculated compressive strength of concrete mixes made with NA, classified in terms of their quality. The physical quality of the A-I aggregates is better than that of A-II, then followed by B-I. The average *k* value is the average of k1 and k2 from Figure 7, according to the (**a**) Bolomey and (**b**) Feret models. The standard deviation for the *k* values was ≈0.25 in both models.

In order to further show the importance of considering the quality of aggregates, the strength of concrete mixes made with various aggregates qualities (A, B, C, and D, as in Table 1) was drawn (Figure 9).

The results calculated with Bolomey's model came from 84 studies [26–79,81–109] of concrete mixes made with 100% coarse NA and 41 other studies [34–36,42,46–48,50,52–54,57,66,67,70,71,76–78, 81,85,87,90,93–95,100,102–105,107,110–115] of concrete made with 100% coarse RA. The *k* values show that, for a 95% confidence interval, the scatter between the mixes made with NA (mostly classed as A in Table 1) was slightly smaller than that of the mixes made with RA (generally classed as B–D) (Figure 9a). This big scatter of the NA mixes allows one important conclusion—it is conceptually established that the quality of RA affects the strength of concrete, and the same is true for NA, but on a smaller scale, because the differences in quality are also smaller. In other words, the quality of the NA needs to be considered to understand the performance of concrete made with them (Figure 9a). Figure 9b was drawn based on the upper and lower boundaries of Figure 9b. This figure confirms the abovementioned facts demonstrated for concrete mixes made with NA only—as the quality of the aggregates (regardless of the source, i.e., NA or RA) decreases (A to D), the calculated/actual strength ratio increases, i.e., existing formulas overestimate the compressive strength the most for lower quality aggregates.

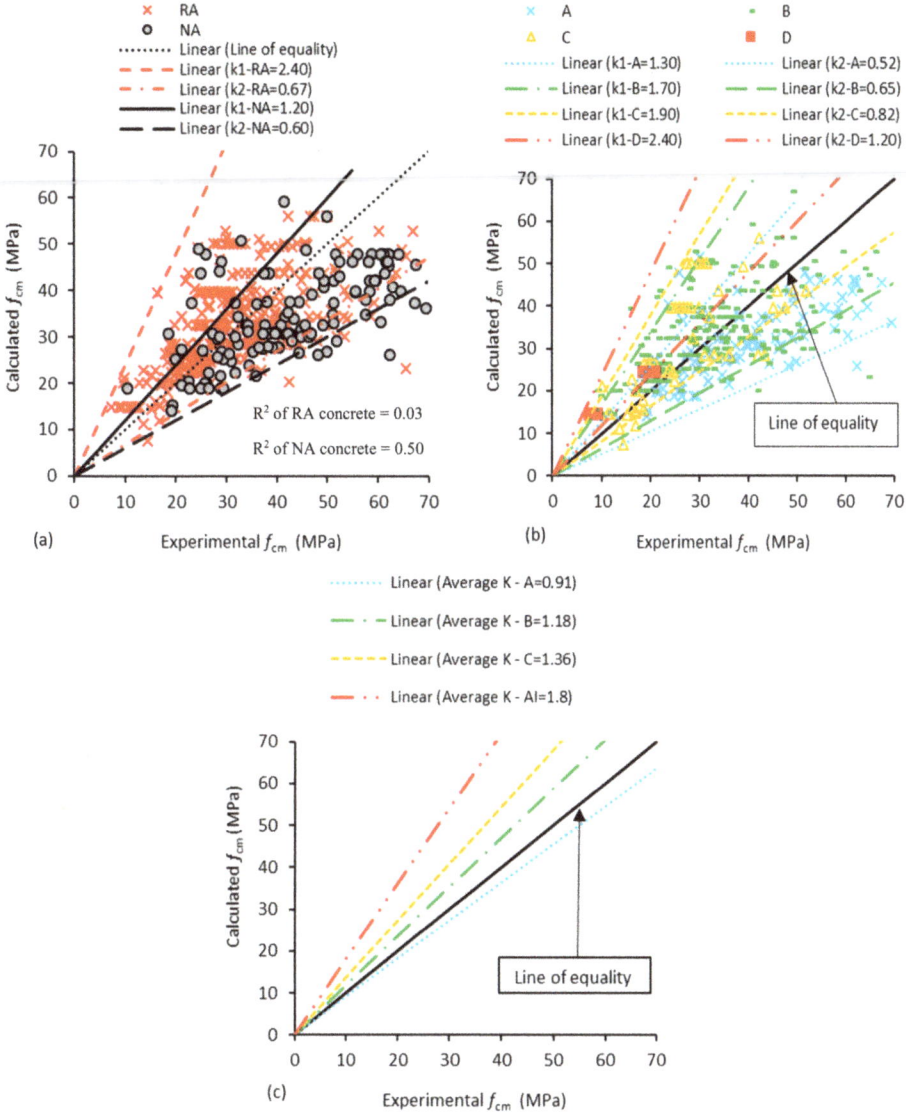

Figure 9. Relationship between the actual and calculated compressive strength of concrete mixes made with 100% NA and RA classified in terms of their quality. The strength was calculated according to Bolomey's model. The physical quality of A aggregates is better than B, followed by C and D. (**a**) *k* values for NA and RA concrete mixes, (**b**) k1 and k2 are the upper and lower boundaries for the 95% confidence intervals, and (**c**) the average *k* value is the average of k1 and k2 from Figure 9b. The standard deviation for the *k* values was ≈0.51 in (**c**).

4. Conclusions

This study allows a better understanding of the compressive strength of concrete made with aggregates of various geological natures (e.g., limestone, basalt, granite, and quartz) and qualities (WA, density, and LA abrasion). The following conclusions can be drawn from this study.

The original (Abrams, Slater, and ACI) and modified (Bolomey and Feret) models cannot accurately estimate the compressive strength of concrete when the characteristics of aggregates are neglected. In other words, the estimated strength of concrete based on the quality of cement paste only is not accurate, and the relationship between the actual and calculated strength for all the original models is poor when only w/c (with and without considering the volume of aggregates to the cement content) is considered as an influencing factor.

The scatter between the calculated and experimental compressive strength of concrete, even if made with 100% NA, is significant. In other words, for the same mix composition (similar cement paste quality), there are significant differences between the results when various geological natures of NA (e.g., limestone, basalt, granite, sandstone) are used in concrete. The same is true when various qualities (WA, density, and LA abrasion) of aggregates are used.

The modified models (Bolomey and Feret) work better than the original models (Abrams, Slater, and ACI) because they consider the strength class of cement as one of the factors. However, Bolomey's model does not consider the ratio of the volume of aggregates to cement paste, and Feret's model does not consider how the aggregates are obtained (e.g., rolled or crushed) as influencing factors.

In most cases, the scatter between concrete mixes measured by the difference between the upper (k1) and lower (k2) boundaries ($k = f_{cm, calculated} / f_{cm, experimental}$) for a 95% confidence interval significantly decreased when they were split based on the geological nature of the aggregates used. The same occurred when the mixes were split based on the quality of their aggregates.

According to both modified models (Bolomey and Ferret), the calculated/actual strength ratio of the concrete mixes made with basalt is higher than that of the mixes with aggregates of other geological natures—in descending order, limestone, quartz, and granite. In terms of the physical characteristics of the aggregates, the calculated/actual strength ratio of concrete increases as the quality (high WA, low density, and LA abrasion) of the aggregates decreases.

In order to further show the importance of considering the physical characteristics of aggregates, the strength of a few examples of concrete made with aggregates of various qualities (A, B, C, and D, according to Table 1), namely incorporating 100% coarse RA or NA, was calculated using Bolomey's model. For a 95% confidence interval, the scatter between the mixes made of NA (mostly classed as A) was slightly smaller than that of the mixes made with RA (generally classed as B–D), but still significant. This means that to estimate the compressive strength of concrete made with either NA or RA, the influence of the geological nature and physical quality cannot be ignored, and this should be included both in the standard formulas and in the codes.

Author Contributions: J.d.B. and R.K. conceived and designed the experiments; R.K. analyzed the data; P.R.d.S. contributed reagents/materials/analysis tools; J.d.B. and R.K. wrote the paper.

Acknowledgments: The authors appreciate the support of CERIS unit from IST-University of Lisbon and the Foundation for Science and Technology of Portugal.

Conflicts of Interest: The authors declare no conflict of interest. The founding sponsors had no role in the design of the study; in the collection, analyses, or interpretation of data; in the writing of the manuscript, and in the decision to publish the results.

References

1. Feret, R. On the compactness of hydraulic mortars (in French–Sur la compacité des mortiers hydrauliques). *Ann. Ponts Chaussées Sér.* **1892**, *7*, 5–164.
2. Abrams, L.D. *Properties of Concrete*, 3rd ed.; Pitman Publishing Ltd.: London, UK, 1919.
3. Powers, T.; Brownyard, T. *Studies of the Physical Properties of Hardened Portland Cement Paste*; American Concrete Institute (ACI): Farmington Hills, MI, USA, 1946; Volume 18, pp. 669–712.
4. Silva, R. Use of Recycled Aggregates from Construction and Demolition Wastes in the Production of Structural Concrete. Ph.D. Thesis, Universidade de Lisboa, Instituto Superior Técnico, Lisbon, Portugal, 2015.
5. Mechling, J.; Lecomte, A.; Diliberto, C. Relation between cement composition and compressive strength of pure pastes. *Cem. Concr. Compos.* **2009**, *31*, 255–262. [CrossRef]

6. Maruyama, I.; Igarashi, G. Cement reaction and resultant physical properties of cement paste. *J. Adv. Concr. Technol.* **2014**, *12*, 200–213. [CrossRef]

7. Silva, R.V.; de Brito, J.; Dhir, R.K. Properties and composition of recycled aggregates from construction and demolition waste suitable for concrete production. *Constr. Build. Mater.* **2014**, *65*, 201–217. [CrossRef]

8. Struble, L.; Skalny, J.; Mindess, S. A review of the cement-aggregate bond. *Cem. Concr. Res.* **1980**, *10*, 277–286. [CrossRef]

9. Brandt, M. *Optimization Methods for Material Design of Cement-Based Composites*; CRC Press. Boca Raton, FL, USA, 2014; 328p.

10. Bolomey, J. Granulation et prevision de la resistance probable des betons. *Travoux* **1935**, *19*, 228–232.

11. Lijuan, K.; Yuanbo, D. Interfacial interaction of aggregate-cement paste in concrete. *J. Wuhan Univ. Technol.-Mater. Sci. Ed.* **2015**, *30*, 117–121. [CrossRef]

12. Kurda, R.; de Brito, J.; Silvestre, J.D. Influence of recycled aggregates and high contents of fly ash on concrete fresh properties. *Cem. Concr. Compos.* **2017**, *84*, 198–213. [CrossRef]

13. De Larrard, F. *Concrete Mixture Proportioning: A Scientific Approach*; CRC Press: London, UK, 1999; p. 448.

14. Ollivier, J.P.; Maso, J.C.; Bourdette, B. Interfacial transition zone in concrete. *Adv. Cem. Based Mater.* **1995**, *2*, 30–38. [CrossRef]

15. Behnood, A.; Behnood, V.; Gharehveran, M.M.; Alyamac, K.E. Prediction of the compressive strength of normal and high-performance concretes using M5P model tree algorithm. *Constr. Build. Mater.* **2017**, *142*, 199–207. [CrossRef]

16. Chidiac, S.E.; Moutassem, F.; Mahmoodzadeh, F. Compressive strength model for concrete. *Mag. Concr. Res.* **2013**, *65*, 557–572. [CrossRef]

17. Hicks, T. *Civil Engineering Formulas*, 2nd ed.; McGraw-Hill Education-Europe: New York, NY, USA, 2009; 416p.

18. Slater, W.A. Relation of 7-day to 28-day compressive strength of mortar and concrete. *J. Proc.* **1926**, *22*, 437–449.

19. ACI 2000-I. *ACI Manual of Concrete Practice 2000, Part 1: Materials and General Properties of Concrete*; American Concrete Institute (ACI): Farmington Hills, MI, USA, 2000.

20. Karni, J. Prediction of compressive strength of concrete. *Mater. Struct.* **1974**, *7*, 197–200. [CrossRef]

21. Popovics, S. New formulas for the prediction of the effect of porosity on concrete strength. *ACI Mater. J.* **1985**, *82*, 136–146.

22. Popovics, S. History of a mathematical model for strength development of Portland cement concrete. *ACI Mater. J.* **1998**, *95*, 593–600.

23. Popovics, S.; Ujhelyi, J. Contribution to the concrete strength versus water-cement ratio relationship. *J. Mater. Civ. Eng.* **2008**, *20*, 459–463. [CrossRef]

24. Adewole, K.; Olutoge, F.; Habib, H. Effect of Nigerian Portland limestone cement grades on concrete compressive strength. *Int. J. Civ. Environ. Struct. Constr. Archit. Eng.* **2014**, *8*, 1199–1202.

25. Faury, J. *Le Béton*; Dunod: Paris, France, 1958. (In French)

26. Ajdukiewicz, A.B.; Kliszczewicz, A.T. Comparative tests of beams and columns made of recycled aggregate concrete and natural aggregate concrete. *J. Adv. Concr. Technol.* **2007**, *5*, 259–273. [CrossRef]

27. Al-Swaidani, A.; Baddoura, M.; Aliyan, S.; Choeb, W. Assesment of alkali resistance of basalt used as concrete aggregates. *SSP-J. Civ. Eng.* **2015**, *10*, 17–27. [CrossRef]

28. Bairagi, N.K.; Ravande, K.; Pareek, V.K. Behaviour of concrete with different proportions of natural and recycled aggregates. *Resour. Conserv. Recyc.* **1993**, *9*, 109–126. [CrossRef]

29. Collins, F.; Sanjayan, J.G. Strength and shrinkage properties of alkali-activated slag concrete containing porous coarse aggregate. *Cem. Concr. Res.* **1999**, *29*, 607–610. [CrossRef]

30. Dilbas, H.; Şimşek, M.; Çakır, Ö. An investigation on mechanical and physical properties of recycled aggregate concrete (RAC) with and without silica fume. *Constr. Build. Mater.* **2014**, *61*, 50–59. [CrossRef]

31. Kishore, I.; Mounika, L.; Prasad, C.; Krishna, B. Experimental study on the use of basalt aggregate in concrete mixes. *SSRG Int. J. Civ. Eng.* **2015**, *2*, 39–42.

32. Mittal, A.; Kaisare, M.; Shetti, R. Experimental study on use of fly ash in concrete. Use of SCC in a pump house at TAPP 3 & 4, Tarapur. *Indian Concr. J.* **2004**, *78*, 30–34.

33. Özturan, T.; Çeçen, C. Effect of coarse aggregate type on mechanical properties of concretes with different strengths. *Cem. Concr. Res.* **1997**, *27*, 165–170. [CrossRef]

34. Paine, K.; Collery, D.; Dhir, R. Strength and deformation characteristics of concrete containing coarse recycled and manufactured aggregates. In Proceedings of the 11th International Conference on Non-Conventional Materials and Technologies (NOCMAT 2009), Bath, UK, 6–9 September 2009; p. 9.

35. Rahal, K. Mechanical properties of concrete with recycled coarse aggregate. *Build. Environ.* **2007**, *42*, 407–415. [CrossRef]

36. Akbarnezhad, A.; Ong, K.C.G.; Zhang, M.H.; Tam, C.T.; Foo, T.W.J. Microwave-assisted beneficiation of recycled concrete aggregates. *Constr. Build. Mater.* **2011**, *25*, 3469–3479. [CrossRef]

37. Barra, M.; Vázquez, E. Properties of concretes with recycled aggregates: Influence of properties of the aggregates and their interpretation. In Proceedings of the International Symposium on Sustainable Construction: Use of Recycled Concrete Aggregate, London, UK, 11–12 November 1998; pp. 19–30.

38. Berndt, M.L. Properties of sustainable concrete containing fly ash, slag and recycled concrete aggregate. *Constr. Build. Mater.* **2009**, *23*, 2606–2613. [CrossRef]

39. Bordelon, A.; Cervantes, V.; Roesler, J.R. Fracture properties of concrete containing recycled concrete aggregates. *Mag. Concr. Res.* **2009**, *61*, 665–670. [CrossRef]

40. Bravo, M.; de Brito, J.; Pontes, J.; Evangelista, L. Performance of concrete made with recycled aggregates from Portuguese CDW recycling plants. *Key Eng. Mater.* **2015**, *634*, 193–205. [CrossRef]

41. Butler, L.; West, J.S.; Tighe, S.L. The effect of recycled concrete aggregate properties on the bond strength between RCA concrete and steel reinforcement. *Cem. Concr. Res.* **2011**, *41*, 1037–1049. [CrossRef]

42. Buyle-Bodin, F.; Hadjieva-Zaharieva, R. Influence of industrially produced recycled aggregates on flow properties of concrete. *Mater. Struct.* **2002**, *35*, 504–509. [CrossRef]

43. Carro-López, D.; González-Fonteboa, B.; de Brito, J.; Martínez-Abella, F.; González-Taboada, I.; Silva, P. Study of the rheology of self-compacting concrete with fine recycled concrete aggregates. *Constr. Build. Mater.* **2015**, *96*, 491–501. [CrossRef]

44. Cartuxo, F.; de Brito, J.; Evangelista, L.; Jiménez, J.R.; Ledesma, E.F. Rheological behaviour of concrete made with fine recycled concrete aggregates—Influence of the superplasticizer. *Constr. Build. Mater.* **2015**, *89*, 36–47. [CrossRef]

45. Casuccio, M.; Torrijos, M.C.; Giaccio, G.; Zerbino, R. Failure mechanism of recycled aggregate concrete. *Constr. Build. Mater.* **2008**, *22*, 1500–1506. [CrossRef]

46. Correia, J.; de Brito, J.; Pereira, A. Effects on concrete durability of using recycled ceramic aggregates. *Mater. Struct.* **2006**, *39*, 169–177. [CrossRef]

47. Dapena, E.; Alaejos, P.; Lobet, A.; Pérez, D. Effect of recycled sand content on characteristics of mortars and concretes. *J. Mater. Civ. Eng.* **2011**, *23*, 414–422. [CrossRef]

48. Debieb, F.; Kenai, S. The use of coarse and fine crushed bricks as aggregate in concrete. *Constr. Build. Mater.* **2008**, *22*, 886–893. [CrossRef]

49. Abd Elhakam, A.; Mohamed, A.E.; Awad, E. Influence of self-healing, mixing method and adding silica fume on mechanical properties of recycled aggregates concrete. *Constr. Build. Mater.* **2012**, *35*, 421–427. [CrossRef]

50. Etxeberria, M.; Vázquez, E.; Marí, A.; Barra, M. Influence of amount of recycled coarse aggregates and production process on properties of recycled aggregate concrete. *Cem. Concr. Res.* **2007**, *37*, 735–742. [CrossRef]

51. Evangelista, L.; de Brito, J. Mechanical behaviour of concrete made with fine recycled concrete aggregates. *Cem. Concr. Compos.* **2007**, *29*, 397–401. [CrossRef]

52. García-González, J.; Barroqueiro, T.; Evangelista, L.; de Brito, J.; de Belie, N.; Pozo, J.M.; Juan-Valdés, A. Fracture energy of coarse recycled aggregate concrete using the wedge splitting test method: Influence of water-reducing admixtures. *Mater. Struct.* **2017**, *50*, 120. [CrossRef]

53. Gómez-Soberón, J.M.V. Porosity of recycled concrete with substitution of recycled concrete aggregate: An experimental study. *Cem. Concr. Res.* **2002**, *32*, 1301–1311. [CrossRef]

54. Gonçalves, A.; Esteves, A.; Vieira, M. Influence of recycled concrete aggregates on concrete durability. In Proceedings of the International RILEM Conference on the Use of Recycled Materials in Buildings and Structures, Barcelona, Spain, 8–11 November 2004; pp. 554–562.

55. Gonzalez-Corominas, A.; Etxeberria, M. Properties of high performance concrete made with recycled fine ceramic and coarse mixed aggregates. *Constr. Build. Mater.* **2014**, *68*, 618–626. [CrossRef]

56. Knights, J. Relative performance of high quality concretes containing recycled aggregates and their use in construction. In Proceedings of the International Symposium on Sustainable Construction: Use of Recycled Concrete Aggregate, London, UK, 11–12 November 1998; pp. 275–286.

57. Kou, S.C.; Poon, C.S.; Chan, D. Influence of fly ash as cement replacement on the properties of recycled aggregate concrete. *J. Mater. Civ. Eng.* **2007**, *19*, 709–717. [CrossRef]

58. Kumar, P.; Dhinakaran, G. Effect of admixed recycled aggregate concrete on properties of fresh and hardened concrete. *J. Mater. Civ. Eng.* **2012**, *24*, 494–498. [CrossRef]

59. Kurda, R. Sustainable Development of Cement-Based Materials: Application to Recycled Aggregates Concrete. Ph.D. Thesis, Universidade de Lisboa, Instituto Superior Técnico, Lisbon, Portugal, 2017.

60. Kwan, W.H.; Ramli, M.; Kam, K.J.; Sulieman, M.Z. Influence of the amount of recycled coarse aggregate in concrete design and durability properties. *Constr. Build. Mater.* **2012**, *26*, 565–573. [CrossRef]

61. Lotfy, A.; Al-Fayez, M. Performance evaluation of structural concrete using controlled quality coarse and fine recycled concrete aggregate. *Cem. Concr. Compos.* **2015**, *61*, 36–43. [CrossRef]

62. Lima, C.; Caggiano, A.; Faella, C.; Martinelli, E.; Pepe, M.; Realfonzo, R. Physical properties and mechanical behaviour of concrete made with recycled aggregates and fly ash. *Constr. Build. Mater.* **2013**, *47*, 547–559. [CrossRef]

63. Muhammad, M.; Abdullah, W.; Abdul-Kadir, M. Post-fire mechanical properties of concrete made with recycled tire rubber as fine aggregate replacement. *Sulaimani J. Eng. Sci.* **2017**, *4*, 74–85. [CrossRef]

64. Pedro, D.; de Brito, J.; Evangelista, L. Influence of the crushing process of recycled aggregates on concrete properties. *Key Eng. Mater.* **2015**, *634*, 151–162. [CrossRef]

65. Pepe, M. A Conceptual Model for Designing Recycled Aggregate Concrete for Structural Applications. Ph.D. Thesis, University of Salerno, Fisciano, Italy, 2015.

66. Poon, C.; Kou, S. Effects of fly ash on mechanical properties of 10-year-old concrete prepared with recycled concrete aggregates. In Proceedings of the 2nd International Conference on Waste Engineering Management (ICWEM 2010), Shanghai, China, 13–15 October 2010; pp. 46–59.

67. Poon, C.; Shui, Z.; Lam, L. Effect of microstructure of ITZ on compressive strength of concrete prepared with recycled aggregates. *Constr. Build. Mater.* **2004**, *18*, 461–468. [CrossRef]

68. Poon, C.S.; Shui, Z.H.; Lam, L.; Fok, H.; Kou, S.C. Influence of moisture states of natural and recycled aggregates on the slump and compressive strength of concrete. *Cem. Concr. Res.* **2004**, *34*, 31–36. [CrossRef]

69. Ravindrarajah, R.; Loo, Y.H.; Tam, C.T. Recycled concrete as fine and coarse aggregates in concrete. *Mag. Concr. Res.* **1987**, *39*, 214–220. [CrossRef]

70. Razaqpur, A.G.; Fathifazl, G.; Isgor, B.; Abbas, A.; Fournier, B.; Foo, S. How to produce high quality concrete mixes with recycled concrete aggregate. In Proceedings of the 2nd International Conference on Waste Engineering Management (ICWEM 2010), Shanghai, China, 13–15 October 2010; pp. 11–35.

71. Ridzuan, A.; Ibrahim, A.; Ismail, A.; Diah, A. Durability performance of recycled aggregate concrete. In Proceedings of the International Conference on Global construction: Ultimate Concrete Opportunities: Achieving Sustainability in Construction, London, UK, 5–6 July 2005; pp. 193–202.

72. Safiuddin, M.; Alengaram, U.; Salam, M.; Jumaat, M.; Jaafar, F.; Saad, H. Properties of high-workability concrete with recycled concrete aggregate. *Mater. Res.* **2011**, *14*, 248–255. [CrossRef]

73. Sarhat, S. An experimental investigation on the viability of using fine concrete recycled aggregate in concrete production. In Proceedings of the International Conference on Sustainable Construction Materials and Technologies, Coventry, UK, 1–13 June 2007; pp. 53–57.

74. Soares, D.; de Brito, J.; Ferreira, J.; Pacheco, J. Use of coarse recycled aggregates from precast concrete rejects: Mechanical and durability performance. *Constr. Build. Mater.* **2014**, *71*, 263–272. [CrossRef]

75. Surya, M.; Rao, V.; Parameswaran, L. Mechanical, durability, and time-dependent properties of recycled aggregate concrete with fly ash. *ACI Mater. J.* **2015**, *112*, 653–662. [CrossRef]

76. Thomas, C.; Setién, J.; Polanco, J.A.; Alaejos, P.; de Juan, M.S. Durability of recycled aggregate concrete. *Constr. Build. Mater.* **2013**, *40*, 1054–1065. [CrossRef]

77. Yang, K.; Chung, H.; Ashour, A. Influence of type and replacement level of recycled aggregates on concrete properties. *ACI Mater. J.* **2008**, *105*, 289–296.

78. Yang, J.; Du, Q.; Bao, Y. Concrete with recycled concrete aggregate and crushed clay bricks. *Constr. Build. Mater.* **2011**, *25*, 1935–1945. [CrossRef]

79. Wang, Z.; Wang, L.; Cui, Z.; Zhou, M. Effect of recycled coarse aggregate on concrete compressive strength. *Trans. Tianjin Univ.* **2011**, *17*, 229–234. [CrossRef]
80. Kurad, R.; Silvestre, J.D.; de Brito, J.; Ahmed, H. Effect of incorporation of high volume of recycled concrete aggregates and fly ash on the strength and global warming potential of concrete. *J. Clean. Prod.* **2017**, *166*, 485–502. [CrossRef]
81. Kenai, S.; Debieb, F.; Azzouz, L. Mechanical properties and durability of concrete made with coarse and fine recycled aggregates. In Proceedings of the International Symposium on Sustainable Concrete Construction, Dundee, Scotland, UK, 9–11 September 2002; pp. 383–392.
82. Akça, K.R.; Çakır, Ö.; İpek, M. Properties of polypropylene fiber reinforced concrete using recycled aggregates. *Constr. Build. Mater.* **2015**, *98*, 620–630. [CrossRef]
83. González, B.; Martínez, F. Shear strength of concrete with recycled aggregates. In Proceedings of the International RILEM Conference on the Use of Recycled Materials in Buildings and Structures, Barcelona, Spain, 8–11 November 2004; pp. 619–628.
84. Medina, C.; de Rojas, M.I.S.; Frías, M. Properties of recycled ceramic aggregate concretes: Water resistance. *Cem. Concr. Compos.* **2013**, *40*, 21–29. [CrossRef]
85. Park, S. *Recycled Concrete Construction Rubble as Aggregate for New Concrete*; Study Report No 86; BRANZ: Wellington, New Zealand, 1999; p. 20.
86. Nagataki, S.; Gokce, A.; Saeki, T.; Hisada, M. Assessment of recycling process induced damage sensitivity of recycled concrete aggregates. *Cem. Concr. Res.* **2004**, *34*, 965–971. [CrossRef]
87. Teranishi, K.; Dosho, Y.; Narikawa, M.; Kikuchi, M. Application of recycled aggregate concrete for structural concrete: Part 3-Production of recycled aggregate by real-scale plant and quality of recycled aggregate concrete. In Proceedings of the International Symposium on Sustainable Construction: Use of Recycled Concrete Aggregate, London, UK, 11–12 November 1998; pp. 143–156.
88. Beltrán, M.G.; Barbudo, A.; Agrela, F.; Galvín, A.P.; Jiménez, J.R. Effect of cement addition on the properties of recycled concretes to reach control concretes strengths. *J. Clean. Prod.* **2014**, *79*, 124–133. [CrossRef]
89. Chen, H.J.; Yen, T.; Chen, K.H. The use of building rubbles in concrete and mortar. *J. Chin. Inst. Eng.* **2003**, *26*, 227–236. [CrossRef]
90. Choi, W.-C.; Yun, H.-D. Compressive behavior of reinforced concrete columns with recycled aggregate under uniaxial loading. *Eng. Struct.* **2012**, *41*, 285–293. [CrossRef]
91. Corinaldesi, V.; Moriconi, G. Recycling of concrete in precast concrete production. In Proceedings of the Sustainable Construction Materials and Technologies, London, UK, 11–13 June 2007; pp. 69–75.
92. Corinaldesi, V.; Moriconi, G. Concrete and mortar performance by using recycled aggregates. In Proceedings of the International Conference on Sustainable Waste Management and Recycling: Construction Demolition Waste, London, UK, 14–15 September 2004; pp. 157–164.
93. Dhir, R.K.; Paine, K.A. *Performance Related Approach to the Use of Recycled Aggregates*; Waste and Resources Action Programme (WRAP) Aggregates Research Programme: Banbury, Oxon, UK, 2007; p. 77.
94. Dhir, R.K.; MLimbachiya, C.; Leelawat, T. Suitability of recycled concrete aggregate for use in bs 5328 designated mixes. *Proc. Inst. Civ. Eng.-Struct. Build.* **1999**, *134*, 257–274. [CrossRef]
95. Dhir, R.K.; Paine, K.A.; O'Leary, S. Use of recycled concrete aggregate in concrete pavement construction: A case study. In Proceedings of the International Symposium on Sustainable Waste Management, Dundee, Scotland, UK, 9–11 September 2003; pp. 373–382.
96. Dolara, E.; Di Niro, G.; Cairns, R. Recycled aggregate concrete prestressed beams. In Proceedings of the International Symposium on Sustainable Construction: Use of Recycled Concrete Aggregate, London, UK, 11–12 November 1998; pp. 255–261.
97. Domingo-Cabo, A.; Lázaro, C.; López-Gayarre, F.; Serrano-López, M.A.; Serna, P.; Castaño-Tabares, J.O. Creep and shrinkage of recycled aggregate concrete. *Constr. Build. Mater.* **2009**, *23*, 2545–2553. [CrossRef]
98. Dosho, Y. Development of a sustainable concrete waste recycling system—Application of recycled aggregate concrete produced by aggregate replacing method. *J. Adv. Concr. Technol.* **2007**, *5*, 27–42. [CrossRef]
99. Fan, C.-C.; Huang, R.; Hwang, H.; Chao, S.-J. Properties of concrete incorporating fine recycled aggregates from crushed concrete wastes. *Constr. Build. Mater.* **2016**, *112*, 708–715. [CrossRef]
100. Juan, M.S.; Gutiérrez, P.A. Influence of recycled aggregate quality on concrete properties. In Proceedings of the International RILEM Conference on the Use of Recycled Materials in Buildings and Structures, Barcelona, Spain, 8–11 November 2004; pp. 545–553.

101. Khatib, J.M. Properties of concrete incorporating fine recycled aggregate. *Cem. Concr. Res.* **2005**, *35*, 763–769. [CrossRef]
102. Kim, S.-W.; Yun, H.-D. Influence of recycled coarse aggregates on the bond behavior of deformed bars in concrete. *Eng. Struct.* **2013**, *48*, 133–143. [CrossRef]
103. Kim, K.; Shin, M.; Cha, S. Combined effects of recycled aggregate and fly ash towards concrete sustainability. *Constr. Build. Mater.* **2013**, *48*, 499–507. [CrossRef]
104. Koulouris, A.; Limbachiya, M.C.; Fried, A.N.; Roberts, J.J. Use of recycled aggregate in concrete application: Case studies. In Proceedings of the International Conference on Sustainable Waste Management and Recycling: Challenges and Opportunities, London, UK, 14–15 September 2004; pp. 245–257.
105. Limbachiya, M.; Meddah, M.S.; Ouchagour, Y. Use of recycled concrete aggregate in fly-ash concrete. *Constr. Build. Mater.* **2012**, *27*, 439–449. [CrossRef]
106. Otsuki, N.; Miyazato, S.-I.; Yodsudjai, W. Influence of recycled aggregate on interfacial transition zone, strength, chloride penetration and carbonation of concrete. *J. Mater. Civ. Eng.* **2003**, *15*, 443–451. [CrossRef]
107. Rao, M.; Bhattacharyya, S.; Barai, S. Influence of recycled aggregate on mechanical properties of concrete. In Proceedings of the 5th Civil Engineering Conference in the Asian Region and Australasian Structural Engineering Conference, Sydney, Australia, 8–12 August 2010; pp. 749–754.
108. Waleed, N.; Canisius, T. *Engineering Properties of Concrete Containing Recycled Aggregates*; Waste and Resources Action Programme: Banbury, UK, 2007; p. 104.
109. Zhou, C.; Chen, Z. Mechanical properties of recycled concrete made with different types of coarse aggregate. *Constr. Build. Mater.* **2017**, *134*, 497–506. [CrossRef]
110. Corinaldesi, V.; Moriconi, G. Influence of mineral additions on the performance of 100% recycled aggregate concrete. *Constr. Build. Mater.* **2009**, *23*, 2869–2876. [CrossRef]
111. Padmini, A.K.; Ramamurthy, K.; Mathews, M.S. Influence of parent concrete on the properties of recycled aggregate concrete. *Constr. Build. Mater.* **2009**, *23*, 829–836. [CrossRef]
112. Yanagi, K.; Kasai, Y.; Kaga, S.; Abe, M. Experimental study on the applicability of recycled aggregate concrete to cast-in-place concrete pile. In *Sustainable Construction: Use of Recycled Concrete Aggregate*; Thomas Telford: London, UK, 2015; pp. 359–370.
113. Kou, S.; Poon, C. Properties of self-compacting concrete prepared with coarse and fine recycled concrete aggregates. *Cem. Concr. Compos.* **2009**, *31*, 622–627. [CrossRef]
114. Lin, Y.-H.; Tyan, Y.-Y.; Chang, T.-P.; Chang, C.-Y. An assessment of optimal mixture for concrete made with recycled concrete aggregates. *Cem. Concr. Res.* **2004**, *34*, 1373–1380. [CrossRef]
115. Pedro, D.; de Brito, J.; Evangelista, L. Mechanical characterization of high performance concrete prepared with recycled aggregates and silica fume from the precast industry. *J. Clean. Prod.* **2017**, *164*, 939–949. [CrossRef]
116. Kurda, R.; Silvestre, J.D.; de Brito, J.; Ahmed, H. Optimizing recycled concrete containing high volume of fly ash in terms of the embodied energy and chloride ion resistance. *J. Clean. Prod.* **2018**. [CrossRef]
117. Kurda, R.; de Brito, J.; Silvestre, J.D. Indirect evaluation of the compressive strength of recycled aggregate concrete with high fly ash ratios. *Mag. Concr. Res.* **2018**, *70*, 204–216. [CrossRef]
118. FIB Bullten 42. *Constitutive Modelling of High Strength/High Performance Concrete*; International Federation for Structural Concrete (FIB): Lausanne, Switzerland, 2008.
119. Gul, M. Effect of cube size on the compressive strength of concrete. *Int. J. Eng. Dev. Res.* **2016**, *4*, 956–959.
120. Neville, A.M. *Properties of Concrete*, 4th ed.; John Wiley & Sons: New York, NY, USA, 1997.
121. Felekoğlu, B.; Türkel, S. Effects of specimen type and dimensions on compressive strength of concrete. *GU J. Sci.* **2005**, *18*, 639–645.
122. Gonnerman, H. Effect of size and shape of test specimen on compressive strength of concrete. *Proc. ASTM* **1925**, *25*, 237–250.
123. Nasser, K.; Al-Manaseer, A. It's time for a change from 6_12 to 3_6 inch cylinders. *ACI Mater. J.* **1987**, *84*, 6–213.
124. Malhotra, V. Are 4 × 8 Inch concrete cylinders as good as 6 × 12 Inch cylinder for quality control concrete? *ACI J.* **1976**, *73*, 33–36.
125. Forstie, D.; Schnormeier, R. Development and Use of 4 × 8 in. Concrete Cylinders in Arizona. *Concr. Int.* **1981**, *3*, 41–45.

126. Carrasquillo, R.; Nilson, A.; State, F. Propeties of high strength concrete subject to short-term loads. *ACI J.* **1981**, *78*, 42–45.

127. Lessard, M.; Aitcin, P. *Testing of High Performance Concrete, High Performance Concrete*; E & FN Spon: London, UK, 1992; pp. 196–213.

128. Halstead, P.E. The early history of Portland cement. *Trans. Newcom. Soc.* **1961**, *34*, 37–54. [CrossRef]

129. Lea, F.; Mason, T. *Cement*; Encyclopædia Britannica, Inc.: Chicago, IL, USA, 2015.

130. Bruhl, R. *Understanding Statistical Analysis and Modeling*; SAGE Publications. Thousand Oaks, CA, USA, 2017, p. 440; ISBN 9781506317373.

131. Dahlquist, G.; Bjorck, A. *Numerical Methods in Scientific Computing: Volume 1*; SIAM: Philadelphia, PA, USA, 2008; p. 717; ISBN 9780898717785.

132. Ingham, J.P. 4-Aggregates. In *Geomaterials under the Microscope*; Academic Press: Boston, MA, USA, 2013; pp. 61–74.

133. Gill, R. *Igneous Rocks and Processes: A Practical Guide*; Wiley-Blackwell: Hoboken, NJ, USA, 2010; p. 438.

applied
sciences

MDPI

Article

A Study on the Properties of Recycled Aggregate Concrete and Its Production Facilities

Jung-Ho Kim [1,*], Jong-Hyun Sung [1], Chan-Soo Jeon [2], Sae-Hyun Lee [2] and Han-Soo Kim [3]

[1] Technology Research Center, Hallaencom, 1170-5, Poseunghyangnamro, Hwaseong-si, Gyeonggi-do 18572, Korea; jonghyun.sung@hallaencom.com

[2] Department of Living and Built Environment Research, Korea Institute of Civil Engineering and Building Technology, 283, Goyang-daero, Ilsanseo-gu, Goyang-si, Gyeonggi-do 10223, Korea; jcsi0815@kict.re.kr (C.-S.J.); shlee@kict.re.kr (S.-H.L.)

[3] Department of Architecture, Konkuk University, 120 Neungdong-ro, Gwangjin-gu, Seoul 05029, Korea; hskim@konkuk.ac.kr

* Correspondence: jungho.kim@hallaencom.com; Tel.: +82-10-5057-5390

Received: 28 March 2019; Accepted: 30 April 2019; Published: 10 May 2019

Abstract: In recent years, the amount of construction waste and recycled aggregate has been increasing every year in Korea. However, as the recycled aggregate is poor quality, it is not used for concrete, and the Korean government has strengthened the quality standards for recycled aggregate for concrete. In this study, research was conducted on the mechanical and durability characteristics of concrete using recycled aggregate, after developing equipment to improve the quality of recycled aggregate to increase the use of recycled aggregate for environmental improvements. The results illustrated improvements in the air volume, slump, compressive strength, freezing and thawing resistance, and drying shrinkage. Furthermore, this study is expected to contribute to the increased use of recycled aggregate in the future.

Keywords: recycled aggregate; concrete; construction waste; mechanical characteristics; durable characteristics

1. Introduction

In the latter part of the 1980s, Korea constructed large-scale housing developments, as well as new cities for building pleasant residential environments and resolving housing issues. Thereafter, during the 2010s, with the drastic increase of old housing that caused re-development and re-construction to become increasingly active, the production of construction waste materials increased yearly. According to the nationwide waste material volume of 2016, which was reported by the Korea Environment Corporation in 2017. The volume of construction waste materials has been on the rise every year (as shown in Figure 1), with the highest value reaching up to approximately 48%. From the figure, the ratio of waste concrete was 62.8% and the waste asphalt concrete was 17.9%, illustrating that the recycling of waste concrete has become an urgent matter [1].

In accordance with the provision of Article 35 of the Construction Waste Recycling Promotion Act, the government established the Recycled Aggregate Quality Standard for promoting the recycling of waste materials from construction, to recommend a more diversified and broadened facilitation of waste concrete. Therefore, in terms of applying recycled aggregate for concrete, there have been a significant number of studies that have been conducted at home and abroad, and recently, there have been active studies on the addition of diversified mixed materials to the recycled aggregate concrete or the mixed use of natural aggregate, to provide recycled aggregate concrete with capabilities equivalent to that of ordinary concrete [2–14]. However, recycled aggregate for use in concrete has not been widely employed in Korea, as the aggregates are of poor quality; they have a high absorption rate and low density, therefore, are mostly used for filling the ground.

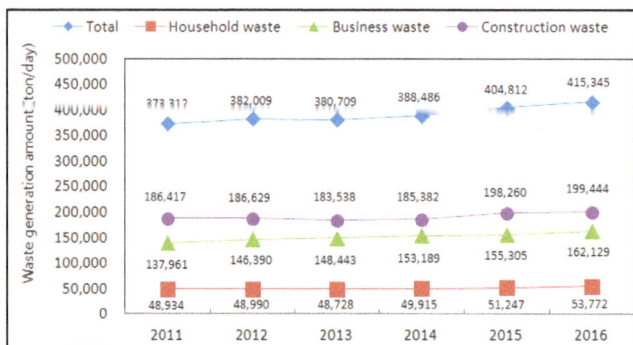

Figure 1. Changes in the waste material volume in Korea [1].

The recycled aggregates are not used in structural buildings as their quality fluctuation is severe. It might only be used in buildings if the cement paste on the surface of the recycled aggregate is removed, but the removal of impurities and fine aggregates is also very important. Previous research, has studied the removal of cement paste on the surface of recycled aggregates but there have been no research on the development of impurities and the aggregate fine powder removal facilities. Therefore, the main purpose of this study was to develop and verify facilities, to activate recycled aggregate for concrete. Theoretically, there exists a technology that improves the quality of recycled aggregate by adding wind speed to the impact method, which removes the impurities.

Previous studies that have pursued the improvement of the quality of recycled aggregates, have mostly been conducted on the removal of surface cement paste, and the removal methods were typified by impact, chemical, and heating methods. Ogawa and Nawa reported that the impact strength increased as the number of impacts increased, and that the compressive strength of the recycled aggregate concrete decreased [15]. Shima and Tatayasiki have reported a technique of removing the cement paste on the surface of the recycled aggregate, by heating and fracturing [16]. Kim, Kiuchi, Juan, and Ismail, as a study of chemical methods, focused on the technique of removing the cement paste adhered to the recycled aggregate, with hydrochloric acid solution [17–20]. Tam also reported cement paste removal with sulfuric acid solution [12]. However, the heating and fracturing and chemical methods have not been utilized due to its spatial and temporal constraints.

In order to promote the use of recycled aggregates, the Korean government has presented a new standard of use with an expanded scope that requires the use of recycled fine aggregate, from the existing recycled coarse aggregate. This standard was in accordance with the laws and regulations of the quality standard of the recycled aggregate. In the event that the standard of use for the recycled aggregate is applied to concrete for a 21–27 MPa structure, the standard to facilitate the use of recycled fine aggregate has been prepared by revising 30% or less (using only recycled coarse aggregate) of the total aggregate capacity, as is also the case for non-structure concrete of less than 21 MPa. In the event that it is used for concrete for a structure of 27 MPa or less, the standard should be used 30% or less (using recycled coarse aggregate and recycled fine aggregate) of the total aggregate capacity, 60% or less (using only recycled coarse aggregate) of the total coarse aggregate capacity, and 30% or less (using only recycled fine aggregate) of the total fine aggregate capacity, and 30% or less (using only recycled coarse aggregate and recycled fine aggregate) of the total aggregate capacity. In addition, the standard value for recycled fine aggregates has been improved from the existing density of 2.2 g/cm^3 and absorption rate of 5%, to a density of 2.3 g/cm^3 and an absorption rate of 4% [21]. This was intended to minimize the quality deviation through an enhancement of quality for the recycled fine aggregate, but the current production technologies for recycled fine aggregates face difficulties in meeting the revised standard.

In this research, to use recycled aggregate in structural buildings, a facility was developed and research was conducted to remove the cement paste attached to the surface of the recycled aggregate, aggregate fine powder, and impurities. The characteristics of concrete using a recycled aggregate type and ratio were studied.

2. Scope and Method of Research

In this research, the quality improvement of the recycled fine aggregate and the characteristics of concrete using recycled aggregate have been studied through the following studies.

(1)   A study on the quality improvement of recycled fine aggregate

Recycled fine aggregate contains significant amounts of organic and inorganic impurities with an inconsistent quality, creating concerns about quality decline of the concrete. Therefore, the production facilities for quality improvement of the recycled fine aggregate were reviewed, and the equipment used to remove impurities and strip the recycled fine aggregate was developed for stable production and quality improvement of the recycled fine aggregate.

(2)   A study on the quality improvement of mixed aggregate

The recycled and natural aggregates have different specific gravities and are injected separately when producing concrete. If the mixer efficiency is low, this creates quality deviations in the concrete. Accordingly, to achieve a stable mixing of the recycled and natural aggregates, aggregate mixture facilities were developed. The densities, absorption rates, and other properties of the mixed aggregate, with various compositions were tested. In addition, the characteristics of the quality deviation of the recycled aggregate concrete, following premixing, were determined, and a premixing facility for the aggregate was developed.

(3)   An empirical study on the mechanical characteristics and durability of concrete using the recycled aggregate

Before and after modification, the recycled fine aggregates, recycled coarse aggregates, and natural aggregates that were mixed in accordance with the replacement rate of the physical and mechanical properties of the concrete that used the recycled fine aggregate, were evaluated. The concrete using the recycled aggregate was mixed with natural aggregate and the target strength was 24 MPa (general strength) or 35 MPa (high strength). Slump, air volume, compressive strength, drying shrinkage, freezing, and thawing resistance and other properties were measured to analyze the physical, mechanical and durability characteristics of the concrete containing recycled fine aggregate, before and after modification.

3. Development of the Technology to Improve the Recycled Fine Aggregate Quality

3.1. Outline

Waste materials collected from the construction fields are required by law to be classified, but due to the difficulty of field management, there is a significant amount of discharge of mixed waste materials. If the fine aggregate is less than 5 mm, it makes it difficult to separate the foreign matters on an aggregate surface. Therefore, there is a limit to the application of existing technologies for improving the density and absorption rate. In order to remove a large amount of impurities, particularly inflammable impurities, the most frequently used sorting method is separation by an air blower (as shown in Figure 2). Furthermore, the density and absorption rate of the recycled fine aggregate change depending on the contents of the mortar attached to the aggregate surface. As the mortar content increases, the density decreases and the absorption rate increases, diminishing the quality, as a result. Therefore, to improve the aggregate quality, the mortar attached to the aggregate surface must be removed as much as possible. In the past, there was no equipment available to remove foreign materials and mortar attached to the aggregate surface, at the same time.

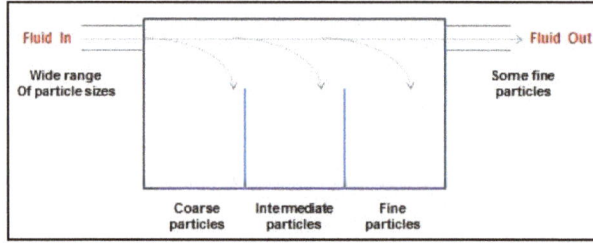

Figure 2. The principle of air-blower impurity sorting.

In general, employing the developed equipment, inflammable impurities with diameters >25 mm are separated through a separate sorting device, such as a trammel. Inflammable impurities in the 10–25 mm range are separated using a screen. Inflammable impurities of 10 mm or less are sorted by ventilation, during transport of the aggregate. Air-flow-type sorting methods have the advantages of being efficient in sorting impurities with low specific gravities; they also employ a simple facility configuration and equipment maintenance method. The purpose of separation selection should be clearly distinguished by locating at the end of processing. However, air-flow-type sorting, can sort out all substances with low specific gravities, including aggregates. Furthermore, this process can scatter well, and if there is a significant flow of aggregate, the sorting efficiency is low. Thus, a method for sorting impurities and for eliminating dust scattering is required, when there is a large amount of aggregates flowing.

Furthermore, the density and absorption rate of the recycled fine aggregate change, depending on the contents of the mortar attached to the aggregate surface. As the mortar content increases, the density decreases, and the absorption rate increases, diminishing the quality.

3.2. Principle of Technology

Technology is needed to resolve the quality decline, due to impurities, by processing the waste materials, implementing technology to remove mortar from the surface of the recycled fine aggregate, and producing recycled fine aggregate with a density of 2.3 g/cm^3 or more, an absorption rate of less than 4%, and organic impurity contents of less than 0.3%. The appearance and perspective drawing of the equipment for stripping impurities and recycled fine aggregates, and the entire production systems, are shown in Figures 3 and 4, respectively.

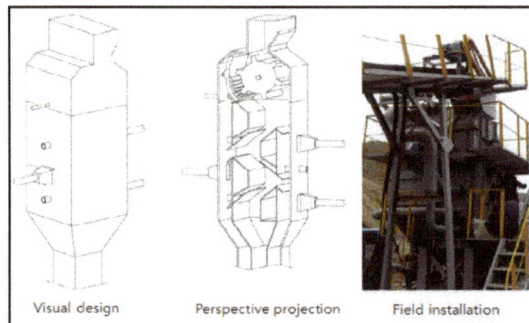

Figure 3. Appearance and perspective drawing of stripping the removal equipment for the impurities of recycled fine aggregate.

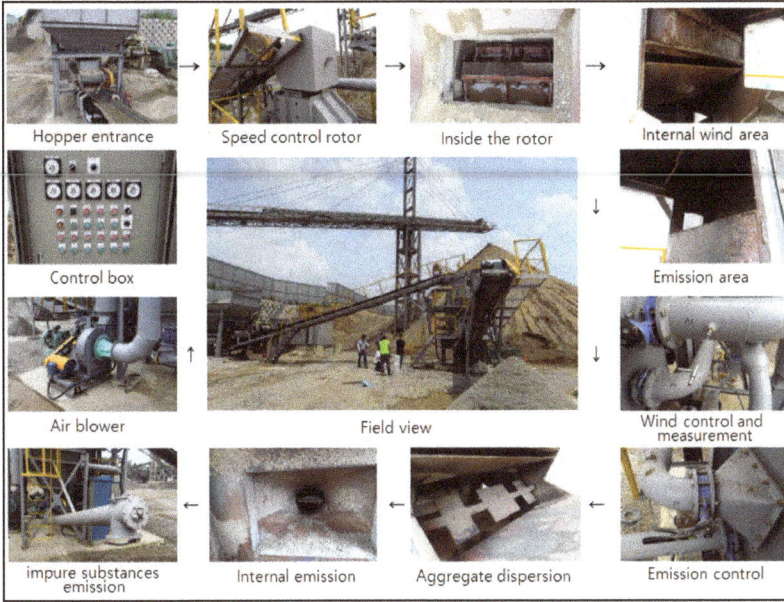

Figure 4. The production process of the technology development scheme.

These facilities are closed and can accommodate a multi-phase flow. The recycled fine aggregate is stripped by the Speed Adjusting Rotor Hammer, and the wind pressure and direction are varied, three times, to remove the impurities.

To separate the aggregate and impurities, a knife-type air-speed supply device and an inhaling device are used to capture the impurities. The wind pressure used to separate the impurities is generated by a recycling fan, with a capacity of 120 m^3/min and the structure between the upper rotation rotor and final discharge phase is enclosed to prevent the scattering of dust. In addition, by adjusting the wind velocity and absorption volume, the system can accommodate various impurities. Furthermore, the Speed Adjusting Rotor Hammer rotational impact device, accelerates the initially supplied aggregate, and the organic impurity is separated by the absorption device in the upper part, due to the wind speed of the rotational force. Mortar on the aggregate surface is removed through the stripping plate. The principle of mortar removal is shown schematically, in Figure 5.

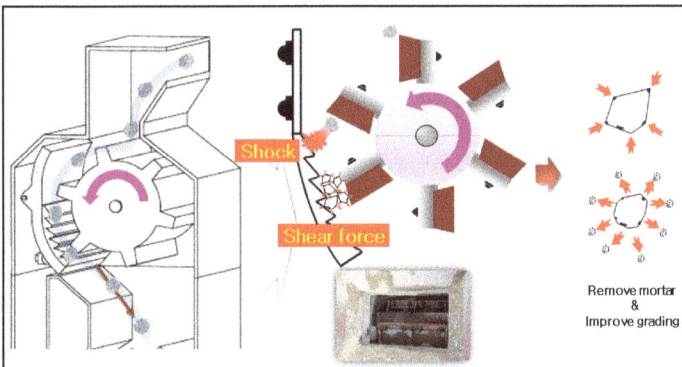

Figure 5. The principle of mortar removal.

3.3. Experiments and Results of the Recycled Fine Aggregate Using Stripping Removal Equipment for the Impurities of Recycled Fine Aggregate

The equipment developed in this study removed cement paste, aggregate particles, and the impurities attached to the surface of recycled aggregate. Removal of the cement paste increased the absolute dry density, decreased the absorption rate, and increased the percentage of the absolute volume of the particle shape. Additionally, through the removal of aggregate particles, the penetration rate of 0.08 mm was reduced and the content of organic and inorganic impurities was reduced through the removal of impurities.

3.3.1. Absolute Dry Density

The absolute dry density could affect the concrete strength if the aggregate density was lower than the Korean Standard requirement of 2.2 g/cm^3 or more. This test was implemented following the KS F 2504 "Method to Test Density and Absorption Rate of Fine Aggregate." The absolute dry density was found to increase by an average of approximately 0.07 g/cm^3. The dry density changed from 2.25 → 2.32 g/cm^3 in the first pass, to 2.25 → 2.33 g/cm^3 in the second pass, and 2.28 → 2.35 g/cm^3 in the third pass.

3.3.2. Absorption Rate

The recycled aggregate had significant amount of mortar attached to the aggregate surface, compared to the natural aggregate. Thus, the recycled aggregate had a greater absorption rate than that of the aggregate used for general concrete. If the recycled aggregate that had been air- or oven-dried was used, the water content of the aggregate was increased, to achieve more loss of slump in the unsolidified concrete, and the pumpability was worsened. Therefore, the KS required an absorption rate of 5% or less. This was measured following the KS F 2504 "Method to Test Density and Absorption Rate of Fine Aggregate." Before and after penetration, the absorption rate decreased by an average of approximately 0.70%. The absorption rate decreased from 5.42% → 4.62% in the first test, 5.35% → 4.68% in the second test, and 5.38% → 4.75% in the third test.

3.3.3. Percentage of Absolute Volume of the Particle Shape

The percentage of absolute volume of the particle shape was used to determine the aggregate particle shape. If the absolute volume percentage was 53% or more, the recycled fine aggregate particles were considered to be adequate. This was measured following the KS F 2527 "Aggregate Crushed for Concrete." Before and after penetration, the absorption rate increased by an average of approximately 0.64%. In the first, second, and third tests, the absolute volume percentage increased from 53.05% → 53.52%, 52.71% → 53.42%, and 52.88% → 53.63%, respectively.

3.3.4. Penetration Rate of 0.08 mm

In the event that there was a significant quantity of small particles with a size of 0.08 mm or less, the quality would be adversely influenced, resulting in, for example, a decline of strength, due to an increase of the unit quantity, when producing ready-mixed concrete. An increase of laitance and drying shrinkage when pouring the concrete, and a decrease in adhesion when joint pouring was applied, which was determined by the Korean Standard to be 7.0% or less. This test was implemented by the KS F 2511 "Method to test chips included in aggregate (penetrating the 0.08 mm sieve)", and as a result of the experiments before and after penetration, it was demonstrated that the chips decreased by an average of approximately 1.90% of 0.08 mm or less, in the first test, from 3.15% → 1.83%, in the second test from 3.65% → 1.27%, and in the third test from 3.39% → 1.39%. This was considered to be due to the air-flow removing the impurity of the development facility that sorted the chips.

3.3.5. Contents of Organic and Inorganic Impurities

Waste concrete has a variety of impurities that are mixed in the recycled aggregate, which was produced by crushing them, resulting in a declined quality. Accordingly, the KS F 2527 "Crushed Aggregate for Concrete", defined the impurity contents (organic and inorganic impurity) of the recycled fine aggregate. This test was implemented by the KS F 2576 "Method to Test the Foreign Substance Contents of Recycled Aggregates", and as a result of the experiment (as shown in Figure 6), before and after penetration, the organic impurity experiment demonstrated a decrease in the organic impurity, for an average of approximately 0.59%, when penetrating in the first test of 1.21% → 0.58%, the second test of 1.47% → 0.82%, and the third test of 1.25% → 0.77%, and as a result of the experiment, before and after penetration, in the inorganic impurity experiment, it demonstrated that a decrease in the contents of the inorganic impurity for an average of approximately 0.46%, in the first test, it showed a decrease from 0.95% → 0.48%, in the second test, from 1.05% → 0.55%, and in the third test, from 0.94% → 0.54%.

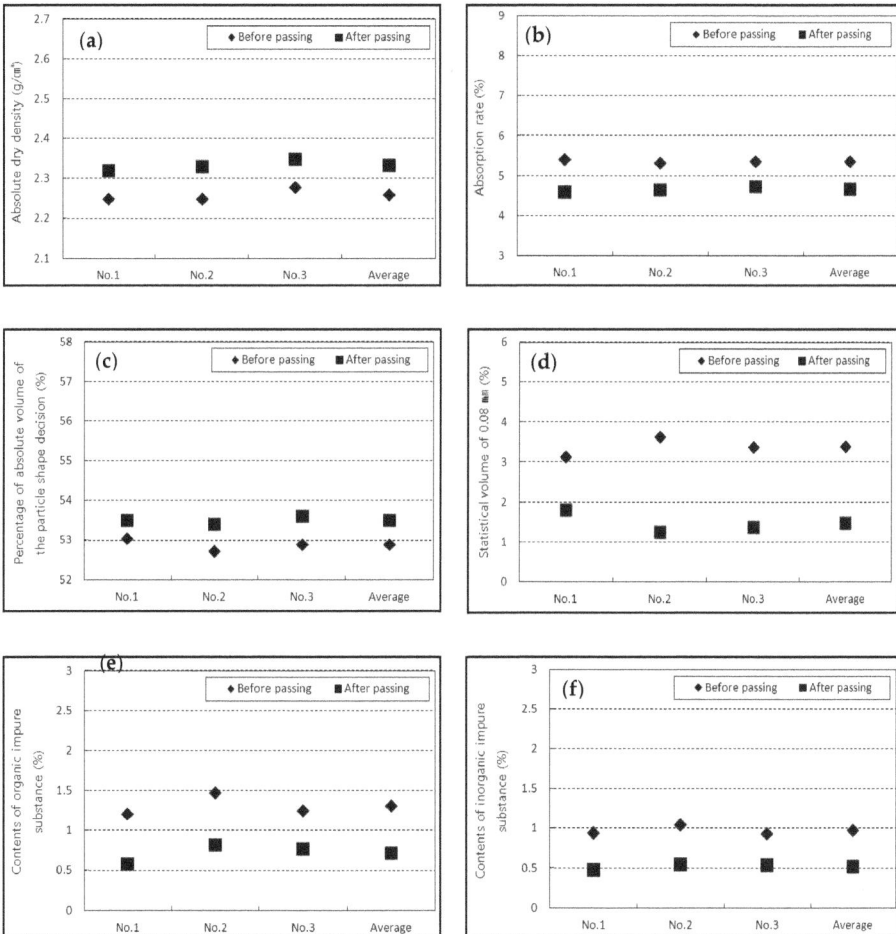

Figure 6. (**a**) Absolute dry density, (**b**) absorption rate, (**c**) percentage of absolute volume of the particle shape, (**d**) statistical volume of 0.08 mm, (**e**) contents of organic impurities, and (**f**) contents of inorganic impurities of recycled fine aggregate, before and after passing.

4. Development of the Technology to Improve the Quality of the Mixed Aggregates

4.1. Outline

Ready-mixed concrete is produced by weighing raw materials to inject into the mixer, at the time of production. However, the mixtures could differ depending on the mixer capability of the batch plant, and if the mixer efficiency is low, it could cause concrete quality deviations. In addition, if the materials with different densities are mixed by simultaneous injection, sufficient mixing is difficult due to the density difference. To improve the mixture capabilities, premixing is an option. According to the literature, if the mixture materials are displaced in a mass of approximately 70% or more, the density difference of cement, fine powder of the blast furnace slag, and fly ash could cause severe quality deviations. When the ready-mixed concrete production is conducted through advanced premixing, the mixing capabilities improve and the quality deviation is moderate [22].

4.2. Principle of Technology

In the Korea Standard, the material age is defined as the aggregate, with a density of 2.5 g/cm^3 or more, recycled coarse aggregate is defined as having a density of 2.5 g/cm^3 or more, and recycled fine aggregate is defined as having a density of 2.2 g/cm^3 or more. A quality deviation is generated, when mixing aggregates of different densities at a content of 70% or more, in the ready-mixed concrete. Technology to appropriately mix such mixtures was developed. The purpose of this was to produce recycled aggregate concrete with a difference of 0.8% or less, in the unit capacity volume of the mortar, and 5% or less in the unit coarse aggregate volume, using the hand mixing capability index defined by KS F 2455 "Test Method of Changes of Mortar and coarse aggregate volume from the Concrete Made with Mixer." The perspective drawing and mixing method of the aggregate mixing equipment are shown in Figure 7.

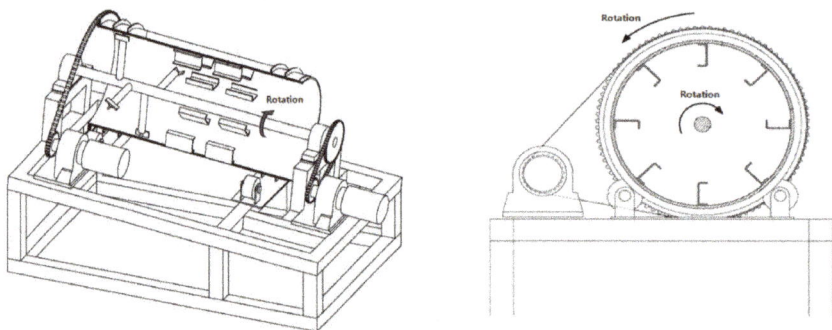

Figure 7. Perspective schematics of the aggregate mixing equipment and mixing method.

Previous aggregate mixing equipment have primarily used gravity methods. For conical-shaped mixers, the aggregate is rotated, allowing the cone to fall simply due to gravity. In this process, the aggregates fall in lumps, while they are attached to the inner wall of the cone-shaped mixer by centrifugal force. As a result, no mixing occurs.

This equipment uses gravity and forced mixing, with a mixing arm around the injection point, to mix the material as it is initially injected. A mixing plate is fixed inside the rotating drum, in a C-shape, for the secondary mixing, to prevent lumping when the material falls due to gravity. In addition, a counterclockwise rotation direction is used for a conical shape and a clockwise direction is used for the central axis. The mixing time is 30 s, and the mixing speed is 20 RPM.

4.3. Experiment and Results of the Mixed Aggregate Quality before and after the Penetration of the Aggregate Mixing Equipment

The mixed aggregate concrete before and after entering the aggregate mixing equipment was examined. The recycled coarse and fine aggregates were displaced by 50% of the entire aggregate volume with ready-mixed concrete after mixing for 45 s. The differences in the unit capacity volume of mortar in the concrete and differences in the unit coarse aggregate volume from the concrete, and the compressive strength under the Korea Standard, were investigated.

The experimental results are presented at Table 1.

Table 1. Physical properties of natural aggregate and recycled aggregate.

Division	Unit Volume Weight (kg/m³)	Absolute Dry Density (g/cm³)	Absorption Rate (%)	Assembly Rate (%)	Organic Impurities (%)
Natural Coarse Aggregate	1518	2.59	1.01	7.02	-
Recycled Coarse Aggregate	1485	2.51	5.85	6.26	0.53
Natural Fine Aggregate	1552	2.61	1.03	2.95	-
Before Modified Recycled Fine Aggregate	1305	2.21	4.89	2.48	0.92
After Modified Recycled Fine Aggregate	1384	2.33	3.62	2.74	0.49

4.3.1. Difference in Unit Capacity Volume of Mortar and Difference in Unit Coarse Aggregate Volume

The difference in the mortar unit capacity volume and the difference in the concrete before and after mixing are shown in Table 2. The difference in the unit capacity volume of mortar, before and after mixing was 16.17 kg at the front part and 15.84 kg at the rear part, for the case before mixing, demonstrating a difference of approximately 2.04%. After mixing, the unit capacity volume of mortar was 16.15 kg in the front part and 16.07 kg in the rear part, exhibiting a difference of approximately 0.49%. In addition, the difference in the unit coarse aggregate volume in the front part of 2.89 kg and the rear part of 2.69 kg was approximately 6.92% before mixing. After mixing, the front part was 2.87 kg and the rear part was 2.75 kg, representing a difference of 4.18%.

Table 2. Experimental results of Unit Volume Weight of Mortar and Unit Coarse Aggregate Amount.

Division	Unit Volume Weight of Mortar			Unit Coarse Aggregate Amount		
	Front (kg)	Back (kg)	Difference Rate (%)	Front (kg)	Back (kg)	Difference Rate (%)
Before	16.17	15.84	2.04	2.89	2.69	6.92
After	16.15	16.07	0.49	2.87	2.75	4.18

4.3.2. Compressive Strength

This concrete compressive strength test was implemented to determine the difference in the compressive strength, before and after mixing (as shown in Figure 8). Before mixing and with a three-day material age, the front portion exhibited compressive strengths in the range of 11.1–12.5 Mpa, with an average of 11.8. In the back portion, the compressive strengths were in the range of 10.4–11.7 Mpa, with an average of 11.0 MPa. Thus, there was a difference in the compressive strengths of a 0.8 MPa between the front and back. After mixing, the front portion had a compressive strength range of 11.5–12.4 Mpa, for an average of 11.9 MPa. In the back portion, the range was 11.9–12.4 MPa, with an average of 12.2 MPa. Thus, there was a 0.3 MPa difference in the compressive strengths, between the front and back. For a material with a seven-day age before mixing, the front portion had a range of 18.1–19.3 Mpa, with an average of 18.6 MPa. In the back portion, the range was 17.1–18.2 MPa, with an average of 17.8 MPa. Thus, there was a 0.8 MPa difference in the compressive strength, between the front and back. After mixing, the front portion had a range of 18.4–19.2 MPa, with an average of 18.8 MPa, and the back portion had a range of 17.0–19.2 MPa, with an average of 18.0 MPa.

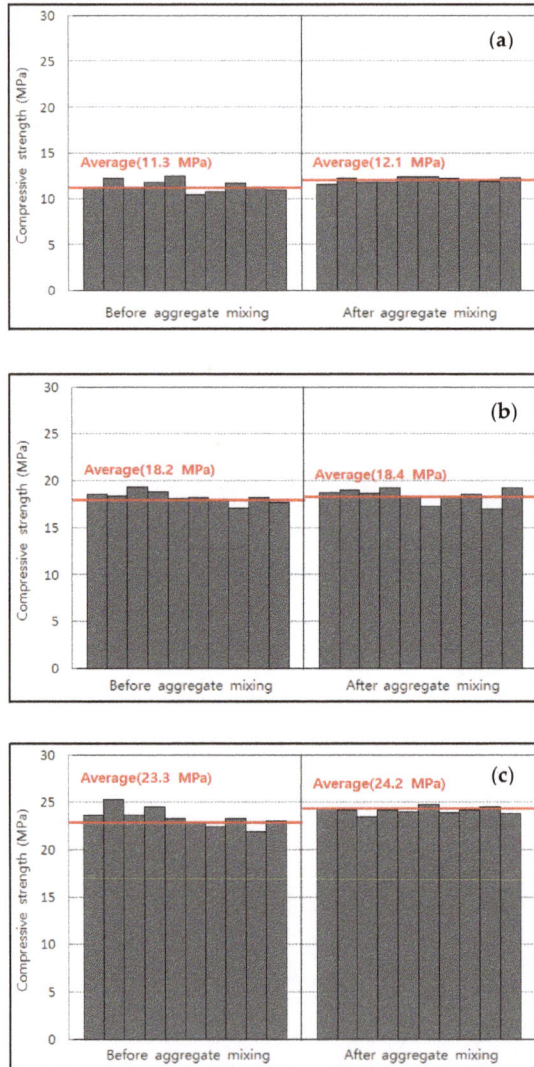

Figure 8. The (**a**) 3-day, (**b**) 7-day, and (**c**) 28-day concrete strength before and after aggregate mixing.

Thus, there was a 0.8 MPa difference in the compressive strengths between the front and back. For the material with a 28 day age before mixing, the front portion had a range of 23.3–25.3 Mpa, with an average of 24.1 MPa, and the back portion had a range of 21.9–23.3 Mpa, with an average of 22.7 MPa.

Thus, there was a 1.4 MPa difference in the compressive strengths between the front and back. After the mixture, the front portion had a range of 23.5–24.4 MPa, with an average of 24.1 MPa, and the back portion had a range of 23.8–24.8 MPa, with an average of 24.2 MPa. Thus, there was a 0.1 MPa difference in the compressive strengths between the front and back.

The differences in the compressive strengths between the samples with three and seven-day material ages had a range of approximately 1.1–1.4 Mpa, before mixing, and approximately 0.7–1.5 Mpa, after mixing. Thus, mixing decreased the difference. The compressive strength was approximately 0.8 Mpa, before mixing, and approximately 0.3 MPa after mixing. However, with a 28-day material age, the deviation of the compressive strength was in the range of approximately 1.4–2.0 Mpa, before mixing, and approximately 0.9–1.0 MPa after mixing, The compressive strength was approximately 1.4 Mpa, before mixing, and approximately 0.1 MPa after mixing, and there was a greater difference for the 28-day material age than for the three-day and seven-day material ages.

5. Mechanical Properties of the Recycled Aggregates Using Concrete and Durability Characteristics

5.1. Experimental Method

The purpose of this research was to expand the use of the recycled aggregate. The recycled aggregates, before and after modifications, were mixed in accordance with the replacement ratio, to evaluate the physical, mechanical, and durability properties of the recycled aggregate concrete. The slump, air volume, compressive strength, freezing and thawing resistance, drying shrinkage and other properties of the concrete, following the recycled aggregate replacement ratio were determined.

5.1.1. Aggregate Replacement Rate

When the quality of the recycled fine aggregate was improved for use as a recycled aggregate, the recycled coarse aggregate was replaced by 0%, 30%, 60%, and 100%, for before and after modifications of the recycled fine aggregate, in order to find out the change in the capability of the recycled aggregate concrete.

5.1.2. Experiment Factor, Level, and Combination Selection

The design standard strength had two levels, the general strength territory of 24 MPa and a high strength territory of 35 MPa, which were generally used for research. Experimental combinations were selected to satisfy the design standard strength, using the natural aggregate for the ready-mixed concrete combination in the preliminary experiment phase.

Concrete mix was determined so that the water to cement ratio (W/C) had a general strength territory of 44.3% and a high strength territory of 39.5%, sand to aggregate ratio (S/A) had a general strength territory of 47.1%, and a high strength territory of 48.0%.

An increase of the water-cement ratio for the recycled aggregate concrete tended to decrease the compressive strength. Similar to ordinary concrete, the aggregate quantity and water-cement ratio were fixed, and a water-reducing agent was used to achieve a target slump of 190 ± 25 mm. In addition, pre-wetting was implemented, prior to the experiment, to suppress the reduction of unit quantity by the air-gap on the recycled aggregate surface of the pastes, and the aggregates were implemented into the experiment under the internal saturation condition of the dried surface. The experiment factor and level of recycled aggregate concrete are shown in Table 3 and the experiment combinations are shown in Table 4.

Table 3. Experimental factors and levels.

Division	Factors	Levels	Symbols
Aggregate Type	Recycled Coarse Aggregate, Recycled Fine Aggregate before Modify, Recycled Fine Aggregate after Modify	3	I, II, III
Recycled Aggregate Replacement Ratio (%)	0, 30, 60, 100	4	1, 2, 3, 4
Target Strength (MPa)	24, 35	2	A, B
Water Cement Ratio (%)	48.5, 44.6	2	-
Fine Aggregate Ratio (%)	47.5, 49.2	2	-

Table 4. Concrete mixing table.

Mix ID	W/C (%)	S/A (%)	Water	Cement	N.F. *	R.B. **	R.A. ***	N.C. ****	R.C. *****
							Sand	Gravel	(Unit Weight kg/m³)
I1II1A						0		974.5	0
I2II1A								682.2	292.4
I3II1A					868.4			389.8	584.7
I4II1A							0	0	974.5
I1II2A	44.3	47.1	146.5	330.5	607.9	260.5			
I1II3A					347.4	521.0			
I1II4A					0	868.4		974.5	0
I1III2A					607.9		260.5		
I1III3A					347.4		521.0		
I1III4A					0		868.4		
I1II1B						0		895.9	
I2II1B					825.6			627.1	268.8
I3II1B								358.4	537.5
I4II1B							0	0	895.9
I1II2B	39.5	48.0	167.9	425.3	577.9	247.7			
I1II3B					330.2	495.4			
I1II4B					0	825.6		895.9	0
I1III2B					577.9		247.7		
I1III3B					330.2	0	495.4		
I1III4B					0		825.6		

* N.F.: Natural fine aggregate; ** R.B.: Recycled fine aggregate before modification; *** R.A.: Recycled fine aggregate after modification; **** N.C.: Natural coarse aggregate; and ***** R.C.: Recycled coarse aggregate.

5.2. Physical and Mechanical Characteristics

Air volume (as shown in Figure 9), slump (as shown in Figure 10), and compressive strength (as shown in Figure 11), test results were as follows.

5.2.1. Air Volume

The air volume increased in, both, the general strength territory and the high strength territory, as the replacement rate of the recycled coarse aggregate increased. The air volume increased slightly by 0.3%–0.5%, up to a replacement rate of 60%, which was not significant, compared to the air volume increase of ordinary concrete. However, at a replacement rate of 100%, the air volume increased by 1.1%–1.7%. In addition, during the replacement of the recycled fine aggregates, before modification, a replacement rate of 30% in the general strength territory and a high strength territory changed the air volume by −0.3%–0.1%, compared to the ordinary concrete. However, at a replacement rate of 60%, the air volume increased by 0.6%–1.6%. At a replacement rate of 100%, the air volume increased significantly by 2.0%–2.2%.

In the replacement of the recycled fine aggregate, after modification, both, the general strength territory and the high strength territory exhibited, decreased in air volume, −1.4% to −0.5%, compared to that of ordinary concrete, at up to 60% of the replacement rate. A similar air volume to that of ordinary concrete was displayed at the replacement rate of 100%. Thus, the recycled aggregate contained a massive number of air-gaps on the mortar attached to the surface and the finite cracks, and the air volume increased as the density decreased, or the rate of use of the recycled aggregate (with a high absorption) rate increased. However, for the recycled fine aggregate, with an oven dried density of 2.33 g/cm^3 and an absorption rate of 4.62%, the air volume did not increase. Therefore, if the recycled coarse aggregate and the recycled fine aggregate, before modifications are used, it is prudent to prepare for changes in the air volume, for the design's standard strength and a continuous concrete quality management. Appropriate air volume management through an independent improvement of the recycled aggregate, is required.

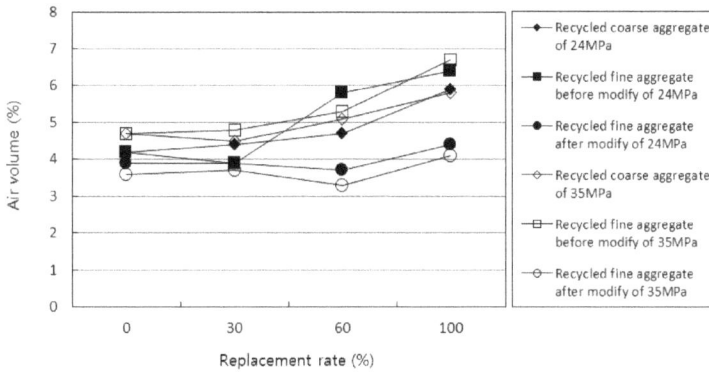

Figure 9. Air volume versus the replacement rate for each type of recycled aggregate for the general strength (24 MPa) and high strength (35 MPa) territory.

5.2.2. Slump

The elapse time of the slump showed that the slump decreased, following the elapse time, with an increase in the replacement rate for the general strength territory and a high strength territory, and up to 60% of the replacement rate. The slump decline was 55–75 mm at 60 min of elapse time, which was not significantly different to that of ordinary concrete at 60 mm. However, at 100% of the replacement rate, the slump decline was 100 mm at 60 min of elapse time. In addition, from the replacement of the recycled fine aggregate before modification, the general strength territory and high strength territory showed slump declines, following the elapse time with an increase in the replacement rate. Up to a 30% replacement rate, the slump decline was 60–70 mm, at 60 min elapse time, which was similar to the 60 mm of the slump decline of the ordinary concrete. However, at a 60% replacement rate, the slump decline was 80–105 mm, and at a 100% replacement rate, it was 115–150 mm. For the replacement of the recycled fine aggregate, after modification, up to 60% of the replacement rate, the slump decline was 50–55 mm at 60 min of elapse time, similar to that of ordinary concrete. At a 100% replacement rate, the slump decline was 60–75 mm, corresponding to a slump reduction of up to 20 mm, from ordinary concrete.

Thus, a replacement rate up to 60% for the recycled coarse aggregate, 30% for the recycled fine aggregate before modification, and 60% for the recycled fine aggregate after modification, were suitable. For any replacement of these limits, a slump reduction with elapse time occurred.

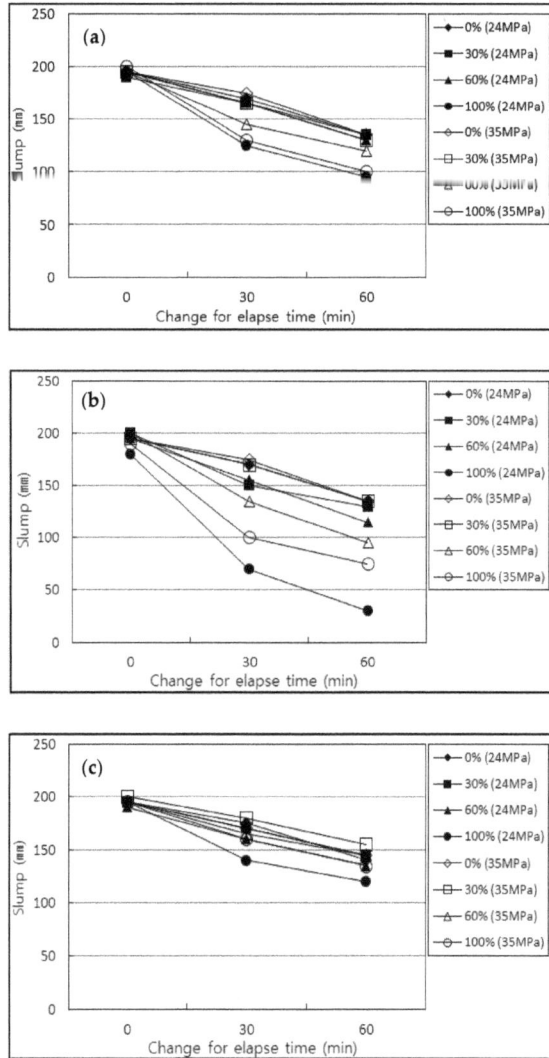

Figure 10. Slump change for strength area and replacement rate. (**a**) Recycled coarse aggregate replacement, (**b**) recycled fine aggregate before modification, and (**c**) recycled fine aggregate after modification.

5.2.3. Compressive Strength

Following the replacement of the recycled coarse aggregate and recycled fine aggregate, before and after modification, the compressive strengths were measured.

As the replacement rate of the recycled coarse aggregate increased, the compressive strength increased up to a 60% replacement rate, but drastically decreased at a 100% replacement rate. In particular, in the general strength territory, at a 60% replacement rate, the compressive strength was 26.3 MPa, which was 1.2 MPa greater than that of ordinary concrete (25.1 MPa). However, at a 100% replacement rate, the compressive strength was 19.8 MPa, which was 5.3 MPa less than that of ordinary concrete. In the high strength territory, at a 60% replacement rate, the compressive strength was 36.2 MPa, which was 0.3 MPa greater than that of ordinary concrete (35.9 MPa). However, at 100%, the compressive strength was 32.2 MPa, which was 3.7 MPa less than that of ordinary concrete. This finding was similar to that of a previous study [23], in which the recycled coarse aggregate had a density of 2.51 g/cm^3 and an absorption rate of 2.85%. Furthermore, the aggregate was slightly stabilized so that the air volume following the replacement was minimal, up to a 60% replacement rate, and increased at 100%. Thus, in their study, there was no decline in the compressive strength for the recycled coarse aggregate, up to 60% of the replacement rate, but the decline of compressive strength was shown at 100% of the replacement rate.

Furthermore, the replacement rate of the recycled coarse aggregate had a strength improvement effect compared to ordinary concrete, by up to 60%, because the compressive strength improved with the increase of the absolute volume percentage in the concrete. The volume percentage increased because of the particle shape improvement, as a relatively round particle shape, with 3.93% of the natural coarse aggregate and 3.48% of the flattening rate of the recycled coarse aggregate. For the use of the recycled fine aggregate, before modification, the compressive strength decreased as the replacement rate increased, and at 30% of the replacement rate, there was a small reduction of 1.0 MPa in the general strength territory and 0.8 MPa in the high strength territory, compared to ordinary concrete. However, at 60% and 100% of the replacement rate, there was a slight decrease in the general strength territory of 3.8 and 8.3 Mpa and the high strength territory of 3.1 and 7.4 Mpa, compared to the ordinary concrete.

Based on previous studies, when the mortar attachment to the surface of the recycled fine aggregate was excessive, the replacement rate of the recycled fine aggregate was adequate, for 30% or less, but in excess of 30%. It was reported that the compressive strength decreased due to interference with the hydration products, following the excessive injection of the mortar particles, without reaction, due to the reduction effect of the water–cement ratio. Furthermore, in this study, up to 30% of the replacement of the recycled fine aggregate, before modification, produced minimal property changes, such as a reduction in the slump elapse time and increase of the air volume. However, 60% or more of the replacement produced property changes, and in particular, the air volume increased. This influenced the compressive strength [24,25]. When recycled fine aggregates after modification were used, at replacement rates of 0%, 30%, 60%, and 100%, the compressive strengths in the general strength territory were 25.7, 25.4, 26.5 and 23.8 MPa, respectively, and 36.1, 36.4, 37.3 and 34.8 MPa in the high strength territory, respectively. In particular, at the replacement rate of 100%, the general strength territory had a value of 1.9 MPa and the high strength territory had a value of 1.3 Mpa, compared to the ordinary concrete, showing a tendency of a slight decline, but the difference was minimal if the recycled fine aggregate with a density of 2.33 g/cm^3, absorption rate of 4.62%, and organic impurity content of 0.49% was used. This material could be used as an alternative material to natural aggregates.

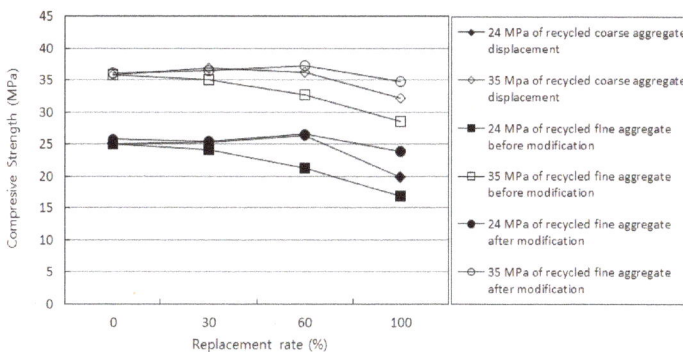

Figure 11. Compressive strength for strength area and replacement rate for recycled coarse aggregate replacement, recycled fine aggregate before modification, and recycled fine aggregate after modification.

5.3. Durability Characteristics

5.3.1. Freezing and Thawing Resistance

Freezing and thawing experiments of the recycled aggregate concrete were conducted according to KS F 2456 "standard test method for resistance of concrete to rapid freezing and thawing" and the experimental results are shown as Figure 12.

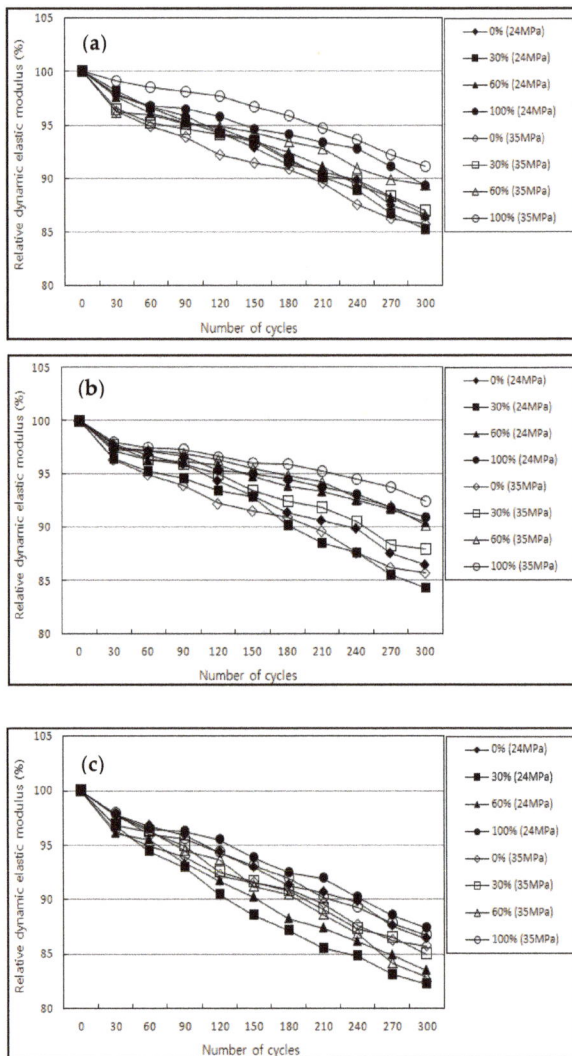

Figure 12. Relative dynamic elastic modulus and number of cycles: (**a**) Recycled coarse aggregate replacement, (**b**) recycled fine aggregate before modification, and (**c**) recycled fine aggregate after modification.

The durability index of ordinary concrete was 86, in the general strength territory and high strength territory. The durability indices of the recycled aggregate concrete that used the recycled coarse aggregate were 85, 87, and 89, in the general strength territory and 87, 89, and 91, in the high strength territory at replacement rates of 30, 60, and 100%, respectively. The durability indices of the recycled aggregate concrete that used recycled fine aggregate after modification, exhibited slightly higher durability indices than that of ordinary concrete, with values of 84, 90, and 91, in the general strength territory and 88, 90, and 92, in the high strength territory for replacement rates of 30, 60, and 100%, respectively. Furthermore, in the recycled aggregate concrete that used the recycled fine aggregate after modification, the durability indices were 82, 84, and 87, in the general strength territory and 85, 83, and 87 in the high strength territory, for replacement rates of 30%, 60%, and 100%, respectively which were slightly lower than those of the recycled coarse aggregate and recycled fine aggregate, before modification. The durability index was thought to have increased due to a buffer effect, caused by the expansion and relaxation of moisture due to freezing and thawing. This is occurs due to a lowering of the air-spacing in the concrete as a result of the entrained air bubbles attached to the recycled aggregate. Furthermore, the amount of air entrained in the concrete containing recycled fine aggregate, after modification, is relatively small. In Figure 13, the relationship between the durability index and air volume of the recycled aggregate concrete is shown. As a result of the analysis, the significance probability was 0.856, so it was considered that there was a slightly high significance probability between the durability index and air volume of the recycled aggregate concrete. Accordingly, to improve the frost resistance, it was necessary to secure an appropriate air volume through combination management, through the use of an AE agent and other factors.

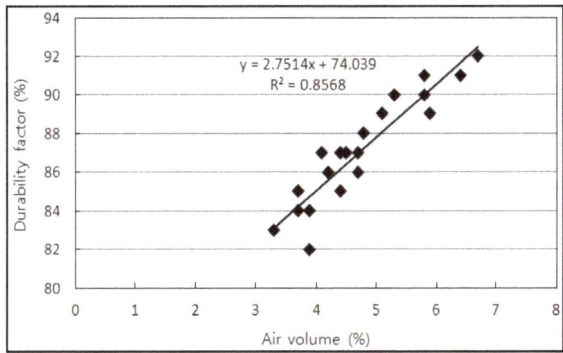

Figure 13. Relationship between durability and air volume.

5.3.2. Drying Shrinkage

Freezing and thawing experimental results are as Figure 14.

After 12 weeks of material aging, the length variation ratio due to the drying shrinkage of the ordinary concrete was 682×10^{-6}, in the general strength territory, and 526×10^{-6}, in the high strength territory. For the replacement rates of the recycled coarse aggregate of 30%, 60%, and 100%, the general strength territory exhibited length variations of 619, 652, and 533×10^{-6}, respectively, which were 4.4%–21.8% smaller than that of ordinary concrete. In the high strength territory, the length variations were 510, 476, and 508×10^{-6}, respectively, which were 3.0%–9.5% less than that of ordinary concrete. For the replacement rates of the recycled fine aggregate before modification of 30%, 60%, and 100%, the length variations in the general strength territory were 658, 417, and 454×10^{-6}, respectively, which were 3.5%–38.9% less than that of ordinary concrete, and in the high strength territory were 498, 371, and 324×10^{-6}, respectively, which were 5.3%–41.4% less than the ordinary concrete. In the replacement of the recycled fine aggregate after modification, for 30%, 60%, and 100% replacement rates, the length variations in the general strength territory were 697, 624, and 630×10^{-6}, respectively, which were -2.2%–8.5% of that of ordinary concrete. The length variations in the high strength territory were 488, 497, and 424×10^{-6}, respectively, which were 5.5%–19.4% smaller than that of ordinary concrete.

Thus, when the recycled fine aggregate before modification was used, the length variations were similar to ordinary concrete, up to a replacement of 30%, but when the replacement was 60% and 100%, the length variation ratios were significantly reduced. After modification, the use of the recycled fine aggregate produced lower length variation ratios if the volume of the recycled aggregate was increased, but there was no significant decrease up to a difference of approximately 20%. Reducing the water–cement ratio reduced the amount of water absorbed into the air-gap of the mortar attached to the recycled aggregate.

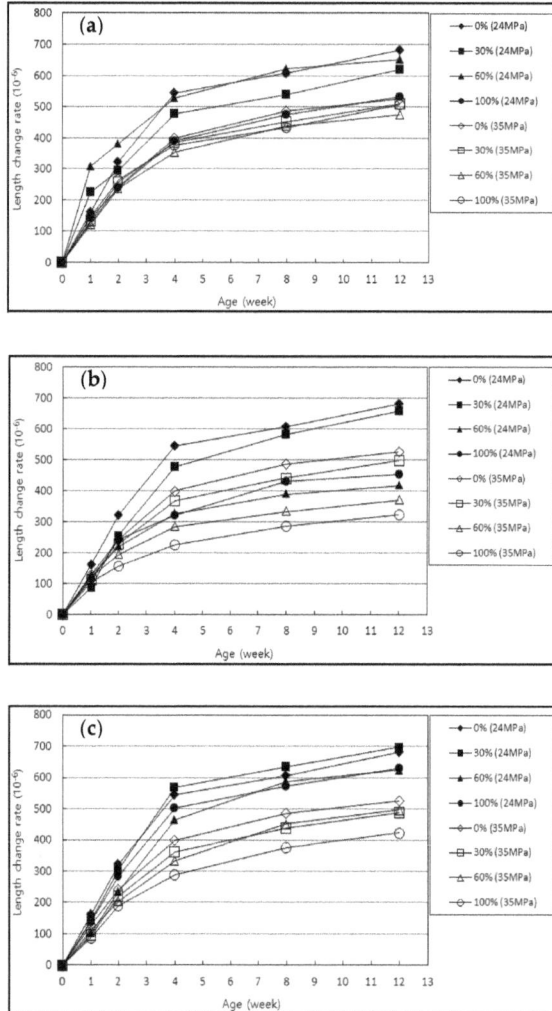

Figure 14. Length change rate according to age: (**a**) Recycled coarse aggregate replacement, (**b**) recycled fine aggregate before modification, and (**c**) recycled fine aggregate after modification.

6. Conclusions

The conclusions of this study are as follows.

- The quality before and after injecting the recycled fine aggregate using the equipment for impurity removal and stripping of recycled fine aggregate was examined. As a result, it was confirmed that the quality of the recycled fine aggregate was improved.
- For the recycled coarse aggregate, the increase in the air volume was minimal up to a replacement rate of 60%, but the air volume was increased by 1.1%–1.7% at a 100% replacement rate. For the recycled fine aggregate before modification, the air volume increased by 0.6%–1.6% at 60% of the replacement rate and 2.0%–2.2% at 100% of the replacement rate. However, there was no change in the air volume of the modified fine aggregate.
- For the recycled coarse aggregate in the slump tests, a replacement rate of 60% was appropriate. For the recycled coarse aggregate before modification, a 30% replacement rate was appropriate. For the recycled fine aggregate, a replacement rate of 60% was appropriate. For higher replacements, the slump decreased as the slump elapse time increased.
- For the recycled coarse aggregate, the compressive strength increased up to a replacement rate of 60% but decreased at a replacement rate of 100%. For the recycled fine aggregate before modification, as the replacement rate increased, the compression strength decreased. For the recycled fine aggregate after modification, a replacement rate of 100% produced similar results to those of ordinary concrete.
- Under the freezing and thawing resistance, the recycled coarse aggregate before modification and recycled fine aggregate before modification, exhibited slightly higher durability indices than that of the fine aggregate after modification. For the recycled fine aggregate after modification, the durability index was similar to that of ordinary concrete. In addition, in the mass reduction rate, the recycled coarse aggregate and the recycled fine aggregate after modification showed similar results to those of the ordinary concrete, for all combinations. However, the mass reduction rate of the recirculating fine aggregate, before modification, was increased from the replacement rate of 60% or more.
- If the recycled fine aggregate was used before modification in the drying shrinkage for up to 30% of replacement, the results were similar to those of ordinary concrete. However, for replacements of 60% or 100%, the length variation ratio significantly decreased. For the recycled coarse aggregate and the recycled fine aggregate after modification, the length variation ratio decreased when the volume of the recycled aggregate increased.

This study was conducted to generate the base data for the stable facilitation and expansion of recycled aggregate, through a concrete quality evaluation, by using the modified recycled fine aggregate and facility development for the stable production of recycled fine aggregate, under the revised quality standards. Recycled fine aggregate can be used with concrete, in an equal mixture, through aggregate pre-mixing before batch plant mixing, to improve the quality by using impurity removal and stripping removal equipment. Furthermore, this study is expected to contribute to the use of recycled aggregate in the future. As a follow-up study, the mechanical and durability characteristics of recycled aggregate concrete with varying density and various absorption rates should be explored and CO_2 emissions from recycled aggregate production should be studied.

Author Contributions: Conceptualization, J.-H.K.; Data curation, J.-H.K.; Formal analysis, J.-H.K., J.-H.S., C.-S.J. and S.-H.L.; Investigation, J.-H.S. and C.-S.J.; Project administration, J.-H.K., S.-H.L. and H.-S.K.; Supervision, H.-S.K.; Validation, H.-S.K.; Writing—original draft, J.-H.K.

Funding: This research was supported by a grant (18SCIP-C120606-03) from the Construction Technology Research Project Program funded by Ministry of Land, Infrastructure, and Transport of the Korean government.

Conflicts of Interest: The authors declare no conflict of interest.

References

1. Ministry of Environment. *National Waste Generation and Disposal Status in 2016*; Ministry of Environment: Seoul, Korea, 2017.
2. Kim, Y.C.; Bae, J.H.; Shin, J.R.; Nam, J.S.; Yeo, J.D.; Gwag, Y.G. *A study on the Mechanical Properties of Concrete According to Substitute Rate of Recycled Aggregate*; Architectural Institute of Korea: Choongju, Korea, 2010; pp. 295–299.
3. Yang, I.H.; Jung, J.Y. Effect of Recycled Coarse Aggregate on Compressive Strength and Mechanical Properties of Concrete. *J. Korea Concr. Inst.* **2016**, *28*, 105–113. [CrossRef]
4. Sim, J.S.; Park, C.W.; Park, S.J.; Kim, Y.J. Characterization of Compressive Strength and Elastic Modulus of Recycled Aggregate Concrete with Respect to Replacement Ratios. *J. Korean Soc. Civ. Eng.* **2006**, *26*, 213–218.
5. Jo, H.D. Studies on the Development of Removal Technique of Attached Mortar of Recycled Aggregate for Concrete. Ph.D. Thesis, Daegu University, Daegu, Korea, 2012.
6. Park, R.S. A Study on the Characteristics of Concrete Using the Surface Coated Recycled Aggregate. Ph.D. Thesis, Chonbuk National University, Jeonju, Korea, 2007.
7. Tam, V.W.; Gao, X.F.; Tam, C.M. Microstructural analysis of recycled aggregate concrete produced from two-stage mixing approach. *Cem. Concr. Res.* **2004**, *35*, 1195–1203. [CrossRef]
8. Etxeberria, M.; Vázquez, E.; Marí, A.; Barra, M. Influence of amount of recycled coarse aggregates and production process on properties of recycled aggregate concrete. *Cem. Concr. Res.* **2007**, *37*, 735–742. [CrossRef]
9. Zaharieva, R.; Franççois, B.B.; Wirquin, E. Frost resistance of recycled aggregate concrete. *Cem. Concr. Res.* **2004**, *34*, 1927–1932. [CrossRef]
10. Kim, Y.; Hanif, A.; Usman, M.; Munir, M.J.; Kazmi, S.M.S.; Kim, S. Slag waste incorporation in high early strength concrete as cement replacement: Environmental impact and influence on hydration; durability attributes. *J. Clean. Prod.* **2018**, *172*, 3056–3065. [CrossRef]
11. Hanif, A.; Kim, Y.; Lu, Z.; Park, C. Early-age behavior of recycled aggregate concrete under steam curing regime. *J. Clean. Prod.* **2017**, *152*, 103–114. [CrossRef]
12. Tam, V.W.Y.; Tam, C.M.; Le, K.N. Removal of cement mortar remains from recycled aggregate using pre-soaking approaches. *Resour. Conserv. Recycl.* **2007**, *50*, 82–101. [CrossRef]
13. Kou, S.C.; Poon, C.S. Properties of concrete prepared with PVA-impregnated recycled concrete aggregates Cem. *Concr. Compos.* **2010**, *32*, 649–654. [CrossRef]
14. Kou, S.C.; Zhan, B.J.; Poon, C.S. Use of a CO2 curing step to improve the properties of concrete prepared with recycled aggregates Cem. *Concr. Compos.* **2014**, *45*, 22–28. [CrossRef]
15. Ogawa, H.; Nawa, T. Improving the Quality of Recycled Fine Aggregates by Selective Removal of Brittleness Defects. *J. Adv. Concr. Technol.* **2012**, *10*, 396. [CrossRef]
16. Shima, H.; Matsuhashi, R.; Yoshida, Y.; Tatayasiki, H. Life cycle Assesment of High Quality Aggregate Recovered from Demolished Concrete with Heat and Rubbing Method. *J. Clin. Investig.* **2001**, *23*, 67–72.
17. Kiuchi, T.; Horiuchi, E. An Experimental Study on Recycle Concrete by using High Quality Recycled Coarse Aggregate. *J. Adv. Concr. Technol.* **2003**, *44*, 42–43.
18. Juan, M.S.; Gutiéerrez, P.A. Study on the Influence of Attached Mortar Content on the Properties of Recycled Concrete Aggregate. *Constr. Build. Mater.* **2009**, *23*, 875–876.
19. Ismail, S.; Ramli, M. Engineering Properties of Treated Recycled Concrete Aggregate (RCA) for Structural Applications. *Constr. Build. Mater.* **2013**, *44*, 473–475. [CrossRef]
20. Kim, Y.; Hanif, A.; Kazmi, S.M.; Munir, M.J.; Park, C. Properties enhancement of recycled aggregate concrete through pretreatment of coarse aggregates—Comparative assessment of assorted techniques. *J. Clean. Prod.* **2018**, *191*, 339–349. [CrossRef]
21. Ministry of Land, Infrastructure and Transport. *Recycled Aggregate Quality Standard*; Ministry of Land Public Notice: Nairobi, Kenya, 2017.
22. Kim, J.H.; Park, Y.S.; Park, J.M.; Youm, G.S.; Jeon, H.G. An Experimental Study on the Concrete Performance Improvement Using Pre-Mixer. *Spring Conf. J. Korea Concr. Inst.* **2011**, 287–288.

23. Kim, B.Y. An Experimental Study on the Durability of Recycled Aggregate Concrete. Ph.D. Thesis, Konkuk University, Seoul, Korea, 2005.

24. Kim, Y.; Hanif, A.; Usman, M.; Park, W. Influence of bonded mortar of recycled concrete aggregates on interfacial characteristics—Porosity assessment based on pore segmentation from backscattered electron image analysis, Constr. *Build. Mater.* **2019**, *212*, 149–163. [CrossRef]

25. Barnhouse, P.W.; Srubar, W.V. Material characterization and hydraulic conductivity modeling of macroporous recycled-aggregate pervious concrete. *Constr. Build. Mater.* **2016**, *110*, 89–97. [CrossRef]

applied
sciences

MDPI

Article

Investigation of the Use of Recycled Concrete Aggregates Originating from a Single Ready-Mix Concrete Plant

Eleftherios Anastasiou *, Michail Papachristoforou, Dimitrios Anesiadis, Konstantinos Zafeiridis and Eirini-Chrysanthi Tsardaka

Laboratory of Building Materials, Department of Civil Engineering, Aristotle University of Thessaloniki, 54124 Thessaloniki, Greece; papchr@civil.auth.gr (M.P.); d.anesiadis@gmail.com (D.A.); konzafi@gmail.com (K.Z.); extsardaka@gmail.com (E.-C.T.)
* Correspondence: elan@civil.auth.gr; Tel.: +30-2310-995787

Received: 2 October 2018; Accepted: 31 October 2018; Published: 3 November 2018

Featured Application: Recycling of the major by-products from a ready-mix concrete plant within the plant as concrete aggregates.

Abstract: The waste produced from ready-mixed concrete (RMC) industries poses an environmental challenge regarding recycling. Three different waste products form RMC plants were investigated for use as recycled aggregates in construction applications. Crushed hardened concrete from test specimens of at least 40 MPa compressive strength (HR) and crushed hardened concrete from returned concrete (CR) were tested for their suitability as concrete aggregates and then used as fine and coarse aggregate in new concrete mixtures. In addition, cement sludge fines (CSF) originating from the washing of concrete trucks were tested for their properties as filler for construction applications. Then, CSF was used at 10% and 20% replacement rates as a cement replacement for mortar production and as an additive for soil stabilization. The results show that, although there is some reduction in the properties of the resulting concrete, both HR and CR can be considered good-quality recycled aggregates, especially when the coarse fraction is used. Furthermore, HR performs considerably better than CR both as coarse and as fine aggregate. CSF seems to be a fine material with good properties as a filler, provided that it is properly crushed and sieved through a 75 μm sieve.

Keywords: ready-mixed concrete; recycled concrete aggregates; returned concrete; concrete sludge fines; soil stabilization

---

1. Introduction

The construction industry has put considerable effort over the past years into becoming more sustainable [1]. The sustainability efforts of the industry involve practices such as energy conservation, waste minimization, and recycling. These practices apply to all aspects of construction activity, including cement and concrete production, and one important sector of concrete production is ready-mixed concrete. It is estimated that ready-mixed concrete production accounts for 350 million tons per year in the EU in 2016 [2], and the waste associated with this production varies greatly depending on local construction practices and is estimated at about 0.5–2% of the total production [3]. The waste streams generated from the ready-mix concrete plants (RMC) also vary, depending on the waste management strategies employed. The literature indicates several waste management strategies for RMC plants: some strategies include the use of the washing-out process for reclaiming the aggregates from the fresh concrete returns [4], the direct use of the fresh concrete returns in downgraded concrete products [5], and water reclamation by pressuring the sedimentation tanks [6].

The main solid wastes generated from RMC are hardened concrete from trial mixes or concrete testing, returned fresh concrete, and sludge fines from the washing of trucks. All of these materials could be used, potentially, in new construction products as recycled aggregates [7–9].

The use of recycled aggregates in concrete is of paramount importance for the sustainability of the construction industry, either seen from the point of view of natural resource scarcity [10] or from the point of view of waste and landfilling minimization [11]. As in most recycling processes, it is not an easy task; recycled aggregates originate from various sources and their quality level is not guaranteed [12]. Typically, they have inferior properties compared to natural aggregates [13,14] and, furthermore, they pose some technical difficulties (contaminants, higher absorption, etc.) [15,16]. However, a great number of researchers have explored the possibilities of using recycled aggregates in concrete in a beneficial way, either by suggesting ways of upgrading the product [17–20] or by categorizing and directing the proper quality of recycling aggregates to a suitable application [21–23].

The present report focuses on three main waste products from RMC, in the form of recycled aggregates from hardened good quality concrete (HR), recycled aggregates from hardened fresh concrete returns (CR), and recycled cement sludge fines from the washing of the concrete trucks (CSF). HR was produced by crushing test concrete specimens, typically used in the quality control of RMC concrete batches. In order to have a good-quality recycled aggregate, concrete specimens with an original compressive strength of 40 MPa or higher were selected for crushing, as it is shown from the literature that the original strength of the concrete influences the properties of the recycled aggregate and of the new concrete [24,25]. CR was also produced from crushing hardened concrete; however, prior to unloading the trucks, it is common that a considerable amount of water is used to ease the placing process. Therefore, CR originated from a lower-strength concrete than HR and was expected to be of lower quality. Kou et al. (2012) suggest that CR aggregates should be used in concrete mixtures with a w/c ratio as low as 0.35. HR and CR were tested for their suitability as concrete aggregates before substituting natural aggregates in test concrete mixtures. CSF has been found to consist mostly of hardened and unhydrated cement grain and fine aggregate [26]. It is a fine material which can be easily ground and sieved to a small size and has the potential to be used as a construction material [27].

In the present report, HR and CR were considered as recycled concrete aggregates (both coarse and fine) and were tested in concrete mixtures. On the other hand, CSF was considered as a filler and was tested separately. Following an investigation on the use of similar fine materials [28–30], and after investigating its physicochemical properties, CSF was tested in the present research as a material for soil stabilization at a 10% and 20% addition rate and as a filler, substituting 10 and 20 wt %. of cement in mortars.

The recycled aggregates from RMC have the benefit of being recycled concrete aggregates (RCA) without any mixed construction and demolition waste inclusions, which renders them acceptable for many standards regarding the use of RCA in concrete [31]. In addition, if they can be recycled on-site in new concrete products, then two other benefits compared to other RCAs arise; they are expected to have less fluctuation in properties compared to RCA from various sources, and there are no transportation costs. The pretreatment required for such aggregates (drying, crushing, sieving and soaking) can easily be carried out within the RMC plant [32,33]. Therefore, the recycling of considerable amounts of RMC waste would render the plant more sustainable and environmentally friendly.

2. Materials and Methods

2.1. Recycled Concrete Aggregates (HR and CR)

Crushed limestone (L) was used as a reference aggregate, as it is the most common aggregate used in concrete production in Greece. Fresh concrete returns (CR) and recycled concrete of high compressive strength (HR) were crushed and sieved in order to achieve the same gradation as limestone aggregates.

The recycled aggregates (HR and CR) were used as natural coarse aggregate replacements or as both fine and coarse aggregate replacements in concrete mixtures. The different fractions of recycled aggregates were examined separately since they typically have different properties, such as different water absorption and different amounts of adhered mortar [34–36]. The replacement ratio of either the fine or coarse fraction was selected to be 100%, in order to determine clearly the effect of aggregate substitution on the properties of the new concrete, although the optimum replacement ratio is usually less than 100% [37,38]. The aggregates under investigation were tested for their physical and chemical suitability for use in concrete, as well as for their durability.

2.1.1. Recycled Aggregate Properties Testing

The aggregate tests that were conducted were the sand equivalent test according to EN 933-8 and methylene blue test according to EN 933-9, in order to determine the quantity and quality of their fines. Grain size distribution was measured according to EN 933-2, in order to simulate similar aggregate gradation to that of the natural aggregates. Density (bulk, dry, saturated surface dry) and water absorption for the saturated-surface dry (SSD) condition were determined in order to perform the concrete mix design. Since water absorption for the recycled aggregates is usually higher than that of natural aggregates, the recycled aggregates were used in the SSD condition in order to have the same effective w/c ration in all mixtures. X-ray diffraction analysis (XRD) was used for the determination of the mineralogical composition, with a PW 1840 Phillips diffractometer (Philips, Amsterdam, Netherlands) and Ionic Chromatography for the determination of the water-soluble sulfate salts content (SO_3wt %), with Dionex CS-1100, Thermoscientific Instruments (ThermoFisher Scientific, Waltham, USA). Regarding the assessment of the durability of the recycled aggregates, two tests were carried out: resistance to fragmentation (Los Angeles test) according to EN 1097-2 and resistance to freezing and thawing according to EN 1367-1.

2.1.2. Design of Concrete Mixtures

Five concrete mixtures were prepared in the laboratory and their terminology is given in Table 1. The reference concrete mixture with limestone aggregates (LL) was a standard C25/30 mixture with maximum aggregate size of 16 mm. All concrete mixtures were produced using 300 kg/m^3 cement CEM II32.5, 0.55 effective water/cement ratio and 1.5 kg/m^3 superplasticizer. Limestone coarse aggregates (4–16 mm fractions) were replaced in mixtures CR55 and HR55 by CR and HR aggregates, respectively. The total amount of aggregates in mixtures CR100 and HR100 consisted of CR and HR, respectively. All the aggregates were added in a saturated-surface-dry condition by estimating the water absorbed by the aggregates and pre-soaking them. Each fraction of each gradation (i.e., 0.25–0.50 mm, 0.50–1.00 mm, etc.) was separately sieved for the three aggregate types used, so that the aggregate gradation curve would be the same for all mixtures. These particle-size distribution curves are shown in Figure 1. The final proportions of all mixtures are presented in Table 2. All specimens cast were cured at 20 °C and 95% RH until testing.

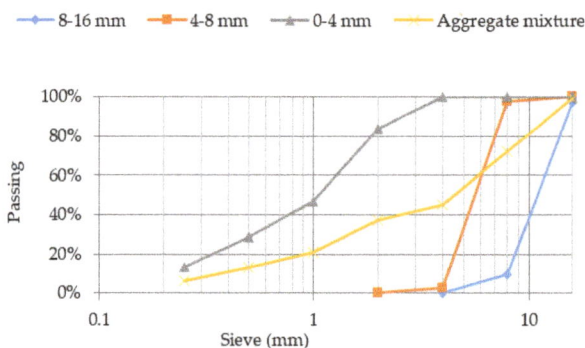

Figure 1. Particle size distribution and aggregate mix gradation curve for all test mixtures.

Table 1. Terminology for all concrete mixtures.

Mixture	Designation
LL	Reference (100% of aggregates limestone)
CR55	Recycled fresh concrete returns as coarse aggregates
CR100	100% of aggregates recycled fresh concrete returns
HR55	Recycled concrete of high compressive strength as coarse aggregates
HR100	100% of aggregates recycled concrete of high compressive strength

Table 2. Proportions of all mixtures in kg/m^3.

Concrete Composition	LL	CR55	CR100	HR55	HR100
Cement CEM II32.5	300	300	300	300	300
Tap water	165	165	165	165	165
Fine aggregate 0–4 mm	842	843	528	843	499
Coarse aggregate 4–8 mm	478	478	300	478	283
Coarse aggregate 8–16 mm	576	525	525	542	542
Superplasticizer	1.5	1.5	1.5	1.5	1.5

2.1.3. Fresh and Hardened Concrete Testing

In order to assess the fresh properties of the test mixtures, the air content, bulk density and consistency by the slump test of fresh concrete were measured according to EN 12350. Hardened concrete was tested for its mechanical properties and durability with the following tests:

- Open porosity according to RILEM CPC11.3 in water under vacuum;
- Seven and 28-day compressive strength test in 150 mm cubes (at least three for each testing age and mix);
- Elastic modulus in 150 × 300 mm cylinders (at least three for each testing age and mix);
- Three-point bending strength in 100 × 100 × 400 mm prisms (at least three for each testing age and mix);
- Drying shrinkage by 100 × 100 × 1000 mm, cured in 20 °C and <50% RH;
- Resistance to chloride ion penetration according to ASTM C 1202-97;
- Freeze–thaw resistance with de-icing salts after 10 cycles according to CEN/TS 12390-9.

2.2. Cement Sludge Fines (CSF)

Cement sludge fines (CSF) were obtained from the same RMC plant as the recycled aggregates, in wet condition. Figures 2 and 3 show the sedimentation tanks and the storage of CSF in the RMC plant, while Figure 4 shows the CSF as received. The measured moisture in CSF was 42% and, therefore, it was left to dry in laboratory conditions for 2–3 days and then oven-dried at 100 °C until a constant mass was reached. After drying, it was crushed and sieved in order to obtain a fine material passing from the 75 μm sieve. The chemical composition of CSF and of the reference CEM I42.5 cement used were determined using atomic absorption spectroscopy, AAnalyst 400, Perkin Elmer. Additionally, their water-soluble salts were determined using ionic chromatography, Dionex, while simultaneous DTA-TGA (differential thermal–thermogravimetric analysis), SDT 2960 TA Instruments, was used for the determination of the calcium carbonate ($CaCO_3$) content of CSF, under an N_2 atmosphere from 10 °C to 1000 °C. The particle size distribution of CSF was determined with a Malvern Mastersizer 2000 analyzer. Furthermore, strength activity tests with lime according to ASTM C593-95 were performed, in order to determine its potential pozzolanicity.

Figure 2. Sedimentation tank for the water-washing of the trucks.

Figure 3. Stockpiled cement sludge fines (CSF) within the ready-mixed concrete (RMC) plant.

Figure 4. CSF as received (with 42% moisture content).

2.2.1. CSF as a Replacement of Cement in Mortars

CSF was considered as a filler material and it was used as 10 and 20 wt %. Portland cement CEM I 42.5 replacement in mortars. All mixtures were prepared using a cement/siliceous river sand ratio of 0.33 and a water/binder ratio of 0.50. The fresh mortars were tested for bulk density and consistency and were cured at 20 °C and 95% RH until testing. Compressive strength was measured in 40 × 40 × 40 mm cubes (at least three for each testing age and mix) at ages of 3, 7 and 28 days. Additionally, X-ray diffraction analysis (XRD) was used for the determination of the mineralogical composition of the mortar with 20% CSF, and the porosity of hardened mortars was measured according to RILEM CPC11.3 in water under vacuum, after the completion of 28 days.

2.2.2. CSF as an Additive for Soil Stabilization

A well-known use of filler materials is for the improvement of poor soils, and CSF was also tested as an additive for soil stabilization. A sample of a low-quality reference soil with a 5% California Bearing Ratio (CBR) value was taken from a metro station excavation site in Thessaloniki, Greece. Two addition rates of CSF in soil were examined: 10 and 20 wt %. The maximum dry density and optimum moisture content according to the modified Proctor test and the CBR coefficient of soils were measured and compared.

3. Results and Discussion

3.1. Recycled Concrete Aggregates (HR and CR)

3.1.1. Recycled Aggregate Test Results

With the sand equivalent test, the calculation of SE value gives an estimation of the percentage of fines (aggregate particles that pass the 0.063 mm sieve). For higher SE values, the percentage of fines decreases. The results of the test are given in Figure 5. The higher concentration of fines was observed for CR, which can be explained by the high w/c ratio of concrete returns, which reduces the mechanical properties of the cement paste attached on the aggregates and thus produces more fines during crushing. On the other hand, in HR aggregates, the fines concentration is relatively low since the hardened cement paste is of high strength.

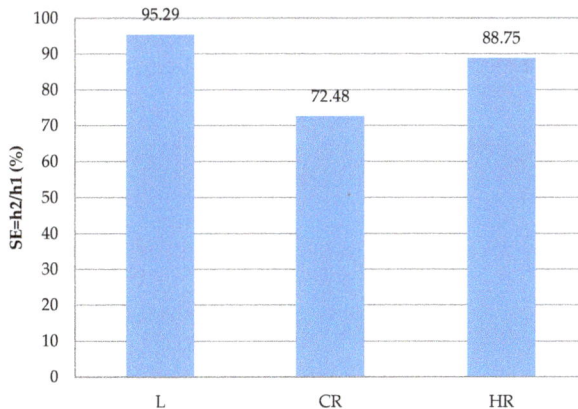

Figure 5. Results of sand equivalent test. L: crushed limestone; CR: crushed hardened concrete from returned concrete; HR: crushed hardened concrete from test specimens of at least 40 MPa compressive strength.

After identifying the content of fines in the recycled aggregates, it was considered necessary to determine if any clay particles exist in these fines and also measure the particle size distribution of aggregates. The results of the methylene blue test that was conducted for this purpose can be seen in Figure 6. It was observed that, for CR aggregates, a higher amount of injected dye solution is required, which is an indication of clay concentration. On the contrary, the concentration of clay in HR fines is insignificant. Since CR is 100% recycled concrete, the existence of clay can be attributed to storing and handling these aggregates in the facility of RMC.

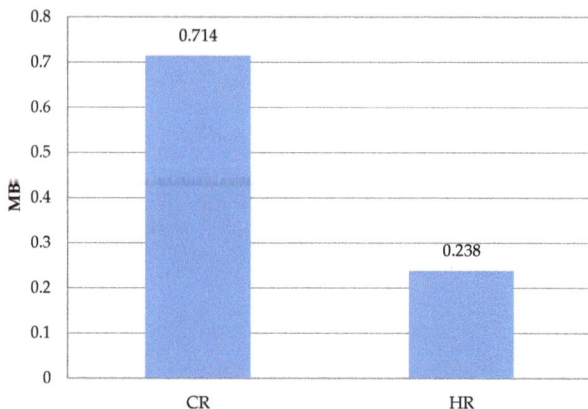

Figure 6. Results of methylene blue test.

Table 3 shows the results of water absorption (WA_{24}), particle bulk density (p_a), particle dry density (p_{rd}), particle saturated-surface-dry density (p_{ssd}) and Los Angeles coefficient (LA) tests. CR seems to absorb more water compared to HR, which is an indication of lower quality. However, the densities of the two aggregates do not differ essentially. The LA coefficient of CR was lower, but comparable to HR, and both aggregates have acceptable resistance to fragmentation according to the EN 12620 categorization. The mineralogical compositions of the raw materials CR and HR are given in Figures 7 and 8. The presence of calcite and quartz in both materials is related to the binder attached on the aggregates, as well as tri-calcium–silicate compounds (C_3S) traces. The C_3S in CR was attributed to unhydrated forms of cement. Additionally, portlandite content (P) originating from the hydration of cement paste was found in HR.

Table 3. Aggregate test results.

	CR Fine	CR Coarse	HR Fine	HR Coarse
WA24 (%)	6.13	5.45	5.88	3.21
p_a (kg/m^3)	1723	2604	1617	2610
p_{rd} (kg/m^3)	1558	2281	1477	2409
p_{ssd} (kg/m^3)	1654	2405	1563	2486
LA	-	31.1%	-	36.5%

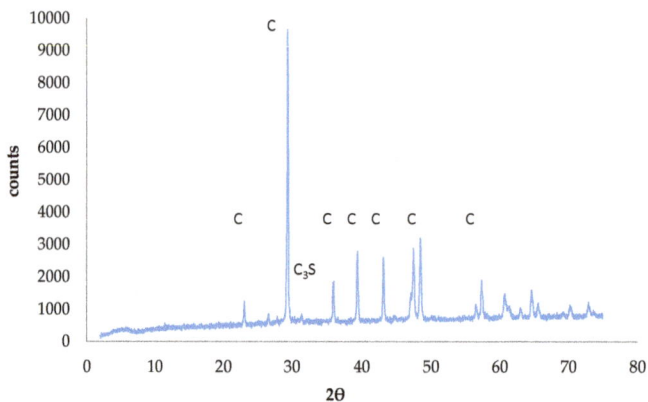

Figure 7. XRD diagram of CR, where C: calcite, Q: quartz and traces of C_3S.

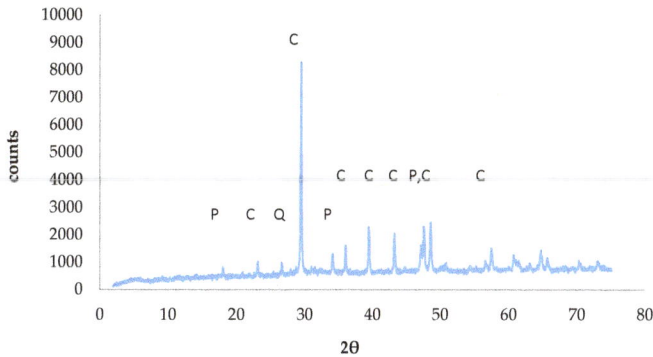

Figure 8. XRD diagram of HR, where C: calcite, Q: quartz and P: portlandite.

The durability test results for CR and HR regarding mass loss (wt %) after 10 cycles of freezing–thawing are given in Figure 9. HR aggregates showed better results compared to CR. The mass loss of both aggregates is less than the upper category limit of 40% according to EN 12620.

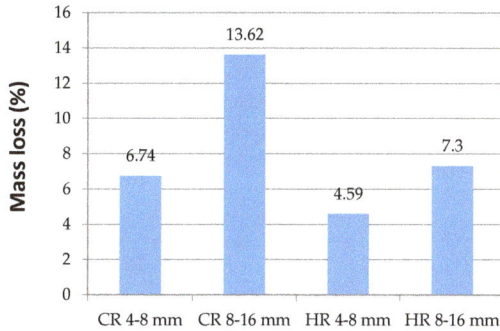

Figure 9. Resistance to freezing and thawing expressed by mass loss.

The sieved representative samples (<0.75 mm) of CR and HR were collected in order to quantify their water-soluble sulfate salts, utilizing ionic chromatography. Sulfate salts were found in a low proportion in the mate and HR, respectively.

Although both CR and HR originate from hardened concrete, with a recycled concrete content of 90% or higher, it seems clear that HR is a higher quality aggregate, as expressed by the water absorption values, the fines quality characterization and the durability testing.

3.1.2. Concrete Test Results

The results of the fresh concrete tests are shown in Table 4. Slump values increased as the rate of recycled aggregate increased, which can be attributed to the method of achieving the SSD condition of the aggregates, with HR aggregates showing better results. The recycled aggregates have been soaked with the amount of water required to reach the SSD state; however, it is not certain if the aggregates have actually absorbed all the water in the presoaking process or if they produced free water during the mixing process, mainly due to their porous nature. It is very probable that they did not remain completely saturated and that an amount of free water was released from the aggregates, resulting in an increased w/c ratio and increased workability. Of course, there is a reduction in the strength of the concrete with recycled aggregates, which is attributable to this excess water. The air content of all mixtures was around 2%, regardless of the type of aggregates used. The density seemed to decrease as the rate of recycled aggregates in the mixture increased, but the reduction is greater, as expected, when the fine recycled aggregates are used.

Table 4. Fresh concrete properties.

	LL	CR55	CR100	HR55	HR100
Air content (%)	?	?	?3	??	?
Slump (cm)	6.5	11	17	15	18
Density (kg/m³)	2418	2318	2165	2378	2157

The results of the tests that were conducted on hardened concrete are given in Table 5, while in Figure 10, various mechanical properties of the recycled-base mixtures (Xi) are compared to the corresponding ones of reference concrete (X_{LL}). HR-based mixtures showed increased mechanical properties compared to those with CR aggregates and were lower compared to reference concrete. The open porosity was similar to that of the reference concrete when coarse recycled aggregates were used, but it increased considerably when the fine recycled aggregate was used. Compared to LL, HR 55 and HR100 showed a reduction of compressive strength of only 11% and 15%, respectively, which indicates that even the finer fraction of HR does not reduce the compressive strength of the produced concrete considerably. The flexural strength and elastic modulus, on the other hand, rely more on the quality of the interfacial transition zone between the aggregate and the cement paste, and the reduction in these values is in the range of 9–13% and 28–38% for HR55 and HR100, respectively. This implies that the bond between the fine aggregates and the cement paste was weaker when fine HR was used. Regarding CR, however, CR 55 and CR100 showed a reduction of compressive strength by 26% and 47%, respectively, which corresponds with the lower properties of CR measured earlier and the same trend applied for flexural strength and elastic modulus.

Table 5. Hardened concrete tests.

		LL	CR55	CR100	HR55	HR100
Open porosity (%)		13	11	21	10	22
Compressive strength	7-day	33	22	18	28	27
(MPa)	28-day	38	28	20	34	33
28-day flexural strength (MPa)		7.03	5.36	4.87	6.10	5.05
28-day elastic Modulus (GPa)		44.7	40.2	25.3	40.8	27.6
28-day drying shrinkage (µstrain)		180	400	590	220	300
Resistance to chloride	Charge passed	2491	3912	8284	2913	9342
Ion penetration	Chloride ion penetrability	MODERATE	MODERATE	HIGH	MODERATE	HIGH
Freeze–thaw mass loss (wt %)		0.422	0.362	0.845	0.412	0.890

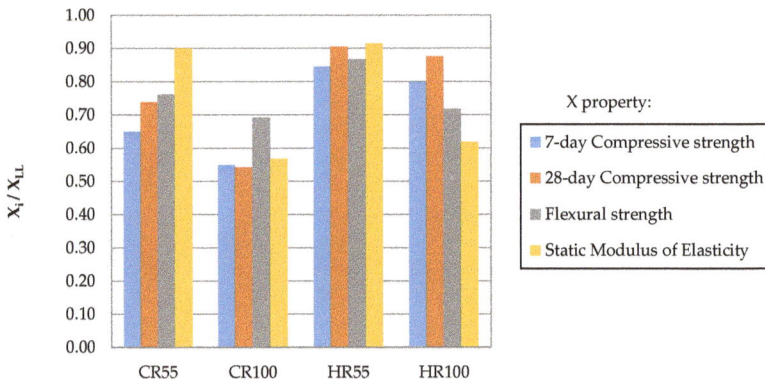

Figure 10. Comparison of various properties of mixtures (X) with the reference concrete (X_{LL}).

Regarding the durability tests, as shown in Table 5, the concretes with coarse CR and HR showed similar chloride ion penetrability to the reference concrete, while the mixtures with both fine and coarse recycled aggregates showed considerably higher penetrability. This can be attributed to the increased porosity of CR100 and HR100, which allowed the chloride ions to enter the concrete mass. Freeze–thaw results serve only as an indication of durability, since more cycles are usually required; however, the tested concretes with only coarse recycled aggregates seem to perform in a similar way to the reference concrete. The drying shrinkage results, as shown in Figure 11, show that HR aggregates, regardless of size, perform better than CR aggregates. In particular, concrete HR55 was closer to the reference concrete, while CR100 showed a distinct increase in shrinkage and, hence, reduced durability, which corresponds to the previous tests.

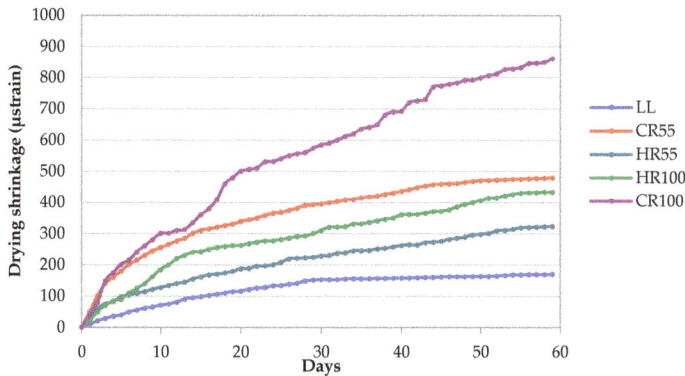

Figure 11. Drying shrinkage of test concretes.

In general, the mechanical properties of concrete decreased as the rate of recycled aggregates increased. However, the mixtures with coarse recycled aggregates (CR55 and HR55) showed only a minimal decrease in both mechanical and durability properties when compared to the reference concrete. Furthermore, there is a clear distinction between the two types of recycled aggregates, as HR performed better than CR in all cases. In the cases of compressive strength and drying shrinkage, concrete with both fine and coarse HR showed better results than concrete with only coarse CR. These results, along with the initial test results on the materials, imply that HR aggregates are of better quality than CR aggregates and should be managed separately within the RMC plant and directed at different applications.

3.2. Cement Sludge Fines (CSF)

The chemical composition of the CSF, OPC (CEM I42.5 N) and soil used are shown in Table 6, while their particle size distribution in comparison to that of CEM I42.5 is shown in Figure 12. CSF consists mostly of hardened cement paste and fine limestone aggregates; therefore, its CaO and SiO_2 contents were not expected to be reactive. The high value of loss on ignition can be attributed to the bonded water proportion and CO_2 quantity that was emitted from the sample, while the DTA-TG results (Figure 13) indicated a small amount portlandite (0.18 wt %) and the presence of a high amount of $CaCO_3$ (67.77 wt %).

Table 6. Chemical composition of CSF and cement I42.5 N.

(wt %)	Na_2O	K_2O	CaO	MgO	Fe_2O_3	Al_2O_3	SiO_2	L.I.%	Cl^-	NO_3^-	SO_3
CSF	0.19	0.21	32.09	1.05	1.26	2.45	36.51	25.22	0.01	<0.01	0.05
CEM I42.5 N	0.57	1.08	66.84	3.91	2.40	3.74	19.55	1.91	0.03	0.02	1.49
Soil	0.90	0.98	8.67	5.39	3.79	6.88	63.14	10.25	0.08	0.02	0.02

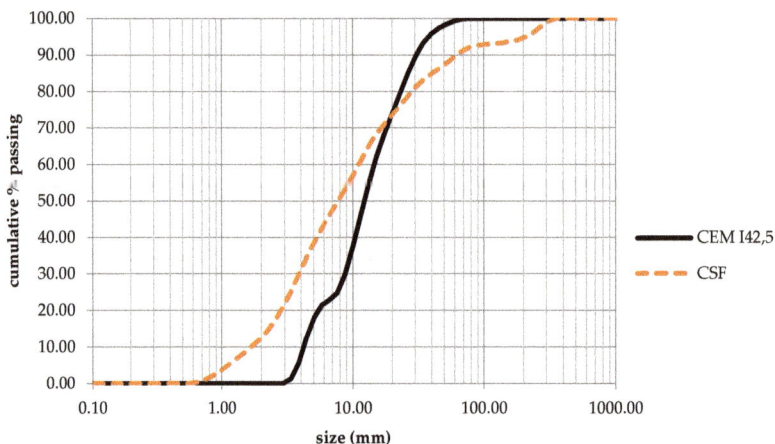

Figure 12. Particle size distribution of CSF and cement.

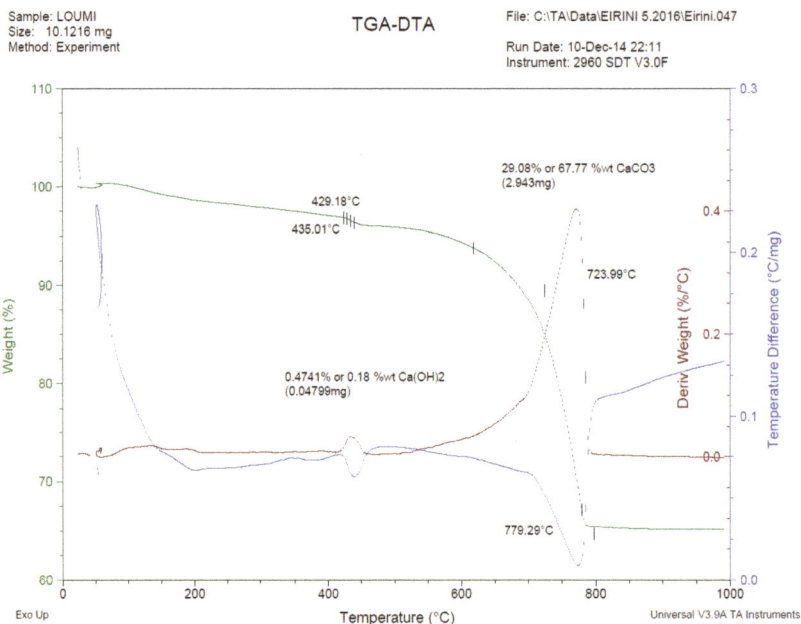

Figure 13. Differential thermal–thermogravimetric analysis (DTA-TGA) analysis of CSF and quantification of calcite content as a percentage by mass of sample.

The XRD diagram (Figure 14) shows that the mineralogical composition of CSF was mainly calcite. Also, an absence of C_3S and traces of C_2S peaks were observed. The latter may indicate full hydration of sludge. Also, a strength development test with lime according to ASTM C593-95 showed minimal strength development at 28 days (<0.50 MPa), which implies that there is no significant pozzolanic activity from CSF. The results of the analytical tests imply that CSF is probably a fine material without any hydraulic or pozzolanic properties. The fact that it can be easily crushed and sieved through the 75 μm implies that CSF could be used as a filler material. Regarding handling and processing, it would be preferable to use CSF in its original wet condition in order to avoid the drying process; however, further research is required in this direction.

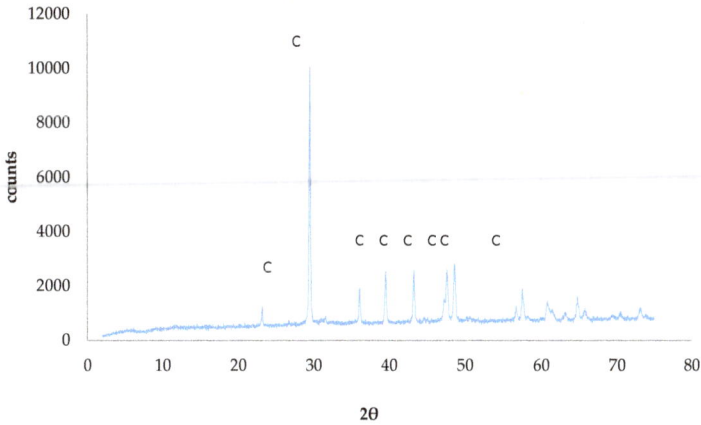

Figure 14. XRD diagram of CSF, where C: calcite.

3.2.1. CSF as a Cement Replacement in Mortars

The results of compressive strength testing are displayed in Figure 15. Mortar with a 10% replacement of cement by CSF reached almost the same strength levels as the reference. The early strength was slightly lower than the reference; however, the 28-day strength was slightly higher compared to reference. The mortar with 20% CSF showed a 7% reduction of strength compared to the reference. These results can be attributed to the filler effect [39], while the XRD results (Figure 16) on 28-day mortar samples with 20% CSF confirm the absence of ettringite. When a filler is added in a certain percentage, it influences the pH value of the binders in the pore solution. If the pH is reduced, this affects the hydration and as a result the compressive strength [40].

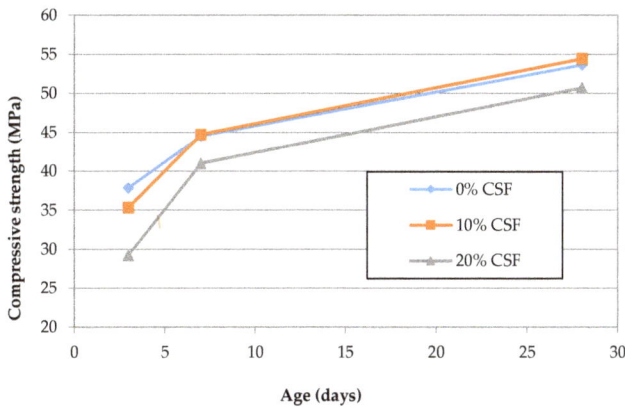

Figure 15. Compressive strength development of mortars.

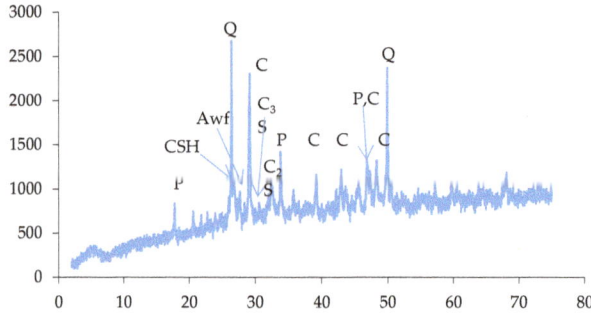

Figure 16. XRD diagram of cement mortar with 80% cement and 20% CSF, where C: calcite, P: portladite, Q: quartz, Awf: Awfillite, CSH: hydrated cement.

The decrease of compressive strength when 20% CSF was added can be related to the calcite content of CSF and its behavior as filler. Similar results have been observed by Nehdi et al. when limestone filler was used for cement substitution in mortars [41]. The porosity of mortar with 10% CSF was 6.6% and of mortar with 20% CSF was 10.3%, showing an increase compared to the reference mortar with no CSF inclusion, which had 3.5% porosity. This increase can be attributed to the porous nature of the hydrated cement mortar present in CSF, in a similar manner to the porous nature of fine recycled aggregates.

3.2.2. CSF for Soil Stabilization

The soil used as a reference had the chemical composition shown in Table 6, while its Atterberg limits were as follows: liquid limit LL: 56, plastic limit PL: 29.4 and plasticity Index PI: 26.6. Overall, it was categorized as poor for subgrade. Figure 17 shows the dry density versus moisture content for soil with 0%, 10% and 20% soil addition. As shown in Table 7, the dry density increased as the CSF rate increased while the required moisture in order to reach the maximum dry density reduced. Both of these properties are desirable for soil stabilization, but the most significant increase is that of the load-bearing capacity of soils, expressed by the CBR value. Furthermore, a higher percentage of CSF increased the CBR value even more. A similar strength increase has been reported (at lower addition percentages) when using lime for soil improvement [42,43].

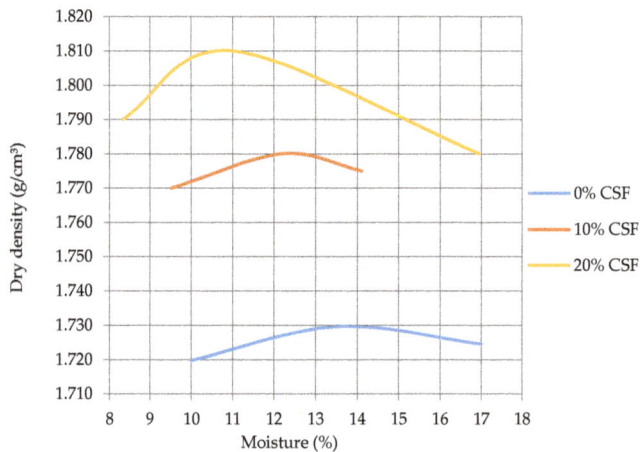

Figure 17. Dry density versus moisture content for soil stabilized with various CSF additions.

Table 7. Soil stabilization test results.

	0% CSF	10% CSF	20% CSF
Maximum dry density (g/cm^3)	1.73	1.78	1.81
Optimum moisture content (%)	14.0	12.4	10.6
CBR (%)	5	12	33

The considerable increase in the CBR value of the soil treated with CSF implies that it can serve as a potential soil stabilization material, provided that it is primarily tested for micropollutants. The higher addition rate required compared to other possible additions (lime, cement) is not a great disadvantage, as CSF is a low-value industrial by-product. Also, Figure 17 shows that that soil with 20% CSF addition was more sensitive to water variation compared to soil without CSF, which needs to be taken account at the design stage.

4. Conclusions

The main conclusions drawn from the research can be summarized as follows:

- The properties of the recycled aggregates produced within a single RMC plant vary greatly, despite the fact that they are recycled concrete aggregates, without any mixed construction materials. Therefore, they should be managed separately and directed to suitable applications;
- All three recycled aggregates tested can be potentially be used in construction applications, after simple processing (crushing, sieving and pre-soaking), indicating that they do not require the energy and cost for transportation outside the RMC plant;
- The recycled aggregates from hardened concrete specimens (HR) had better properties compared to recycled aggregates from concrete returns (CR) and the use of their fine fraction could also be considered for structural concrete production;
- The durability properties of concrete with HR and CR were slightly reduced when 100% of the coarse fraction was used and considerably reduced when 100% of the fine and coarse fraction was used. Future research could determine the optimum amount of natural aggregate replacement, as in all recycled aggregates;
- CSF seems to be a good-quality fine material, provided that it is crushed and sieved through the 75 μm sieve;
- CSF could potentially be used as a filler material in cement mortar production and as an additive for soil stabilization.

Author Contributions: Conceptualization, E.A. and M.P.; methodology, E.A.; validation, M.P. and E.-C.T.; formal analysis, E.-C.T.; investigation, D.A. and K.Z.; data curation, M.P., D.A., K.Z. and E.-C.T.; writing—original draft preparation, E.A. and M.P.; writing—review and editing, E.A.**Funding:** This research received no external funding.

Acknowledgments: The authors would like to thank G. Moraitis and Interbeton S.A. for their assistance in providing information and materials for the present research.

Conflicts of Interest: The authors declare no conflict of interest.

References

1. Behera, M.; Bhattacharyya, S.K.; Minocha, A.K.; Deoliya, R.; Maiti, S. Recycled aggregate from C&D waste & its use in concrete—A breakthrough towards sustainability in construction sector: A review. *Constr. Build. Mater.* **2014**, *68*, 501–516. [CrossRef]
2. European Ready Mixed Concrete Organization. Ready-Mixed Concrete Industry Statistics—Year 2016. Available online: http://ermco.eu/new/wp-content/uploads/2018/07/ERMCO-Statistics-2016-18.02.04.pdf (accessed on 28 September 2018).

3.   Sealey, B.J.; Hill, G.J.; Phillips, P.S. Review of Strategy for Recycling and Reuse of Waste Materials. In *Recycling and Reuse of Sewage Sludge, Proceedings of the International Symposium Organised by the Concrete Technology Unit and Held at the University of Dundee, Scotland, UK, 19–20 March 2001*; Dhir, R.K., Limbachiya, M.K., McCarthy, M.J., Eds.; Thomas Telford: London, UK, 2001; pp. 325–336.

4.   Vieira, L.D.B.P.; de Figueiredo, A.D. Evaluation of concrete recycling system efficiency for ready-mix concrete plants. *Waste Manag.* **2016**, *56*, 337–351. [CrossRef] [PubMed]

5.   Xuan, D.; Poon, C.S.; Zheng, W. Management and sustainable utilization of processing wastes from ready-mixed concrete plants in construction: A review. *Resour. Conserv. Recycl.* **2018**, *136*, 238–247. [CrossRef]

6.   Xuan, D.; Zhan, B.; Poon, C.S.; Zheng, W. Innovative reuse of concrete slurry waste from ready-mixed concrete plants in construction products. *J. Hazard. Mater.* **2016**, *312*, 65–72. [CrossRef] [PubMed]

7.   Pedro, D.; De Brito, J.; Evangelista, L. Influence of the use of recycled concrete aggregates from different sources on structural concrete. *Constr. Build. Mater.* **2014**, *71*, 141–151. [CrossRef]

8.   Kou, S.C.; Zhan, B.J.; Poon, C.S. Feasibility study of using recycled fresh concrete waste as coarse aggregates in concrete. *Constr. Build. Mater.* **2012**, *28*, 549–556. [CrossRef]

9.   Audo, M.; Mahieux, P.Y.; Turcry, P. Utilization of sludge from ready-mixed concrete plants as a substitute for limestone fillers. *Constr. Build. Mater.* **2016**, *112*, 790–799. [CrossRef]

10.   Poon, C.S.; Chan, D. The use of recycled aggregate in concrete in Hong Kong. *Resour. Conserv. Recycl.* **2007**, *50*, 293–305. [CrossRef]

11.   Rao, A.; Jha, K.N.; Misra, S. Use of aggregates from recycled construction and demolition waste in concrete. *Resour. Conserv. Recycl.* **2007**, *50*, 71–81. [CrossRef]

12.   Yehia, S.; Abdelfatah, A. Examining the Variability of Recycled Concrete Aggregate Properties. In Proceedings of the International Conference on Civil, Architecture and Sustainable Development (CASD-2016), London, UK, 1–2 December 2016; pp. 57–60.

13.   Katz, A. Properties of concrete made with recycled aggregate from partially hydrated old concrete. *Cem. Concr. Res.* **2003**, *33*, 703–711. [CrossRef]

14.   Rahal, K. Mechanical properties of concrete with recycled coarse aggregate. *Build. Environ.* **2007**, *42*, 407–415. [CrossRef]

15.   Sagoe-Crentsil, K.K.; Brown, T.; Taylor, A.H. Performance of concrete made with commercially produced coarse recycled concrete aggregate. *Cem. Concr. Res.* **2001**, *31*, 707–712. [CrossRef]

16.   Poon, C.S.; Shui, Z.H.; Lam, L. Effect of microstructure of ITZ on compressive strength of concrete prepared with recycled aggregates. *Constr. Build. Mater.* **2004**, *18*, 461–468. [CrossRef]

17.   Shi, C.; Li, Y.; Zhang, J.; Li, W.; Chong, L.; Xie, Z. Performance enhancement of recycled concrete aggregate—A review. *J. Clean. Prod.* **2016**, *112*, 466–472. [CrossRef]

18.   De Juan, M.S.; Gutiérrez, P.A. Study on the influence of attached mortar content on the properties of recycled concrete aggregate. *Constr. Build. Mater.* **2009**, *23*, 872–877. [CrossRef]

19.   Dimitriou, G.; Savva, P.; Petrou, M.F. Enhancing mechanical and durability properties of recycled aggregate concrete. *Constr. Build. Mater.* **2018**, *158*, 228–235. [CrossRef]

20.   Pepe, M.; Toledo Filho, R.D.; Koenders, E.A.; Martinelli, E. Alternative processing procedures for recycled aggregates in structural concrete. *Constr. Build. Mater.* **2014**, *69*, 124–132. [CrossRef]

21.   Silva, R.V.; De Brito, J.; Dhir, R.K. Availability and processing of recycled aggregates within the construction and demolition supply chain: A review. *J. Clean. Prod.* **2017**, *143*, 598–614. [CrossRef]

22.   Oikonomou, N.D. Recycled concrete aggregates. *Cem. Concr. Compos.* **2005**, *27*, 315–318. [CrossRef]

23.   Agrela, F.; De Juan, M.S.; Ayuso, J.; Geraldes, V.L.; Jiménez, J.R. Limiting properties in the characterisation of mixed recycled aggregates for use in the manufacture of concrete. *Constr. Build. Mater.* **2011**, *25*, 3950–3955. [CrossRef]

24.   Padmini, A.K.; Ramamurthy, K.; Mathews, M.S. Influence of parent concrete on the properties of recycled aggregate concrete. *Constr. Build. Mater.* **2009**, *23*, 829–836. [CrossRef]

25.   Yang, K.H.; Chung, H.S.; Ashour, A. Influence of type and replacement level of recycled aggregates on concrete properties. *ACI Mater. J.* **2008**, *3*, 289–296. [CrossRef]

26.   Audo, M.; Mahieux, P.Y.; Turcry, P.; Chateau, L.; Churlaud, C. Characterization of ready-mixed concrete plants sludge and incorporation into mortars: Origin of pollutants, environmental characterization and impacts on mortars characteristics. *J. Clean. Prod.* **2018**, *183*, 153–161. [CrossRef]

27. Chini, S.A.; Mbwambo, W.J. Environmentally friendly solutions for the disposal of concrete wash water from ready mixed concrete operations. In Proceedings of the CIB W89 Beijing International Conference, Beijing, China, 21–24 October 1996; pp. 21–24.

28. Mashaly, A.O.; El-Kaliouby, B.A.; Shalaby, B.N.; El-Gohary, A.M.; Rashwan, M.A. Effects of marble sludge incorporation on the properties of cement composites and concrete paving blocks. *J. Clean. Prod.* **2016**, *112*, 731–741. [CrossRef]

29. Galetakis, M.; Soultana, A. A review on the utilisation of quarry and ornamental stone industry fine by-products in the construction sector. *Constr. Build. Mater.* **2016**, *102*, 769–781. [CrossRef]

30. Cardoso, R.; Silva, R.V.; de Brito, J.; Dhir, R. Use of recycled aggregates from construction and demolition waste in geotechnical applications: A literature review. *Waste Manag.* **2016**, *49*, 131–145. [CrossRef] [PubMed]

31. De Brito, J.; Saikia, N. *Recycled Aggregate in Concrete: Use of Industrial, Construction and Demolition Waste*; Springer: London, UK, 2012; ISBN 9781447145394.

32. Kazaz, A.; Ulubeyli, S.; Er, B.; Arslan, V.; Atici, M.; Arslan, A. Fresh ready-mixed concrete waste in construction projects: A planning approach. *Organ. Technol. Manag. Constr. Int. J.* **2015**, *7*, 1280–1288. [CrossRef]

33. Ferreira, L.; De Brito, J.; Barra, M. Influence of the pre-saturation of recycled coarse concrete aggregates on concrete properties. *Mag. Concr. Res.* **2011**, *63*, 617–627. [CrossRef]

34. Zhao, Z.; Remond, S.; Damidot, D.; Xu, W. Influence of fine recycled concrete aggregates on the properties of mortars. *Constr. Build. Mater.* **2015**, *81*, 179–186. [CrossRef]

35. Evangelista, L.; Guedes, M.; De Brito, J.; Ferro, A.C.; Pereira, M.F. Physical, chemical and mineralogical properties of fine recycled aggregates made from concrete waste. *Constr. Build. Mater.* **2015**, *86*, 178–188. [CrossRef]

36. Khatib, J.M. Properties of concrete incorporating fine recycled aggregate. *Cem. Concr. Res.* **2005**, *35*, 763–769. [CrossRef]

37. Etxeberria, M.; Vázquez, E.; Marí, A.; Barra, M. Influence of amount of recycled coarse aggregates and production process on properties of recycled aggregate concrete. *Cem. Concr. Res.* **2007**, *37*, 735–742. [CrossRef]

38. Evangelista, L.; De Brito, J. Mechanical behaviour of concrete made with fine recycled concrete aggregates. *Cem. Concr. Compos.* **2007**, *29*, 397–401. [CrossRef]

39. Berodier, E.; Scrivener, K. Understanding the Filler Effect on the Nucleation and Growth of C-S-H. *J. Am. Ceram. Soc.* **2014**, *97*, 3764–3773. [CrossRef]

40. De Weerdt, K.; Haha, M.B.; Le Saout, G.; Kjellsen, K.O.; Justnes, H.; Lothenbach, B. Hydration mechanisms of ternary Portland cements containing limestone powder and fly ash. *Cem. Concr. Res.* **2011**, *41*, 279–291. [CrossRef]

41. Nehdi, M.; Mindess, S.; Aïtcin, P.C. Optimization of high strength limestone filler cement mortars. *Cem. Concr. Res.* **1996**, *26*, 883–893. [CrossRef]

42. Manasseh, J.; Olufemi, A.I. Effect of lime on some geotechnical properties of Igumale shale. *Electron. J. Geotech. Eng.* **2008**, *13*, 1–12.

43. Amu, O.O.; Fajobi, A.B.; Oke, B.O. Effect of eggshell powder on the stabilizing potential of lime on an expansive clay soil. *J. Appl. Sci.* **2005**, *5*, 1474–1478. [CrossRef]

applied
sciences

MDPI

Article

Effect of Different Types of Recycled Concrete Aggregates on Equivalent Concrete Strength and Drying Shrinkage Properties

Sungchul Yang

School of Architectural Engineering, Hongik University, Sejong 30016, Korea; scyang@hongik.ac.kr;
Tel.: +82-10-2523-2665

Received: 10 October 2018; Accepted: 5 November 2018; Published: 8 November 2018

Featured Application: Use of coarse RCA produced from old PC concrete is promising for structural concrete mix, which is proportioned by the modified equivalent mortar volume method, especially for a bottom layer of two-lift paving concrete. This concrete mix exhibits equivalent strength and drying shrinkage properties.

Abstract: Residual mortar attached to recycled concrete aggregate (RCA) always leads to a decrease in Young's modulus and an increase in the drying shrinkage of RCA concrete, mainly due to an increase of total mortar volume. To overcome this inherent problem, the modified and equivalent mortar volume (EMV) methods were proposed by researchers. Despite the comparable test results, both models are still subject to the slump loss problem. Thus, under the same W/C (water to cement ratio) ratio and slump condition, this study assessed the influence of the modified EMV mix method on RCA concrete properties. A total of six mixes were proportioned using the modified EMV method with three different RCAs. Test results show that the concrete mixed with RCA produced from old PC concrete sleepers exhibited compressive strength, Young's modulus, and flexural strength values within 2% variation, equivalent to those values of the companion natural aggregate concrete. In other mixes, compressive strength was found to decrease to 11–20%. It was observed that for 100% replacement of RCA mix, Young's modulus increased to 10% and drying shrinkage increased to 8% only, while for 50% replacement of RCA mix, Young's modulus decreased to 8% and drying shrinkage dropped to 4%.

Keywords: recycled concrete; aggregate; mixture proportioning

---

1. Introduction

There is a general consensus in the literature that recycled concrete aggregate (RCA) is more porous and heterogeneous than natural aggregate. High-quality RCA can be obtained from waste concrete via a crushing process. This usually entails three to seven steps, including the elimination of foreign substances, rebar, and residual mortar (RM). During the crushing process, the RM quantity adhering to the RCA is altered. The primary properties adversely influenced by the adhered RM are density, absorption, etc. [1]. In particular, adhering RM in RCA results in greater porosity and consequently greater water absorption of RCAs [2], where the porosity is represented by the water absorption [3]. Usually, RCA with lower strength resulted in higher porosity in the aggregate and a newly made interfacial transition zone (ITZ). This, in turn, influences the properties of the concrete that is produced afterward. As a result of the increase in the RM, the physical concrete characteristics are impacted, including the compressive strength, Young's modulus, flexural strength, drying shrinkage, thermal expansion coefficient, freeze-thaw resistance, etc. [4–7]. Thus, many researchers have carried out various experimental studies on the use of RCA, such as using RCA source derived from precast concretes, the two-lift paving method, modification of mixing processes, new mix design approaches, etc. [8–12].

One ideal way of acquiring high-quality RCA is to derive it from precast concrete [13–15] or concrete sleepers. The main advantage of retaining RCA from sleepers is the possibility of producing reliable products and reducing sorting costs. Furthermore, several research groups have investigated the material properties of high performance RCA concrete railway sleepers and found that, in comparison to the use of ordinary concrete, adequate material properties could be obtained [16–18].

According to Federal Highway Administration (FHWA) data [7], the use of RCA on bases of new pavements is currently allowed in 41 state departments of transportation (DOT) in the USA. Among these states, 11 states use RCA for paving concrete mix. Moreover, two-lift concrete pavements have been successfully constructed in the USA with the bottom layer containing low-quality recycled aggregate concrete [19–21]. It was pointed out by Shi et al. [22] that two-lift construction using recycled materials in the bottom lift can have the highest positive impacts from a social and environmental perspective.

Tam and Tam [23] proposed a two-stage mixing approach (TSMA), while Sicakova and Urban [24] suggested a triple- mixing procedure to manufacture strong and durable RCA concrete. TSMA divides the mixing process into two parts: initial mixing of all the aggregates and half of the required water and final concrete mixing with the other half of the required water and cement. It was observed that the strength, shrinkage, creep and permeability properties of RCA concrete were enhanced by adopting TSMA [23,25]. The triple-mixing process divides the mixing process into three parts: coating coarse aggregates mixed by application of additive and a certain amount of the water required for coating, adding cement with fine aggregate, and final concrete mixing with the remaining water and plasticizer. It was observed that the density, compressive and splitting tensile strength, and water absorption capacity properties were improved.

New modified mix proportioning methods for producing RCA concrete have been proposed by a few researchers [4,5,26]. The equivalent coarse aggregate mass (ECAM) method was proposed by Gupta et al. [26]. The main concept is that the attached mortar is treated as part of the sand. Test results show that the compressive strength values of the RAC mixes (up to 50% of RCA replacement ratios) were comparable with the compressive strength of the control natural concrete mixes, while slump decreased with the addition of RCAs. The original equivalent mortar volume (EMV) method proposed by Abbas [4] and Fathifazl [5] is considered effective for structural concrete mixes, typically with about 800 kg/m^3 of fine aggregate. However, the characteristics of the EMV method lead to lower fine aggregate amounts, resulting in a rough mix and slump loss [5,13], but these are acceptable for paving concrete. The low slump problem may be overcome in paving concrete by forcibly vibrating the pavement surface with a slipform paver to finish it [13]. In response to these issues, Kim et al. [27] proposed a revised EMV method. It was assumed that some part of the RM volume fraction may be arithmetically considered as that of the original virgin aggregate (OVA), whereas the other part is considered as total mortar (TM). Figure 1 shows the concepts of different mix designs such as the modified EMV as well as traditional mixture designs. Note that TCA denotes the total coarse aggregate volume. Looking at the conventional mix design in Figure 1b, it can be seen that the TM volume of RCA concrete is greater than the TM volume of natural aggregate (NA) concrete, shown in Figure 1a. It is shown in Figure 1b that RCA is the sum of RM and OVA (equal to TCA). Therefore, the traditional RCA concrete mix design yields TM volume increase, which successively influences material properties. Figure 1c illustrates the unique characteristics of the EMV method. As explained before, the TM volume in the RCA concrete shown in Figure 1c, which is considered as the sum of the new mortar (NM) and RM volumes, is equal to the TM volume of NA in the traditional concrete shown in Figure 1a [8]. The NM volume in Figure 1c decreased in proportion to the RM amount. In the modified EMV model shown in Figure 1d, RM adhering to RCA acts as aggregate in the fresh state concrete, and later hardens as mortar. Now, the volume of RM of RCA concrete is treated by the mortar volume fraction (RM$_a$) and the other aggregate volume fraction (RM$_b$). Additional explanation of the revised EMV concept can be found in the reference [13].

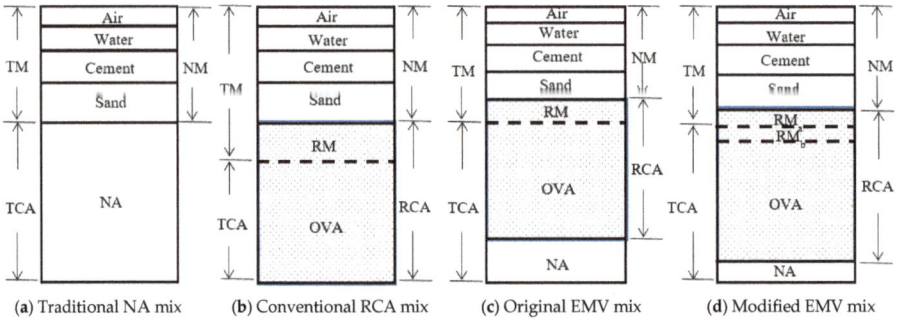

Figure 1. Schematic diagrams of different mix designs; TM: Total Mortar; TCA: Total Coarse Aggregate; NA: Natural Aggregate; NM: New Mortar; RM: Residual Mortar; OVA: Original Virgin Aggregate; RCA: Recycled Concrete Aggregate.

Previous studies have mainly focused on obtaining test results for the elastic modulus [5,8,12,13,27,28] and drying shrinkage [28–30] of RCA concrete mixes with EMV mixes, similar to those of mixes made with NCA or RCA by the traditional mix proportioning method. It should be noted that previous studies on mixes made using the revised EMV method did not consider improvement in compressive strength. Test results illustrated that RCA concretes made with the revised EMV mix method did not always yield compressive strengths that were similar to those of the control mixes [5,8,12,13,27,28].

Despite test results comparable to those for the modified EMV mix designs, the model is still subject to the slump loss problem. It should be noted that previous test results were obtained from EMV mixes, which showed lower slump values with variation of air content, in comparison to the control NA concrete mix proportioned from the ACI mix design. Therefore, this study sought to assess the influence of the revised EMV mix proportion method on the mechanical strength and drying shrinkage properties of RCA concretes, where the same W/C ratio, and slump and air content were applied. This experimental study used three grades of RCAs with different water absorption ratios, where two different RCAs (with water absorption ratio of 3.82% and 6.61%) were crushed from the same source of old concrete. The third RCA was produced from old PC concrete sleepers.

2. Experimental Program

This experimental study used recycled concrete aggregates (RCAs) produced from two different sources in South Korea. The 'RA' aggregate and 'RB' aggregate were crushed with a maximum size of 20 mm RCA from the same source of old concrete at 'I' recycling plant in South Korea. Meanwhile, the 'RR' aggregate was produced with a maximum size of 20 mm RCA from old railway concrete sleepers. The specific gravity [31], absorption ratio [31], LA abrasion coefficient [32], and residual mortar content (RMC) of the RCAs properties were tested [33,34] and are presented in Table 1. Test results showed that the average absorption ratio of 'RA', 'RB', and 'RR' was 3.82%, 6.61%, and 4.53%, respectively. It should be noted that all RCAs did not meet the specified Korean standards (KS) for concrete with respect to 2.5 as the specific gravity and 3.0% as water absorption, except for the 'RA' aggregate, which marginally satisfied the specific gravity standard with a value of 2.52 [35–38].

The RMC values for three RCAs were determined by the thermal treatment suggested by Juan and Gutierrez. [33]. Recycled aggregate samples were prepared and dried in a muffle furnace (DF-5 model made from Daeheung Science in South Korea) at 500 °C for two hours. The sample was then immersed in cold water. Extra mortar that still remained may be removed by the sudden cooling.

Table 1. RCA material properties; RA: Type A recycled aggregate; RB: Type B recycled aggregate; RR: Recycled aggregate derived from Railway concrete sleepers; KS: Korean standards.

Test Items	RA	RB	RR	KS Specification [38]
Specific gravity	2.52	2.34	2.48	>2.5
Absorption ratio (%)	3.82	6.61	4.53	<3.0
LA abrasion coef. (%) [1]	-	-	32.2	<25(paving), 40(others)
RMC [2,3]	25.0 [2]	46.8 [2]	39.9 [2] (40.1 [3])	-

[1] tested by reference [32], [2] RMC test results from thermal treatment of reference [33], [3] RMC test results from chemical treatment of reference [34].

In addition, for the 'RR' aggregate, the RMC value was evaluated by the chemical treatment method recommended by Akbarnezhad et al. [34]. Recycled specimens are prepared in a 2 L beaker and a 3 M H_2SO_4 solution was added, where the volume was five times higher than the RCA sample volume. Finally, the samples were washed with a 4-mm sieve to detach the tangled mortar, and the washed and oven-dried RCA sample weights were weighed to evaluate RMC contents.

The RMC was evaluated by using the following equation [8]:

$$RMC\ (\%) = (W_{RCA} - W_{OVA})/W_{RCA} \tag{1}$$

where W_{RCA} is the first oven-dried RCA sample weight and W_{OVA} is the final oven-dried OVA weight after removal of the RM. It is surprising from the test results of the 'RR' sample in Table 1 that the average RMC value of 39.9% acquired from the thermal treatment method was very close to the RMC value of 40.1% obtained from the chemical treatment method.

Natural river sand was incorporated as fine aggregate. Aggregate test results showed specific gravity with a value of 2.60 and absorption ratio with a value of 0.95%. Figure 2a shows that the particle size of the river sand is well distributed along the midpoint of the lower and upper limit of the gradation test requirement in Korean standards. Crushed granite was used as natural coarse aggregate (NA) with the specific gravity of 2.71 and the water absorption ratio of 0.37. Table 2 tabulates the material properties of natural aggregates. Figure 2b shows the aggregate gradation results, satisfying Korean standards.

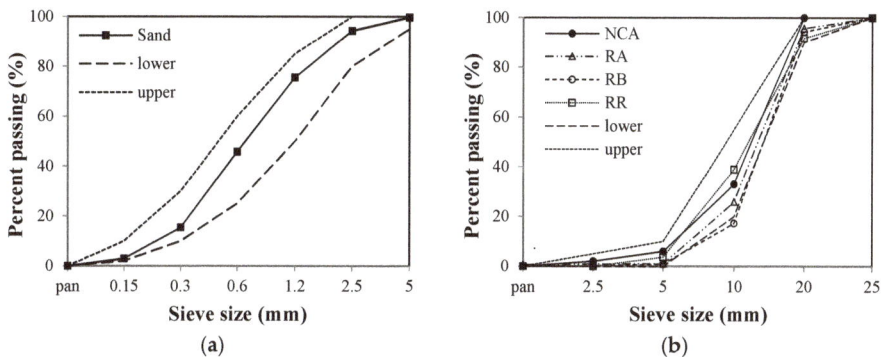

Figure 2. Aggregate gradation: (**a**) fine aggregate and (**b**) coarse aggregates.

Table 2. Material properties of natural aggregate.

Test Items	Specific Gravity	Absorption Ratio (%)
Fine aggregate	2.60	0.95
Natural coarse aggregate	2.71	0.37

3. Experimental Tests

3.1. Mix Design

Six mixtures of 35 MPa grade concrete have been studied. The notations are footnoted in Table 3. Based on the traditional mix design (CNC mix) as the reference mix, five other concrete mixtures were mixed with the modified EMV. Figure 3 shows cement and sand reductions, and coarse aggregate additions in the modified EMV mix designs, compared to those of the reference mix design with NCA. The ERA1 mix proportioned according to the modified EMV method with S = 1 (the original EMV method) results in a 28.3% decrease of cement and 28.2% of sand, but an increase of total coarse aggregate of 25.0%.

Table 3. Mix proportion of concrete per 1 m^3; W/C: Water to cement ratio; FA: Fine aggregate; NCA: Natural coarse aggregate, a: Total aggregate (FA+NCA+RCA).

Number	Mix-ID [1]	W/C	FA/a	RCA, %	S	Materials (kg)					
						W	C	FA	NCA	RCA	Admixture [2] (%)
I	CNC	0.39	45.0	0	-	187	480	742	907	0	0.5
II	ERA1	0.39	32.4	100	1	146	374	579	0	1210	0.7
III	ERA2	0.39	38.7	100	2	165	424	656	0	1037	0.5
IV	ERB1	0.39	34.3	50	1	152	390	604	622	537	0.65
V	ERB2	0.39	39.5	50	2	168	431	666	546	472	0.5
VI	ERR3	0.39	38.2	100	3	163	418	646	0	1046	0.5

[1] Firstly, C represent conventional mix and E as EMV mix. Secondly, N denotes natural coarse aggregate and RA, RB, and RR denote RCAs explained in Table 1. Thirdly, 1–3 is calculate from RMa/RM. [2] Superplasticizer was adopted as admixture.

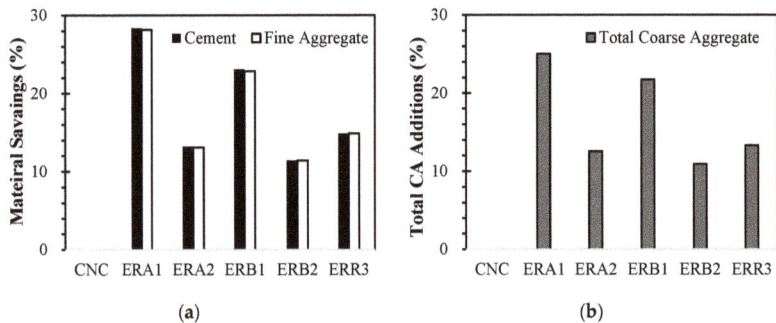

Figure 3. Material quantity change: (**a**) savings in cement and fine aggregate and (**b**) additions in total coarse aggregate.

Sixty-liter volume capacity of pan mixer was utilized at the laboratory located in Hongik University of South Korea. The superplasticizer in the mixing water was thoroughly dispersed, before the addition of water. Portland Type I cement was subsequently added and the mixer was operated for approximately one minute and thirty seconds. Then, the remaining water was added while the pan mixer was operating and the concrete was mixed for another two minutes.

3.2. Test Preparation

All mechanical strengths were the average of three specimens. The compressive strength and Young's modulus were tested according to ASTM C 39 [38] and ASTM C 469 [39], respectively, at 7 and 28 days. The flexural strength and split tensile strength of each mixture were tested at 28 days only.

Drying shrinkage tests were measured with a dial gauge suggested by the KS standard [40], which is similar to ASTM C 157 [41]. Rectangular samples of 100 × 100 × 400 mm were used.

The shrinkage strain was evaluated by an absolute digimatic indicator (ID-S112 model made from Mitutoyo in Japan) with an 0.001 mm resolution. The samples were maintained in an environmental chamber (20 °C and 60% Relative Humidity).

4. Experimental Test Results

4.1. Results of Slump, Air Content and Density

Figure 4 shows the test results of concrete properties at the fresh state such as slump, air content. The mixtures slumps ranged between 140–155 mm and are depicted in Figure 4a. It was explained by Fathifazl [5] that, primarily because of higher water content, concrete mix with NCA resulted in bigger slump, comparing to the EMV mixture, and this usually results in slump loss. In this study, however, the slump loss was adjusted by the superplasticizer.

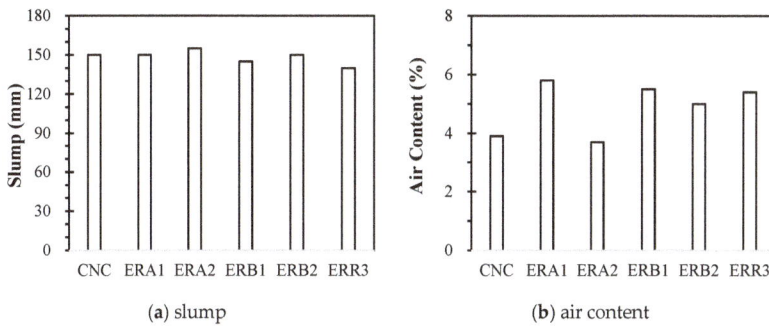

(a) slump   (b) air content

Figure 4. Test values of slump and air content by different mix design methods.

The air contents values of the mixtures ranged between 3.7–5.8% depicted in Figure 4b. Air content of the control mix was 3.9%, whereas that of the modified EMV mixtures ranged from 5.0 to 5.8%, except for the ERA2 mix. It was noted in a previous study [28] that higher air content in the modified EMV mixes may be a result of entrapped air because of its rough mixture and smaller mortar amount. Nonetheless, it appears that the ERA2 mix with a smaller admixture amount, compared to the ERA1 mix, results in the lowest air content value.

Figure 5 shows the density variation of each mixture at the fresh state and hardened states, compared to the CNC mixture. Figure 5 shows that as concrete hardened, its density showed little gain. Figure 5 indicates that the relative densities at the hardened state of the EMV mixtures (excluding 92.8% of ERA2) ranged from 95.6–97.6, compared to the CNC mix. The density of the ERA2 mix dropped about 3% more in both the fresh and hardened states. It is suspected that specimens of the ERA2 were inappropriately made.

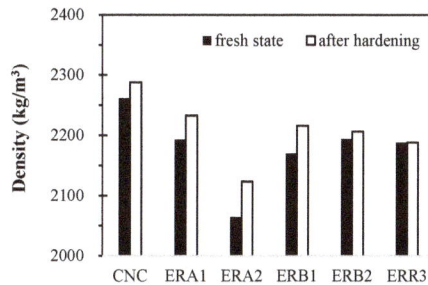

Figure 5. Test values of density by different mix design methods.

4.2. Compressive Strength

Figure 6a presents the average compressive strengths of concrete samples at 7 days and 28 days. In addition, the relative strength ratio of the ERA series, ERB series, and ERR mix are shown in Figure 6b–d, compared to the compressive strength of the control specimens (CNC).

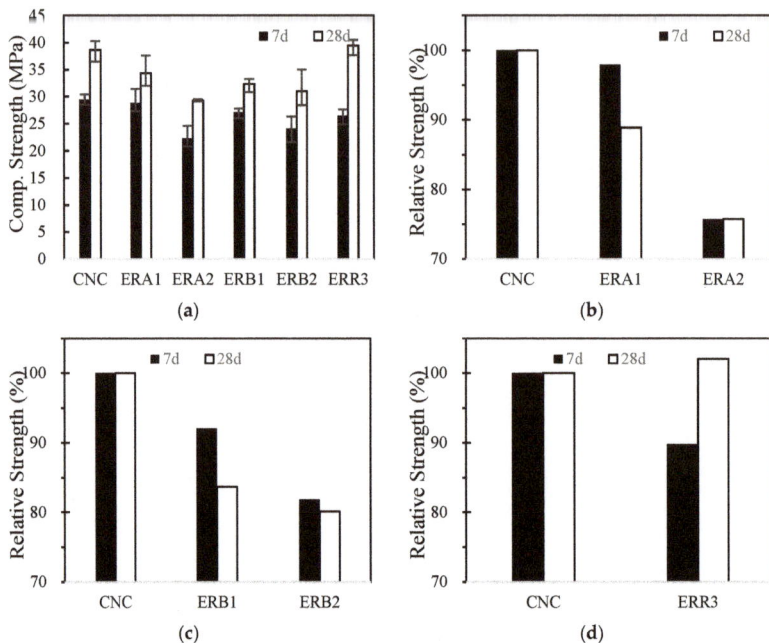

Figure 6. Compressive strength results of concrete samples (**a**) strength, (**b**) relative value of ERA series, (**c**) relative value of ERB series, and (**d**) relative value of ERR mix.

In Figure 6b,c, the relative strengths of the ERA concrete series and ERB concrete series are compared to the CNC concrete, respectively. Test results showed that the compressive strengths decreased by 24% in the ERA2 mix (S = 2), compared to the control CNC mix and ERA1 mix (S = 1). It should be noted that the water-cement ratio, slump, and air content were kept to be almost the same in these mixes in Table 3 and Figure 4a,b, thus not affecting compressive strengths. However, the FA/a ratio (see Table 3) of ERA1 and ERA2 is 32.4% and 38.7%, respectively, whereas that of ERB1 and ERB2 is 34.3% and 39.5%. Because of variation of the FA/a ratio, the compressive strength was ERA1 > ERA2 and ERB1 > ERB2. Obviously, poor water absorption of the RA (3.82%) and RB (6.61%) aggregates resulted in a decrease in the compressive strength, whereas the NA aggregate showed good water absorption (0.37%).

Comparing the test results between the ERA series and ERB series in 'S = 1', the compressive strength was ERA1 > ERB1. Once again, the same water-cement ratio, slump, and air content were employed. A lower FA/a ratio of ERA1 might account for greater compressive strength gain. Ho et al. [42] found an increase of compressive strength of RCA concrete at early ages of 3 days and 7 days with greater replacement of RCA up to 100%. They asserted that the mortar strength of this concrete was likely superior to that of the control mixes, due to the reduction of the effective W/C ratio in the RCA mixes. An increase of compressive strength of RCA concrete with up to 30% replacement of RCA was also reported by Paul et al. [43], while equivalent compressive strength of RCA concrete was attained with 50% replacement and 100% replacement of RCA by Paul [44].

Next, in 'S = 2', however, the compressive strength was ERA2 < ERB2. Very poor density of ERA2 in Figure 6, which might be ascribed to it having the lowest density (see Figure 5), yielded the converse test result.

Test results in Figure 6d indicates that the compressive strength of the ERR mix is superior to that of the control CNC mix. It is due to the high quality of the RCA manufactured from the PC concrete sleepers. This was explained in the author's former study [28]: RCA produced from wastes of precast structural concrete is of high quality and clean.

4.3. Young's Modulus

Figure 7 shows the average Young's modulus of concrete samples at 7 days and 28 days with the traditional mixture and the modified EMV mixtures.

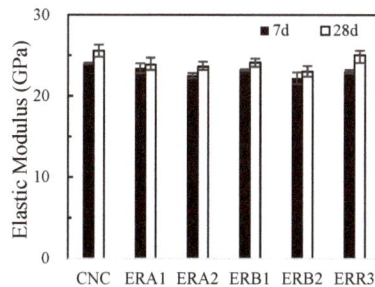

Figure 7. Young's modulus results of concrete samples.

Figure 7 shows that all the other RCA mixes, regardless of having different mix designs, yielded 2–7% decreases in the Young's modulus value at seven days. At 28 days, this gap is enlarged to range from 6–10%. Comparing the relative compressive strength drop of 11–20% in Figure 6b,c, the 6–10% decrease in the relative modulus of the modified EMV concrete mixes might be the result of the characteristic in the modified EMV method. It was explained by Fathifazl et al. [8] that modulus is proportional to the volume of total mortar; however, strength is mostly dependent on the strength of the mortar and the ITZ. Ho et al. [42] reported equivalent modulus test results of RCA concretes with replacement levels of 30%, 50%, and 100% of RCAs. It was pointed out that the modulus of RCA is lower than that of NA and, consequently, due to the porous nature of RCA, the difference in the modulus of RCA and hardened cement paste will be smaller than that of NA and hardened cement paste.

Meanwhile, the ERR mix, which contained high-quality RCA produced from PC concrete sleepers, resulted in only a 2% decrease in the elastic modulus value, compared to the control CNC mix. It is very clear that the strength decrease of concrete produced with RCA is due to the porous interface transition zone (ITZ) surface. Further studies are needed to enhance the strengthening of ITZs and their surroundings by using minerals such as silica fume, slag or ashes.

4.4. Flexural Strength

Figure 8a presents the average flexural strength of concrete samples at 28 days with the traditional mixture and the modified EMV mixtures. In addition, the relative strength ratios of the ERA series, ERB series, and ERR mix are depicted in Figure 8b–d, respectively, compared to the flexural strength of the control specimens (CNC). Compared to the remarkable drop in the compressive strength of as much as 24% in the ERA2 mix, the difference in the flexural strength was reduced to 16% in the ERA2 mix. Two studies explained [45,46] that the flexural strength of RCA concrete is not greatly affected by the presence of RCA, compared to flexural behavior of NA concrete because of the interfacial bond and better mechanical interlocking resulting from rough-textured as well as angular RCA. This trend also can be explained by the relation between flexural strength (modulus of rupture: MOR) and compressive strength. Price [47] suggested from his tests that the MOR of concrete to compressive strength is 14.5% with a compressive strength of 27.6 MPa and 12.8% with 41.4 MPa. However, test results in the ERB series show that the MOR to compression is about 18%, as shown in Table 4. Therefore, either ERB1 or ERB2 mix may be preferred for concrete pavements, where the pavement is loaded in bending.

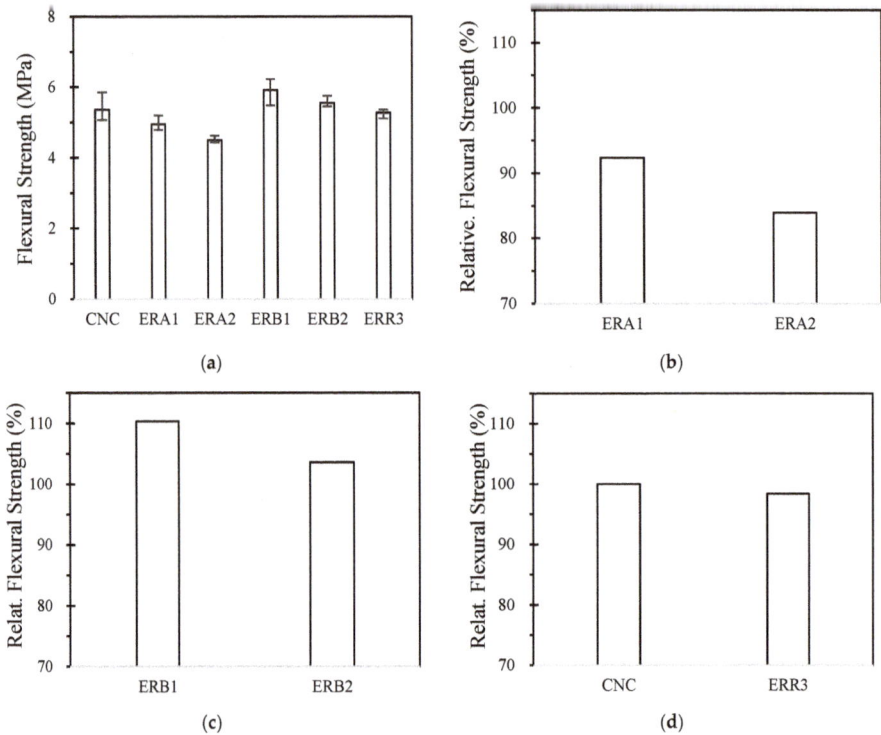

Figure 8. Flexural strength results of concrete samples (**a**) strength, (**b**) relative value of ERA series, (**c**) relative value of ERB series, and (**d**) relative value of ERR mix.

Table 4. Relation between compressive, flexural, and split tensile strength of concrete; Mix-Id: mixture identification.

Mix-Id	Average Strength of Concrete (MPa)			Ratio (%)	
	Compression	Flexure	Split Tension	Flexure to Compression	Split Tension to Compression
CNC	38.7	5.36	3.72	13.9	9.6
ERA1	34.4	4.95	3.40	14.4	9.9
ERA2	29.3	4.50	3.05	15.3	10.4
ERB1	32.4	5.91	3.47	18.3	10.7
ERB2	31.0	5.55	3.31	17.9	10.7
ERR3	39.5	5.27	3.30	13.4	8.4

It is seen in Figure 8b,c that at the same water-cement ratio, slump, and air content, EMV mixtures with S = 1 (ERA1 and ERB1) have higher flexural strength than those with S = 2 (ERA2 and ERB2). The lower FA/a ratio of the ERA1 and ERB1 mixes, compared to the ERA2 and ERB2 mixes, contributed to higher flexural strength.

Next, comparing test results between the ERA series (Figure 8b) and ERB series (Figure 8c), it was observed that flexural strength is decreased as RCA replacement content is increased. Hundred percent of RCA was replaced in the ERA series, while only 50% RCA was replaced in the ERB series. Thus, the relative strength value of the ERB series (104–110%) is far greater than the ERA series (84–92%). It was pointed out by Tripura et al. [48] that failure through old residual mortar (here in ERA series) might result in lower flexural strength and a more irregular failure pattern.

Test results in Figure 8d indicates that the flexural strength of the ERR mix is similar to that of the control CNC mix. The high quality of the RCA, which was produced from PC concrete sleepers, may be one of the main reasons for this.

4.5. Split Tensile Strength

Figure 9a shows the average split tensile strength of concrete samples at 28 days. In addition, the relative strength ratios of the ERA series, ERB series, and ERR mix are shown in Figure 9b–d, respectively, compared to the split tensile strength of the control specimens (CNC). In contrast with the relation of the MOR to compressive strength, the split tension to compressive strength lineally follows the compressive strength trend. In Price's study [47], the split tension of concrete to compressive strength was suggested to be 8.5% with a compressive strength of 4000 psi (27.6 MPa) and 7.7% with 6000 psi (41.4 MPa). Test results in Table 4 shows that the split tension to compression ranges from 9.6–10.7%, except for the ERR3 with a value of 8.4%. Overall, except for the ERR3 mix, a slightly higher split tensile strength values than expected were obtained.

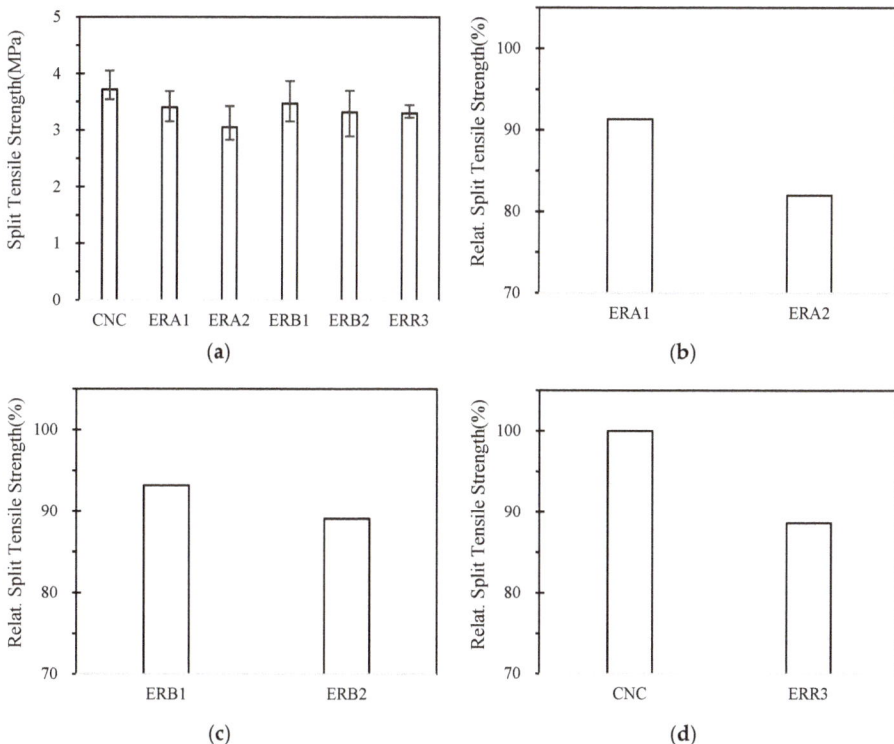

Figure 9. Split tensile strength results of concrete samples (**a**) strength, (**b**) relative value of ERA series, (**c**) relative value of ERB series, and (**d**) relative value of ERR mix.

From Figure 9b,c, the relative split tensile strengths of the ERA concrete series and ERB concrete series were compared to the CNC concrete, respectively. As discussed before, due to variation in the FA/a ratio, split tensile strength was ERA1 > ERA2 and ERB1 > ERB2. Poor water absorption of the RA

(3.82%) and RB (6.61%) aggregates resulted in a decrease of 9–11% (except for 18% of the ERA2) in the split tensile strength results, in contrast to the good water absorption of the NA aggregate (0.37%). It is plainly seen that the strength gap in the split tension narrowed relative to the reference mix, compared to the values of 11–20% (except for 24% of the ERA2) in the compressive strength results.

Comparing the test results between the ERA series and ERB series, except for the ERA2 mix, very similar tensile strength trends were observed. Once again, very poor density of ERA2 in Figure 9 might result in it having the lowest tensile strength.

Test results in Figure 9d shows that the ERR3 mix exhibited 90% of the relative tensile strength to the control mix. According to Price's interpretation of the relation between the tensile strength of concrete to compression, other mixes (9.6–10.7%) had somewhat greater tensile strength gains, in comparison to the compressive strength gains, than the ERR3 mix (8.4%), as tabulated in Table 4.

4.6. Drying Shrinkage

Test results of drying shrinkage are shown in Figure 10a, and their relative drying shrinkage to that of the reference specimen CNC is shown in Figure 10. Finally, at 50 days, the shrinkage strain of the control specimen CNC was 851 μ m/m. Compared to that of the control mix, the drying shrinkage values of the ERA1, ERA2, ERB1, ERB2, and ERR mix were 814 μ m/m, 922 μ mm/mm, 919 μ mm/mm, 1057 μ mm/mm, and 871 μ mm/mm, respectively, indicating roughly a 4% decrease, 8% increase, 8% increase, 24% increase, and 2% increase.

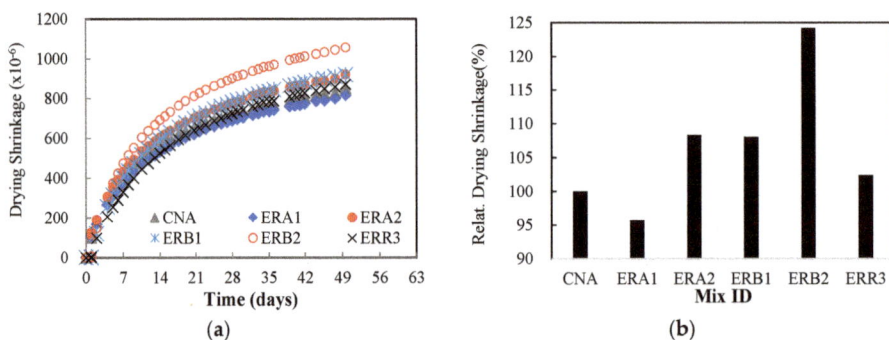

Figure 10. Test results of drying shrinkage of concrete samples (a) shrinkage, (b) relative value.

Using the ACI equation [49], the drying shrinkage difference between the ERA1 and ERA2 mixes can be analyzed. Influencing factors are air content, slump, cement contents, and fine aggregate ratio. Due to the very similar test results for the slump and air content in all mixes, it can be inferred by the ACI 209 equation that combined correction factors from slump and air content do not affect the final drying shrinkage values of all mixes. However, correction factors for fine aggregate ratios of ERA1, ERA2, ERB1, ERB2, and ERR3 dramatically varied at 0.75, 0.84, 0.78, 0.85, and 0.83, respectively. Similarly, cement content changes the correction factor but only slightly. Thus, the drying shrinkage values by the ACI equation were expected to range in the order of ERA1 < ERB1 < ERR3 < ERA2 < ERB2. In a very similar manner, the test results were ERA1 < ERR3 < (ERB1, ERA2) < ERB2. The second best drying shrinkage value from the ERR3 mix may be attributed to be the good quality of the RCAs manufactured from the PC concrete sleepers. Except for the ERB2 mix, the modified EMV mixes are viable against the drying shrinkage, compared to the reference CNC mix.

5. Conclusions

This study assessed the influence of the revised EMV mix proportion method on mechanical strength and drying shrinkage properties of RCA concretes, which are proportioned to have the same W/C ratio, slump and air content. This experimental study used three grades of RCAs with different water absorption ratios, where two different RCAs (water absorption ratio of 3.82% and 6.61%) were crushed from the same source of old concrete. The third RCA was produced from old PC concrete sleepers.

Six mixes were studied for typical structural concrete where the control concrete contained natural coarse aggregate mixed according to the traditional ACI method and the others were prepared with the revised EMV method. From this study, the following conclusions are drawn. Here the test results of ERA2 mix were excluded in the concluding discussion due to low mechanical strength values, which may be ascribed to its low density.

(1) Due to the nature of a lower slump problem that often occurs in the modified EMV method, all the mixture slump values were controlled by using a superplasticizer to range between 140–155 mm and air contents values with 3.7–5.8%.
(2) Except for the split tensile strength, test results showed that the ERR mix with 100% RCA replacement, which was produced from old PC sleepers mixed by the revised EMV mix method, exhibited equivalent compressive strength, Young's modulus, and flexural strength values to the companion reference mix of natural aggregate. In addition, the relative drying shrinkage increased only 2% to the companion control mix.
(3) In other mixes (except for ERR mix), compared to the drop of 11–20% of the compressive strength, the modulus of the modified EMV mixes resulted in only a 6–10% decrease to the companion control mix, which is the result of the characteristic in the revised EMV method.
(4) In the modified EMV mixes with RCA replacement of 100%, the flexural strength of concretes decreased by 8–16%. However, with 50% replacement of RCA mixes, the strength increased by 4–10% and thus may be preferred for concrete pavements, which are loaded in bending.
(5) Although a 7–11% decrease was observed in the modified EMV mixes, the split tension to compressive strength of concrete lineally follows the compressive strength trend.
(6) At 50 days, test results revealed that drying shrinkage of the modified RMV mixes with RCA exhibited a 4% drop to only an 8% increase. There was one, except for the ERB2 mix with a 24% increase, and this might be affected by relatively higher fine aggregate ratio and cement content, compared to the other EMV mixes.

Further studies should be carried out to enhance strengthening ITZs of RCA and their surroundings by using minerals such as silica fume, slag, or bottom ashes. The equivalency of mechanical strength properties then may be more clearly acquired by the revised EMV proportioning method with any RCA source.

Funding: This research was funded by the National Research Foundation (2018 Korea Grant of the Korean Government) for a project on "Structural Performance of Reinforced Concrete Members made with Revised Equivalent Volume Mix Proportioning Method (No. 2016R1A2B4007932)".

Conflicts of Interest: The author declares no conflict of interest.

References

1. Tegguer, A. Determining the water absorption of recycled aggregates utilizing hydrostatic weighing approach. *Constr. Build. Mater.* **2012**, *27*, 113–116.
2. Pauw, P.; Thomas, P.; Vyncke, J.; Desmyter, J. Shrinkage and creep of concrete with recycled materials as coarse aggregates. Sustainable Construction: Use of recycled concrete aggregate. In Proceedings of the International Symposium, London, UK, 11–12 November 1998; pp. 213–225.
3. Maultzch, M.; Mellmann, G. Properties of large scale processed building rubble with respect to the reuse as aggregate in concrete. Sustainable Construction: Use of recycled concrete aggregate. In Proceedings of the International Symposium, London, UK, 11–12 November 1998; pp. 99–107.
4. Abbas, A. Durability of Green Concrete as a Structural Material. Ph.D. Thesis, Carleton University, Ottawa, ON, Canada, March 2007.

5. Fathifazl, G. Structural Performance of Steel Reinforced Recycled Concrete Members. Ph.D. Thesis, Carleton University, Ottawa, ON, Canada, January 2008.

6. Snyder, M. Recycling concrete pavements. In Proceedings of the ACPA Pennsylvania Chapter Presentation, Harrisburg, PA, USA, 27 January 2010.

7. FHWA. Recycled Concrete Aggregate Federal Highway Administration National Review. Available online: http://www.fhwa.dot.gov/Pavement/recycling/rca.cfm (accessed on 27 June 2017).

8. Fathifazl, G.; Abbas, A.; Razaqpur, A.G.; Isgor, O.D.; Fournire, B.; Foo, S. New mixture proportioning method for concrete made with coarse recycled concrete aggregate. *J. Mater. Civ. Eng. ASCE* **2009**, *21*, 601–611. [CrossRef]

9. Abbas, A.; Fathifazl, G.; Isgor, O.B.; Razaqpur, A.G.; Fournire, B.; Foo, S. Durability of recycled aggregate concrete designed with equivalent mortar volume method. *Cem. Concr. Compos.* **2009**, *31*, 555–563. [CrossRef]

10. Fathifazl, G.; Razaqpur, A.G.; Isgor, O.B.; Abbas, A.; Fournire, B.; Foo, S. Flexural performance of steel-reinforced recycled concrete beams. *ACI Struct. J.* **2009**, *106*, 858–867.

11. Fathifazl, G.; Razaqpur, A.G.; Isgor, O.B.; Abbas, A.; Fournire, B.; Foo, S. Shear capacity evaluation of steel reinforced recycled concrete (RRC) beams. *Eng. Struct.* **2011**, *33*, 1025–1033. [CrossRef]

12. Mathew, P.; Baby, V.; Sahoo, D.K.; Joseph, G. Manually recycled coarse aggregate from concrete waste—A sustainable substitute for customary coarse aggregate. *Am. J. Eng. Res.* **2013**, *3*, 34–38.

13. Yang, S.; Lee, H. Mechanical properties of recycled aggregate concrete proportioned with modified equivalent mortar volume method for paving applications. *Constr. Build. Mater.* **2017**, *136*, 9–17. [CrossRef]

14. Desai, S. Sustainable concrete construction: An engineer's views. In Proceedings of the Technical Meeting, Institution of Structural Engineers: South Eastern Countries Branch, London, UK, 2 November 2010.

15. Thomas, C.; Setien, J.; Polanco, J. Structural recycled aggregate concrete made with precast wastes. *Constr. Build. Mater.* **2016**, *124*, 536–546. [CrossRef]

16. Ajdukiewicz, A.; Kliszczewicz, A. Influence of recycled aggregates on mechanical properties of HS/HPC. *Cem. Concr. Compos.* **2000**, *24*, 269–279. [CrossRef]

17. Gomez-Soberon, J. Porosity of recycled concrete with substitution of recycled concrete aggregate-an experimental study. *Cem. Concr. Res.* **2002**, *32*, 1301–1311. [CrossRef]

18. Kou, S.; Poon, C. Effect of the quality of parent concrete on the properties of high performance recycled aggregate concrete. *Constr. Build. Mater.* **2015**, *77*, 501–508. [CrossRef]

19. Hall, K.; Dawood, D.; Vanikar, S.; Tally, R., Jr.; Cackler, T.; Correa, A.; Deem, P.; Duit, J.; Geary, G.; Gisi, A.; et al. *Long-Life Concrete Pavement in Europe and Canada*; FHWA-PL-07-027; Federal Highway Administration: Alexandria, VA, USA, 2007; p. 80.

20. Hu, J.; Siddiqui, M.S.; Fowler, D.W.; Whitney, D. Two-lift concrete paving-case studies and reviews from sustainability, cost effectiveness and construction perspectives. In Proceedings of the 93rd Annual Transportation Research Board Meeting, Washington, DC, USA, 12–16 January 2014.

21. Gerhardt, T. Two-lift Paving: Contractor's Perspective. In *National Open for House Two Lift Concrete Paving*; FHWA: Austin, TX, USA, 2013.

22. Shi, X.; Mukhopadhyay, A.; Zollinger, D. Sustainability assessment for portland cement concrete pavement containing reclaimed asphalt pavement aggregates. *J. Clean. Prod.* **2018**, *192*, 569–581. [CrossRef]

23. Tam, V.; Tam, C. Assessment of durability of recycled aggregate concrete produced by two-stage mixing approach. *J. Mater Sci.* **2007**, *42*, 3592–3602. [CrossRef]

24. Sicakova, A.; Urban, K. The influence of discharge time, kind of additive and kind of aggregate on the properties of three-stage mixed concrete. *Sustainability* **2018**, *10*, 3862. [CrossRef]

25. Tam, V.; Gao, X.; Tam, C. Microstructural analysis of recycled aggregate concrete produced from two-stage mixing approach. *Cem. Concr. Res.* **2005**, *35*, 1195–1203. [CrossRef]

26. Gupta, P.; Khaudhair, Z.; Ahuja, A. A new method for proportioning recycled concrete. *Struct. Concr.* **2016**, *4*, 677–687. [CrossRef]

27. Kim, N.; Kim, J.; Yang, S. Mechanical strength properties of RCA concrete made by a modified EMV method. *Sustainability* **2016**, *8*, 924. [CrossRef]

28. Yang, S.; Lim, Y. Mechanical strength and drying shrinkage properties of RCA concretes produced from old railway concrete sleepers using by a modified EMV method. *Constr. Build. Mater.* **2018**, *185*, 499–507. [CrossRef]

29. Yang, S.; Lee, H. Freeze-thaw resistance and drying shrinkage of recycled aggregate concrete proportioned by the modified equivalent mortar volume method. *Int. J. Concr. Struct. Mater.* **2017**, *11*, 617–626. [CrossRef]

30. Fathifazl, G.; Razaqpur, A.G.; Isgor, O.B.; Abbas, A.; Fournire, B.; Foo, S. Creep and drying shrinkage characteristics of concrete produced with coarse recycled concrete. *Cem. Concr. Compos.* **2011**, *33*, 1026–1037. [CrossRef]

31. *ASTM C127: Standard Test Method for Relative Density (Specific Gravity) and Absorption of Coarse Aggregate*; ASTM International: West Conshohocken, PA, USA, 2015.

32. *ASTM C131: Standard Test Method for Resistance to Degradation of Small-Size Coarse Aggregate by Abrasion and Impact in the Los Angeles Machine*; ASTM International: West Conshohocken, PA, USA, 2014.

33. Juan, M.S.; Gutierrez, P.A. Study on the influence of attached mortar content on the properties of recycled concrete aggregate. *Constr. Build. Mater.* **2009**, *23*, 872–877. [CrossRef]

34. Akbarnezhad, A.; Ong, K.C.; Zhang, M.H.; Tam, C.T. Acid treatment technique for determining the mortar content of recycled concrete aggregates. *J. Test. Eval.* **2013**, *41*, 1–10. [CrossRef]

35. Korea Expressway Corporation Research Institute. *Highway Construction Guide Specification*; Korea Expressway Corporation: Gyungbuk, Korea, 2011. (In Korean)

36. Ministry of Land, Infrastructure and Transportation. *Concrete Structure Specification*; Ministry of Land, Infrastructure and Transportation: Sejong, Korea, 2009. (In Korean)

37. Incheon International Airport Corporation. *Concrete Construction Guidelines*; Incheon International Airport Corporation: Incheon, Korea, 2012. (In Korean)

38. *ASTM C39: Standard Test Method for Compressive Strength of Cylindrical Concrete Specimens*; ASTM International: West Conshohocken, PA, USA, 2016.

39. *ASTM C469: Standard Test Method for Static Modulus of Elasticity and Poisson's Ratio of Concrete in Compression*; ASTM International: West Conshohocken, PA, USA, 2014.

40. *KS F 2424: Standard Test Method for Length Change of Mortar and Concrete*; KATS: Seoul, Korea, 2015.

41. *ASTM C157: Standard Test Method for Length Change of Hardened Hydraulic-Cement Mortar and Concrete*; ASTM International: West Conshohocken, PA, USA, 2012.

42. Ho, N.; Lee, Y.; Lim, W.; Zayed, T.; Chew, K.; Low, G.; Ting, S. Efficient Utilization of Recycled Concrete Aggregate in Structural Concrete. *J. Mater. Civ. Eng.* **2013**, *25*, 318–327. [CrossRef]

43. Paul, S.; Panda, B.; Garg, A. A novel approach in modelling of concrete made with recycled aggregates. *Measurement* **2018**, *115*, 64–72. [CrossRef]

44. Paul, S. Data on optimum recycle aggregate content in production of new structural concrete. *Data Brief* **2017**, *15*, 987–992. [CrossRef] [PubMed]

45. Brito, J.; Saikia, N. *Recycled Aggregate in Concrete, Use of Industrial, Construction and Demolished Waste*; Springer: London, UK, 2013; pp. 379–426.

46. Salam, M.; Jumaat, M.; Jaafar, F.; Saad, H. Properties of high-workability concrete with recycled concrete aggregate. *Mater. Res.* **2011**, *14*, 248–255.

47. Price, W.H. Factors influencing concrete strength. *J. ACI Proc.* **1951**, *47*, 429.

48. Tripura, D.D.; Raj, S.; Mohammad, S.; Das, R. Suitability of recycled aggregate as a replacement for natural aggregate in construction. In *Journal of the ACI Conference*; SP-326; American Concrete Institute: Farmington Hills, MI, USA, 2018; pp. 1–10.

49. ACI Manual of Concrete Practice 209R-92. *Prediction of Creep, Shrinkage, and Temperature Effects in Concrete Structures*; American Concrete Institute: Farmington Hills, MI, USA, 1997; pp. 1–47.

![applied sciences logo]

applied
sciences

MDPI

Article

Analysis and Modelling of Shrinkage and Creep of Reactive Powder Concrete

Pang Chen [1], Wenzhong Zheng [1,2,3,*], Ying Wang [1] and Wei Chang [1]

[1]   School of Civil Engineering, Harbin Institute of Technology, Harbin 150090, China;
     hitchenpang@126.com (P.C.); wangying888@hit.edu.cn (Y.W.); CW1558083@163.com (W.C.)
[2]   Key Lab of Structures Dynamic Behaviour and Control of the Ministry of Education, Harbin Institute of
     Technology, Harbin 150090, China
[3]   Key Lab Smart Prevention and Mitigation of Civil Engineering Disasters of the Ministry of Industry and
     Information Technology, Harbin Institute of Technology, Harbin 150090, China
*   Correspondence: Zhengwenzhong@hit.edu.cn; Tel.: +86-132-846-32829

Received: 8 April 2018; Accepted: 4 May 2018; Published: 5 May 2018

Abstract: The objective of this study was to examine the shrinkage and creep of reactive power concrete (RPC) with different steel fibre contents (0%, 1% and 2% by volume). A total of 37 RPC specimens were prepared and tested for compression strength, elastic modulus, shrinkage, and creep. In addition, different axial stress ratios (0.2, 0.3 and 0.4) were used in the creep tests. Furthermore, the accuracy of the ACI 209-82 model, CEB-FIP 90 model, B3 model, and GL 2000 model for predicting the shrinkage and creep of RPC was evaluated and new numerical shrinkage and creep models were developed. The experimental results revealed that the compressive strength and elastic modulus increase with increasing steel fibre content. The shrinkage and creep decreased with increasing addition of steel fibre from 0% to 2%. A good linear relationship was found between the axial stress ratios and creep strain. All four existing models were unable to accurately predict the shrinkage and creep of RPC. A good agreement between the experimental results and proposed shrinkage and creep numerical models was observed. Therefore, it is suggested that the proposed shrinkage and creep models can be used to calculate the shrinkage and creep of RPC.

Keywords: reactive power concrete; shrinkage; creep; steel fibre; model

1. Introduction

Reactive powder concrete (RPC) is a type of ultra-high performance concrete which has been developed in the last two decades [1,2]. RPC has superior compressive and tensile strengths, which can significantly reduce the dead load of structures [3,4]. Thus, it is especially suitable for long-span bridge decks, thin-plate structures, and field-cast joints for precast bridge decks [5]. It possesses superior energy absorption owing to the addition of steel fibre [6], provides good structural reliability, and has excellent durability, leading to a long service life. It is also almost impermeable, which almost entirely prevents carbonation and penetration of chlorides and sulphates, thereby making it suitable for use in harsh climatic conditions such as freeze-thaw or coastal areas [7]. However, the high cement content (usually as high as 800–1000 kg/m^3) affects production costs and increases the risk of shrinkage [8–12]. Furthermore, the addition of silica fume increases the risk of shrinkage and creep of RPC [13–16].

Shrinkage and creep can significantly affect the long-term characteristics of concrete. In large scale structures, shrinkage and creep can increase the width of cracks and structural deformation. It can also cause stress loss in the prestressed reinforcement of prestressed components [17]. Therefore, it is important to predict and monitor the shrinkage and creep of concrete. In recent years, many studies on the shrinkage and creep of normal strength concrete have been carried out, and mature theories and models have been developed [18–20]. However, there are relatively few studies on the shrinkage and creep of RPC. To produce RPC, it is necessary to minimize the aggregate size, increase the paste/aggregate ratio, decrease the water/binder ratio, add silica fume and steel fibre, and use steam-heat curing. All these measures make the shrinkage and creep characteristics of RPC different from those of normal strength concrete.

Peiliang Shen et al. investigated the autogenous shrinkage of steam-heat cured RPC [21], and the results showed that three factors (i.e., steel fibre, silica fume, and aggregate size) have significant effects on the autogenous shrinkage of RPC. Furthermore, the autogenous shrinkage mainly occurred during the stream-heat curing period. Ehsan Ghafari et al. studied the effects of different supplementary cementitious materials (i.e., silica fume, fly ash, and ground granulated blast furnace slag) on the autogenous shrinkage of RPC, and developed an autogenous shrinkage model for RPC [16]. Shamsad Ahmad et al. conducted studies on the shrinkage of RPC with water curing for 3 days, and observed that the shrinkage increases with increasing water/binder ratio, cement content, and silica fume content [22]. Nguyen found that adding rice husk ash to RPC can reduce the autogenous shrinkage of RPC significantly [23]. Moreover, several other measures such as using expansive additives, shrinkage reducing admixtures, or coarser cement particles, and improving curing conditions were identified to decrease the autogenous shrinkage of RPC [24–27]. Mo Jinchuan et al. demonstrated that both the autogenous and drying shrinkage of RPC increase with raising granulated blast furnace slag (GBFS) [28]. C.M. Tam et al. studied the dry shrinkage of RPC with water curing for 28 days, and found drying shrinkage increases with increasing water/binder ratio and superplasticizer dosage [6]. A. Cwirzen suggested that adding coarse aggregate to RPC can reduce its drying shrinkage and creep [29]. A. Graybeal found that high compressive stress (axial stress of higher than 0.60) on relatively low strength RPC cause significant short-term creep [30].

To the best of the authors' knowledge, previous studies on the shrinkage and creep of RPC have mainly focused on the autogenous shrinkage of RPC. The effects of RPC components on autogenous shrinkage have been investigated. In addition, some measures have been purposed to decrease the autogenous shrinkage of RPC. In present times, RPC is widely used in prefabricated components. For commercial RPC, steam-heat curing is typically used in order to accelerate the hydration process and strength development [31], but there are relatively few studies on the shrinkage of RPC after steam-heat curing and creep in RPC. Furthermore, there is no RPC shrinkage and creep model to predict shrinkage and creep of RPC at present, most of the existing shrinkage and creep models were developed from experimental data fitting, and the results of these models are mostly compared to data related to normal strength concrete. Hence, these models are suitable for normal strength concrete rather than RPC [32]. The applicability of the existing shrinkage and creep models for RPC needs to be further verified.

In this study, the shrinkage after steam-heat curing and creep of RPC have been investigated. To this end, a total of 18 samples were prepared for compressive strength and elastic modulus tests, 9 samples for shrinkage tests, and 10 samples for creep tests. In addition, the influence of steel fibre on the compressive strength, elastic modulus, shrinkage, and creep of RPC was investigated. The effect of axial stress on the creep of RPC was also examined. Finally, the shrinkage and creep of RPC predicted by widely used models were compared with the experimental data to verify the applicability of the models for RPC, and new shrinkage and creep numerical models were developed for RPC.

2. Materials and Mix Proportions

2.1. Raw Materials

Type I Portland cement (PC), silica fume (SF), water, quartz sand, quartz powder, steel fibre (ST), and superplasticizer (SP) were used to produce the RPC samples. The Type I cement conforms to Chinese National Standard GB175-2007 [33], and has a specific gravity of 3.20. The silica fume is an extremely fine material with a particle size smaller than 0.1 μm. The specific surface area and dry bulk density of the silica fume are 18.4 m^2/g and 0.65 kg, respectively. The chemical components of the cement and silica fume are given in Table 1. Quartz sand with a particle size distribution in the range of 0.01–0.50 mm was used as an aggregate, and quartz powder with particle sizes from 10–45 µm was used as a micro-filler. The superplasticizer is a polycarboxylic acid water reducer with a specific gravity of 1.20, allowing for water reduction of up to 25% to achieve the target workability. The mixing water is tap water from Harbin, China. The diameter and length of the steel fibre are 0.22 mm and 13 mm, respectively. The tensile strength of the steel fibre is greater than 2850 MPa. The steel fibre used in this study is shown in Figure 1.

Figure 1. Steel fibre used in this study.

Table 1. Chemical composition of cement and silica fume.

Constituent	Cement (%)	Silica Fume (%)
CaO	63.37	0.39
SiO$_2$	22.08	95.11
Al$_2$O$_3$	5.72	0.43
Fe$_2$O$_3$	3.05	0.42
K$_2$O	0.43	0.48
MgO	2.02	0.17
Na$_2$O	0.19	0.19
Equivalent alkalis (Na$_2$O + 0.658K$_2$O)	0.33	0.64
SO$_3$	2.10	0.28
Loss on ignition	0.71	1.89

2.2. RPC Mixtures and Curing Conditions

37 RPC samples were prepared for tests of the compressive strength, elastic modulus, shrinkage, and creep. The W/B (water-binder ratio) for all RPC specimens is 0.20. Steel fibre contents of 0%, 1% and 2% by volume of mixture were used. The specific mix proportions are listed in Table 2. The numbers in Table 2 represent mass ratios of the RPC mixtures.

Table 2. RPC sample mixtures for experiments.

W/B	Cement	SF	Quartz Sand	Quartz Powder	SP	ST [1] (%)
0.20	1.00	0.30	0.70	0.35	0.024	0
0.20	1.00	0.30	0.70	0.35	0.024	1
0.20	1.00	0.30	0.70	0.35	0.024	2

[1] Volume percentage.

The mixing procedure affects the material properties of RPC [34], and the manufacturing method for RPC is different from that of normal strength concrete. To minimize the impact of the mixing process, all the mixtures were made using the same planetary mixer. Based on previous studies [28,35], the mixing procedure is as follows:

(I)   The cement, silica fume, quartz sand, and quartz powder are mixed in a dry state for approximately 2 min at a low speed of approximately 140 rpm. During this dry mixing process, steel fibre was added to the mixtures.

(II) 50% of the water and 50% of the superplasticizer were gradually added to the mixtures, and the mixtures were stirred for 3 min at a high speed of approximately 280 rpm.

(III) The remaining 50% of the water and the superplasticizer were added to the mixtures, which were then stirred again for 3 min at a high speed of approximately 280 rpm.

(IV) After mixing, the RPC was poured into moulds and vibrated until fully consolidated.

After 24 h, the RPC samples were removed from the moulds, and the samples were cured in a special curing box at a temperature of 90 °C and relative humidity (RH) of greater than 95% for 48 h. The samples were then moved to a testing room with a temperature of 20 °C and a relative humidity (RH) of 60% until the experiments were performed.

3. Experiments

3.1. Compressive Strength and Elastic Modulus Tests

The compressive strength and elastic modulus of RPC samples after 28 days were determined experimentally. According to Chinese standard test methods for mechanical properties of concrete [36], $100 \times 100 \times 300$ mm prisms were loaded uniaxially. The tests were conducted at a loading rate of approximately 0.5 MPa/s. Three samples for the compressive strength test and an additional three samples for the elastic modulus test were tested at each steel fibre content.

3.2. Shrinkage Tests of RPC

For the shrinkage test, $100 \times 100 \times 400$ mm prisms of RPC were cast. After steam-heat curing, all the RPC samples were placed in the testing room at a controlled temperature of 20 °C and a relative humidity of 60%, and the shrinkage tests were conducted immediately. The shrinkage test was designed according to the Chinese standard for test methods of long-term performance and durability of concrete [37]. Two dial gauges with a gauge length of 200 mm were placed on opposite sides of the sample, as shown in Figure 2. For each steel fibre content, the shrinkage was measured for three samples, and the average of the three measured shrinkage strains was reported.

Figure 2. Shrinkage tests of RPC.

3.3. Creep Tests of RPC

For the creep tests, RPC samples were cast in 100 × 100 × 400 mm prisms. After curing the RPC, the samples were all placed in the testing room for 25 days before beginning the creep tests. A total of 10 RPC specimens (in 5 groups) were tested, and the experimental parameters are outlined in Table 3. Self-resisting loading frames were used to conduct the tests, as shown in Figure 3. In each group, two of the same specimens were stacked one on top of the other. A load was applied using an oil-pressure jack at the scheduled loading time, monitored with a pressure sensor, and sustained with the bolts of the pressing rods for the scheduled load duration to determine the compressive creep strain. Two dial gauges were set on opposite sides of the sample with a gauge length of 200 mm. For each self-resisting loading frame, the total strains of the two samples were measured, and their average was reported. The creep strain was determined by subtracting the instantaneous strain and the shrinkage strain from the total strain.

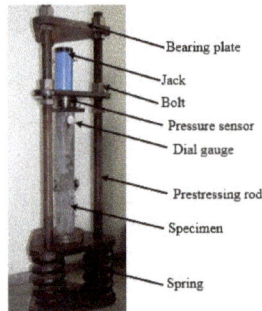

Figure 3. Creep tests of RPC.

Table 3. RPC creep experiment parameters.

Group	t_0 (day)	Δt (day)	f_c (MPa)	η	ST%
I		330	120	0.2	0
II		329	120	0.3	0
III	28	329	120	0.4	0
IV		329	134	0.3	1
V		329	142	0.3	2

Note: t_0 is the RPC age at loading, Δt is the load duration, f_c is the compressive strength at loading, and η is the axial stress ratio, defined as the ratio of the loading stress to the compressive strength.

4. Results and Discussion

4.1. Compressive Strength and Elastic Modulus of RPC

The compressive strength and elastic modulus were determined for RPC samples after 28 days of aging. As shown in Figure 4, the compressive strength and elastic modulus of the RPC samples increase with increasing steel fibre content. The compressive strength increases from 120 MPa to 142 MPa, and the elastic modulus increases from 44.7 GPa to 48.0 GPa as the steel fibre content increases from 0% to 2%. The compressive strengths of the RPC samples with 1% and 2% steel fibre content are 11.9% and 18.3% higher, respectively, than that of RPC samples without steel fibre. The elastic modulus of the RPC samples with 1% and 2% steel fibre content are 4.4% and 7.3% greater, respectively, than that of RPC samples without steel fibre. The strength of RPC is far greater than that of normal strength concrete, which is usually in the range of 20–80 MPa. RPC has a dense structure, which is attributed to the low W/B, hydration of the cement, and the pozzolanic effect of the SF [38]. As a result, the RPC has an ultra-high strength. The steel fibre plays the role of a bridge and dowel in the RPC samples. Increasing steel fibre content can make more fibres to sustain the load, which decreases the stress between fibres and matrix and restricts the development of microcracks and transverse deformation. Thus, the compressive strength and the elastic modulus increase with increasing steel fibre content [39,40].

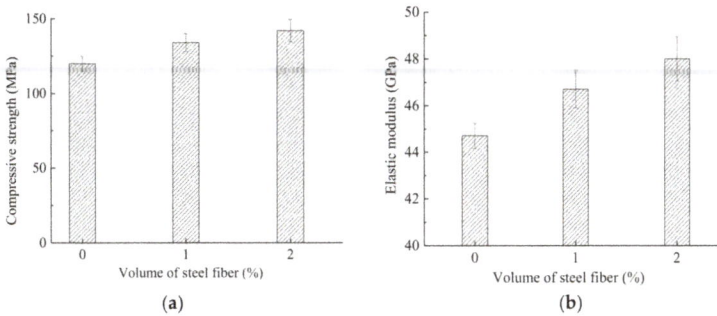

Figure 4. Compressive strength and elastic modulus of the RPC samples: (**a**) Compressive strength; (**b**) Elastic modulus.

4.2. Shrinkage of RPC

Figure 5 shows the variation in the shrinkage strain of the RPC samples with age. At the early ages, the shrinkage strain increases at a fast rate, and the shrinkage reaches a steady state at approximately 200 days. However, the shrinkage of RPC is much smaller than that of normal strength concrete, which is usually on the order of several hundred micro strain or more. With the improved uniformity and minimized pore size of the RPC, the escape of moisture is inhibited, capillary stress is reduced, and as a result, the shrinkage is also reduced. The shrinkage decreases with increasing steel fibre content. The final shrinkage of the RPC samples with 1% and 2% steel fibre content is 10.6% and 15.0% less, respectively, than that of the RPC samples without steel fibre. This is similar to the effect of steel fibre on shrinkage in normal strength concrete [41,42]. The positive effect of steel fibre for reducing the shrinkage of RPC can be explained as follows:

(I)   During the process of mixing the steel fibre into the RPC matrix, some micron-scale water films are formed on the steel fibre surface. As a result, calcium hydroxide crystals form directly on the surface of the fibre and grow with no constraints, which forms a loose reticular structure at the interface of the steel fibre and RPC matrix, thereby reducing shrinkage [43].

(II)   The distribution of steel fibres presents a three-dimensional random state. The steel fibres cross and overlap to form a skeleton, which can hinder the development of free shrinkage owing to the high elastic modulus of the steel fibre [44,45].

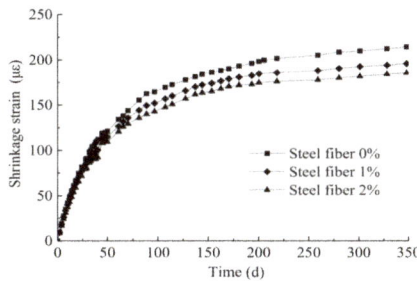

Figure 5. Shrinkage strain of the RPC samples with age.

4.3. Creep of RPC

4.3.1. Effect of the Axial Stress Ratio on the Creep of RPC

The stress-dependent strain is the sum of the instantaneous strain and the creep strain, the specific creep is the creep strain caused by unit stress, and the creep coefficient is the ratio of the creep strain to instantaneous strain at loading time. Figure 6 shows the variations in the stress-dependent strain, specific creep, and creep coefficient of the RPC samples subjected to different stress ratios with time. The stress-dependent strain and its growth rate both increase with increasing axial stress ratio. However, the specific creep and the creep coefficient of RPC samples subjected to different axial stress ratios are almost the same. This indicates that the creep of RPC is linear when the axial stress ratio is less than 0.4, which is consistent with the behaviour of normal strength concrete.

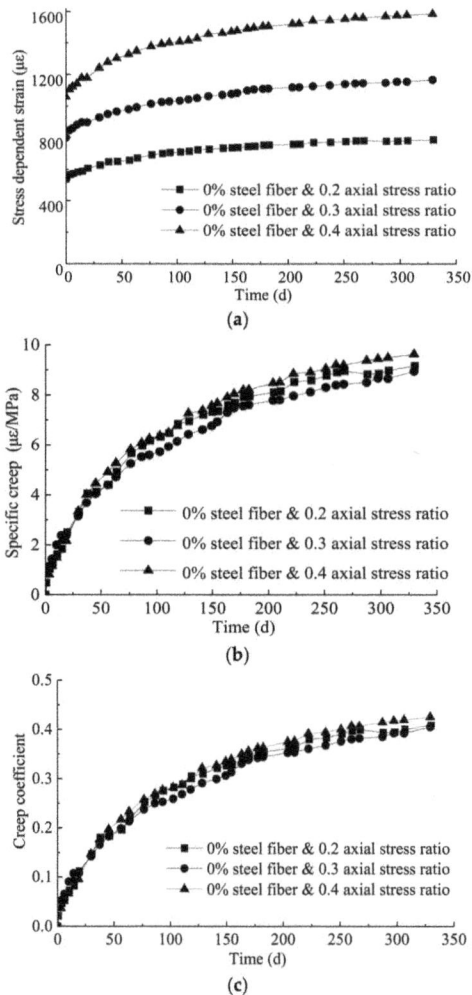

Figure 6. Effect of the axial stress ratio on the stress dependent strain, specific creep, and creep coefficient of RPC: (**a**) Stress-dependent strain; (**b**) Specific creep; (**c**) Creep coefficient.

4.3.2. Effect of Steel Fibre on Creep of RPC

Figure 7 shows variations in the stress-dependent strain, specific creep, and creep coefficient of RPC samples containing different steel fibre contents with time. In Figure 7a, the stress-dependent strain increases with increasing steel fibre content. This is because at the same axial stress ratio, the samples with more steel fibre content bear greater stress, causing more stress-dependent strain. As shown in Figure 7b,c, during the later period after loading, the specific creep and creep coefficient of the RPC samples decrease with increasing steel fibre content. For the samples without steel fibre, the ultimate specific creep is 8.95 $\mu\varepsilon$/MPa. However, for the samples with 1% and 2% steel fibre content, the ultimate specific creep is 7.37 and 6.65 $\mu\varepsilon$/MPa, respectively. The ultimate specific creep of the RPC samples with 1% and 2% steel fibre content are thus 17.7% and 25.7% less, respectively, than that of the RPC samples without steel fibre. For the samples without steel fibre, the ultimate creep coefficient is 0.40. However, for the samples with 1% and 2% steel fibre content, the ultimate creep coefficients are 0.35 and 0.32, respectively. This means that the ultimate creep coefficients of RPC samples with 1% and 2% steel fibre content are 12.5% and 20.0% smaller, respectively, than that of the RPC samples without steel fibre. The positive effect of steel fibre for reducing creep, particularly during the later period after loading, can be explained as follows:

(I) Even within the linear creep range, when the RPC carries a compressive load, microcracks can form and develop gradually inside the RPC due to the inhomogeneity of the RPC matrix. As the microcracks emerge during the later stage, the steel fibres passing through the microcracks can prevent the microcracks from developing further. Hence, the creep strain is reduced.

(II) During the early period after loading, the steel fibre and RPC matrix produce a section slip. As a result, the inhibition effect of the steel fibre on creep is not significant. On the other hand, during the later period after loading, the slip between the steel fibre and RPC matrix tends to be stable. Hence, the ability of steel fibre to inhibit creep gradually appears.

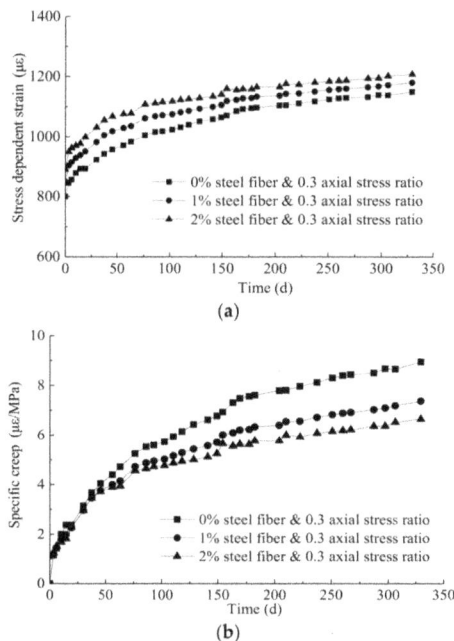

Figure 7. *Cont.*

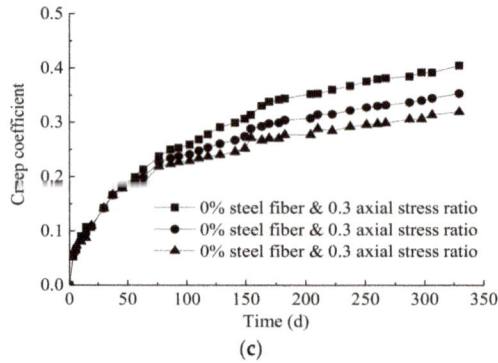

Figure 7. Effect of the steel fibre content on the stress-dependent strain, specific creep, and creep coefficient of RPC: (**a**) Stress-dependent strain; (**b**) Specific creep; (**c**) Creep coefficient.

4.4. Comparison of Shrinkage and Creep of RPC with Existing Models

Shrinkage and creep have complex mechanisms involving many interrelated factors, and there is no single theory which can fully explain these mechanisms. Thus, experimental studies are essential as a basis for shrinkage and creep models. The ACI 209-82 model [46] is the current standard code model recommended by the American Concrete Institute, and is accepted by building codes in the United States. The CEB-FIP 90 model [47] is recommended by the CEB-FIP model code 1990 (Euro-International Committee for Concrete and the International Federation for Prestressing). The B3 model [20] is based on consolidation, has a clear physical background, and considers most of the internal and external factors that affect shrinkage and creep, and thus has a greater accuracy. The GL 2000 model [48] is a modified Atlanta 97 model, influenced by the CEB-FIP 90 model, and was developed to correct the negative relaxation at early loading ages. These four models are the most commonly used shrinkage and creep models for normal strength concrete.

The experimental shrinkage and creep results for RPC in this study were compared with the predicted values from the four models, as shown in Figures 8 and 9. The CEB-FIP 90 model [48] underestimates the shrinkage strain of RPC. However, the ACI 209-82 model [47], B3 model [20], and GL 2000 model [48] overestimate the shrinkage strain of RPC. Moreover, all four models overestimate the creep strain of RPC. From Figures 8 and 9, it can be seen that the shrinkage and creep strain predicted by all four models are inconsistent with the experimental results. The application scopes of the four models are listed in Table 4, from which it can be seen that the four models are mainly focused on compressive strengths at 28 days from 16–90 MPa, water-binder ratios greater than 0.35, and cement contents of less than 719 kg/m^3. They also ignore the influence of the silica fume. However, RPC has an ultra-high strength and dense microstructure due to its low water-binder ratio (often less than 0.2), high cementitious material content compared to normal strength concrete, replacement of aggregate with quartz sand and quartz powder, addition of steel fibre to improve the strength and ductility, and replacement of a portion of the cement with silica fume, which can act as a filler material and participate in the pozzolanic reaction, leading to the production of additional C-S-H gel. Therefore, the four models have errors when predicting the shrinkage and creep of RPC.

Table 4. Application scope of the existing models.

Parameters	ACI 209-82	CEB-FIP 90	B3	GL 2000
f_{c28} (MPa)	-	20–90	17.2–69	16–82
Cement content (kg/m^3)		-	160–719	-
W/B	-	-	0.35–0.85	0.40–0.60
Relative humidity (%)	40–100	40–100	40–100	20–100
t_c (Moist cured)	≥7 days	≤14 days	-	≥1 day
t_c (Steam cured)	≥1 day	≤14 days	-	≥1 day
η	≤0.4	≤0.4	≤0.45	≤0.4

Note: f_{c28} is the compressive strength at 28 days, t_c is the curing age.

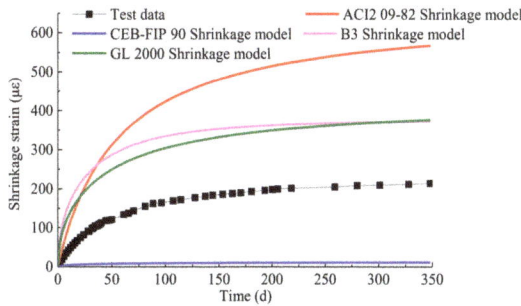

Figure 8. Comparison of the experimental shrinkage strain results for RPC with shrinkage model predictions.

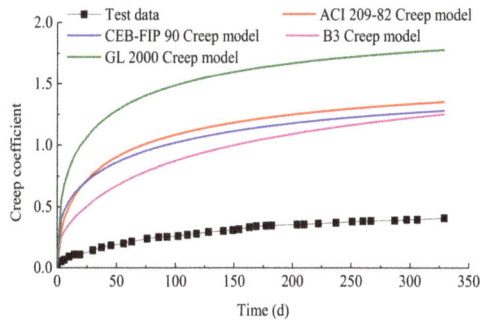

Figure 9. Comparison of the experimental creep strain results for RPC with creep model predictions.

4.5. Shrinkage and Creep Model of RPC

The existing shrinkage and creep models cannot accurately predict the shrinkage and creep of RPC, so it is necessary to develop new simple numerical models to predict the shrinkage and creep of RPC. Based on the ACI 209-82 shrinkage model [46], the general equations for predicting shrinkage of concrete are as follows:

$$\varepsilon_{sh,c}(t, t_c) = F_{sh}(t)\varepsilon_{shu,c} \tag{1}$$

$$F_{sh}(t) = \frac{t - t_c}{b + t - t_c} \tag{2}$$

In these equations, $\varepsilon_{sh,c}(t, t_c)$ is the shrinkage strain of concrete at time, t (days), with curing time t_c (days); $F_{sh}(t)$ is the time function of shrinkage; $\varepsilon_{shu,c}$ is the ultimate shrinkage strain; and b is constant which influences the rate of shrinkage with time. By applying regression fitting to the shrinkage experimental data, a simple shrinkage model (Equations (3) and (4)) for RPC have been developed. In Equation (3), the ultimate shrinkage strain is 246 $\mu\varepsilon$ corresponding to 780 $\mu\varepsilon$ in normal strength concrete. The effects of steel fibre content on the ultimate value of shrinkage is considered by introducing coefficient k_{ss}. A comparison of the proposed shrinkage model predictions and the experimental shrinkage results is shown in Figure 11. The correlation coefficients (R^2) for all the shrinkage samples are greater than 0.991. Therefore, the new shrinkage model can be used to predict the shrinkage of RPC after steam-heat curing with steel fibre content less than 2%.

$$\varepsilon_{sh}(t, t_c) = 246 k_{ss} \frac{(t - t_c)}{(t - t_c) + 47} \tag{3}$$

$$k_{ss} = 0.025\rho_s^2 - 0.115\rho_s + 1 \tag{4}$$

where $\varepsilon_{sh}(t, t_c)$ is the shrinkage strain at time, t, with curing time, t_c; ρ_s is the volume fraction of steel fibre.

Based on the ACI 209-82 creep model [47], the general equations for predicting creep of concrete can be written as:

$$\varphi(t, t_0) = F_c(t)\varphi_{cu,c} \tag{5}$$

$$F_c(t) = \frac{(t - t_0)^A}{B + (t - t_0)^A} \tag{6}$$

In these equations, $\varphi_c(t, t_c)$ is the creep coefficient of concrete at time, t (days), loading at time t_0 (days); $F_c(t)$ is the time function of creep; $\varphi_{cu,c}$ is the ultimate creep strain; A and B are constant. By applying regression fitting to the creep experimental data, a simple creep model (Equations (7)–(9)) for RPC have been developed. In Equation (7), the ultimate creep coefficient is 0.82 corresponding to 2.35 in normal strength concrete. The steel fibre can influence both the ultimate value and the rate of creep. Therefore, the effects of fiber content on the ultimate value and development rate of creep are considered by coefficients k_{sc} and α, respectively. A comparison between the predictions of the proposed creep model and the experimental creep results is shown in Figure 10. The correlation coefficients (R^2) for all the creep samples are greater than 0.986. Therefore, it can be inferred that the new creep model agrees well with the experimental data, and the new model can be used to predict the creep of RPC after steam-heat curing, with the volume fraction of steel fibre below 2% and axial stress ratio less than 0.4.

$$\varphi(t, t_0) = 0.82 k_{sc} \frac{(t - t_0)^{0.61}}{(t - t_0)^{0.61} + 36\alpha} \tag{7}$$

$$k_{sc} = 0.045\rho_s^2 - 0.28\rho_s + 1 \tag{8}$$

$$\alpha = 1 - 0.25\rho_s \tag{9}$$

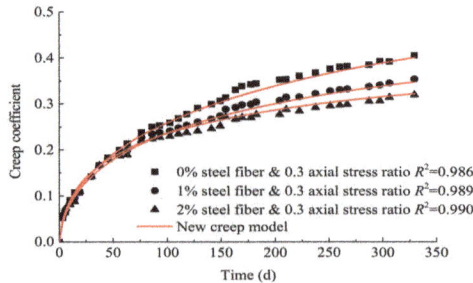

Figure 10. Comparison of the proposed creep model predictions with experimental creep coefficient results for the RPC samples.

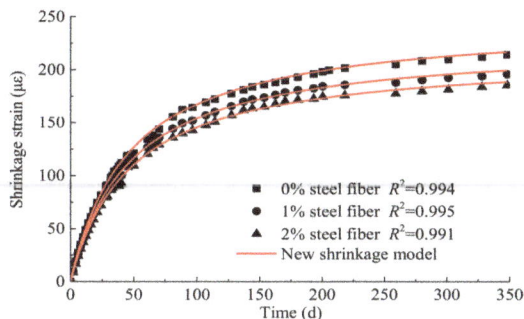

Figure 11. Comparison of the proposed shrinkage model predictions with experimental shrinkage strain results for the RPC samples.

5. Conclusions

In this study, the compressive strength and the elastic modulus of RPC were tested. The shrinkage after steam-heat curing and the creep of RPC were investigated. The effect of steel fibre on shrinkage and creep, and the influence of the axial stress ratio on creep were discussed. Then, a comparison was made between the experimental results and predictions obtained from existing shrinkage and creep models. Finally, a regression analysis of the experimental shrinkage and creep results for RPC was carried out. The conclusions of this study are as follows:

(1)   The compressive strength of RPC is obviously higher than normal strength concrete due to its dense microstructure. The compressive strength and the modulus elastic increase with increasing steel fibre content, as the streel fibre restricts the development of microcracks and transverse deformation.

(2)   The shrinkage of RPC is much smaller than normal strength concrete with the improved uniformity and the narrowed pore size. The shrinkage decreases with increasing steel fibre content due to the micro-scale water films formed on the steel fibre surface and skeleton formed by the cross and overlap of steel fibre.

(3)   The creep of RPC decreases with increasing steel fibre content which is obvious during the later period after loading, as steel fibre can prevent the development of microcracks (emerging mainly during the later stage). Besides, the slip between the steel fibre and RPC matrix tends to be stable in the later period, the ability of steel fibre to inhibit creep gradually appears.

(4)   For axial stress ratios of less than 0.4, the creep strain of RPC varies linearly with the axial stress (RPC is in the linear creep stage).

(5)   The shrinkage strains of RPC predicted by the ACI 209-82 model, B3 model, and GL 2000 model are significantly greater than the corresponding experimental results. However, the shrinkage strains of RPC predicted by the CEB-FIP 90 model are significantly smaller than the experimental results. Moreover, all four models overestimate the creep strain of RPC. Thus, these models cannot be used for predicting the shrinkage and creep of RPC.

(6)   Simple shrinkage and creep models for RPC which consider the influence of steel fibre have been developed.

Author Contributions: Wenzhong Zheng and Ying Wang conceived and designed the experiments; Pang Chen and Wei Chang performed the experiments. All the authors analysed the data and contributed to writing the paper.

Acknowledgments: The authors gratefully acknowledge the financial support of the National Science Foundation of China (Project No. 51678190, No. 51478142).

Conflicts of Interest: The authors declare no conflict of interest.

References

1. Richard, P.; Cheyrezy, M. Composition of reactive powder concretes. *Cem. Concr. Res.* **1995**, *25*, 1501–1511. [CrossRef]
2. Richard, P.; Cheyrezy, M. Reactive powder concretes with high ductility and 200–800 MPa compressive strength. *ACI* **1994**, *114*, 507–518.
3. Cheyrezy, M.; Maret, V.; Frouin, L. Microstructural analysis of RPC (Reactive Powder Concrete). *Cem. Concr. Res.* **1995**, *25*, 1491–1500. [CrossRef]
4. Wang, R.; Gao, X.J. Relationship between Flowability, Entrapped Air Content and Strength of UHPC Mixtures Containing Different Dosage of Steel Fiber. *Appl. Sci.* **2016**, *6*, 216. [CrossRef]
5. Blais, P.Y.; Couture, M. Precast, Prestressed Pedestrian Bridge—World's First Reactive Powder Concrete Structure. *PCI. J.* **1999**, *44*, 60–71. [CrossRef]
6. Tam, C.M.; Tam, V.W.Y.; Ng, K.M. Assessing drying shrinkage and water permeability of reactive powder concrete produced in Hong Kong. *Constr. Build. Mater.* **2013**, *26*, 79–89. [CrossRef]
7. Zheng, W.Z.; Lu, X.Y. Literature review of reactive powder concrete. *J. Build. Struct.* **2015**, *36*, 44–58. (In Chinese)
8. Yazıcı, H.; Yardımcı, M.Y.; Aydın, S.; Karabulut, A.S. Mechanical properties of reactive powder concrete containing mineral admixtures under different curing regimes. *Const. Build. Mater.* **2009**, *23*, 1223–1231. [CrossRef]
9. Su, A.; Qin, L.; Zhang, S.; Zhang, J.; Li, Z. Effects of Shrinkage Reducing Agent and Expansive Admixture on the Volume Deformation of Ultrahigh Performance Concrete. *Adv. Mater. Sci. Eng.* **2017**, *5*, 1–7.
10. Morin, V.; Tenoudji, F.C.; Feylessoufi, A.; Richard, P. Superplasticizer effects on setting and structuration mechanisms of ultrahigh-performance concrete. *Cem. Concr. Res.* **2001**, *31*, 63–71. [CrossRef]
11. Feylessoufi, A.; Tenoudji, F.C.; Morin, V.; Richard, P. Early ages shrinkage mechanisms of ultra-high-performance cement-based materials. *Cem. Concr. Res.* **2001**, *31*, 1573–1579. [CrossRef]
12. Soliman, A. Effect of partially hydrated cementitious materials on early-age shrinkage of ultra-high-performance concrete. *Mag. Con. Res.* **2013**, *65*, 1147–1154. [CrossRef]
13. Maruyama, I.; Teramoto, A. Temperature dependence of autogenous shrinkage of silica fume cement pastes with a very low water–binder ratio. *Cem. Concr. Res.* **2013**, *50*, 41–50. [CrossRef]
14. Jensen, O.M.; Hansen, P.E. Autogenous Relative Humidity Change in Silica Fume Modified Cement Paste. *Aci. Mater. J.* **1996**, *7*, 539–543. [CrossRef]
15. Yang, Y.; Sato, R.; Kawai, K. Autogenous shrinkage of high-strength concrete containing silica fume under drying at early ages. *Cem. Concr. Res.* **2005**, *35*, 449–456. [CrossRef]
16. Ghafari, E.; Ghahari, S.A.; Costa, H.; Júlio, E.; Portugal, A.; Durães, L. Effect of supplementary cementitious materials on autogenous shrinkage of ultra-high performance concrete. *Const. Build. Mater.* **2016**, *127*, 43–49. [CrossRef]
17. Yazdizadeh, Z.; Marzouk, H.; Hadianfard, M.A. Monitoring of concrete shrinkage and creep using Fiber Bragg Grating sensors. *Const. Build. Mater.* **2017**, *137*, 505–512. [CrossRef]
18. Baant, Z.P.; Prasannan, S. Solidification theory for aging creep. *Cem. Concr. Res.* **1988**, *18*, 923–932. [CrossRef]
19. Bazant, Z.P. Prediction of concrete creep and shrinkage: past, present and future. *Nucl. Eng. Des.* **2001**, *203*, 27–38. [CrossRef]
20. Bažant, Z.P.; Baweja, S. Creep and shrinkage prediction model for analysis and design of concrete structures—model B 3. *Mater. Struct.* **1995**, *28*, 357–365.
21. Shen, P.; Lu, L.; He, Y.; Rao, M.; Fu, Z.; Wang, F.; Hu, S. Experimental investigation on the autogenous shrinkage of steam cured ultra-high performance concrete. *Const. Build. Mater.* **2018**, *162*, 512–522. [CrossRef]
22. Ahmad, S.; Zubair, A.; Maslehuddin, M. Effect of the key mixture mixture parameters on shrinkage of reactive powder concrete. *Sci. World. J.* **2014**, *6*. [CrossRef] [PubMed]
23. Tuan, N.V.; Ye, G.; Breugel, K.V. Effect of rice husk ash on autogenous shrinkage of ultra high performance concrete. In Proceedings of the International Rilem Conference on Use of Superabsorbent Polymers and Other New Additives in Concrete, Lyngby, Denmark, 15–18 August 2010.
24. Bentz, D.P.; Peltz, M.A. Reducing Thermal and Autogenous Shrinkage Contributions to Early-Age Cracking. *Aci. Mater. J.* **2008**, *105*, 414–420.

25. Nmai, C.K.; Tomita, R.; Hondo, F.; Buffenbarger, J. Shrinkage-reducing admixtures. *Int. J. Concr. Struct.* **1998**, *20*, 31–37.

26. Soliman, A.M.; Nehdi, M.L. Effects of shrinkage reducing admixture and wollastonite microfiber on early-age behavior of ultra-high performance concrete. *Cem. Concr. Comp.* **2014**, *46*, 81–89. [CrossRef]

27. Soliman, A.M.; Nehdi, M.L. Effect of drying conditions on autogenous shrinkage in ultra-high performance concrete at early-age. *Mater. Struct.* **2011**, *44*, 879–899. [CrossRef]

28. Mo, J.; Ou, Z.; Zhao, X.; Wang, Y. Influence of superabsorbent polymer on shrinkage properties of reactive powder concrete blended with granulated blast furnace slag. *Const. Build. Mater.* **2017**, *146*, 283–296. [CrossRef]

29. Cwirzen, A.; Penttala, V.; Vornanen, C. Reactive powder based concretes: mechanical properties, durability and hybrid use with OPC. *Cem. Concr. Res.* **2008**, *38*, 1217–1226. [CrossRef]

30. Graybeal, A. Characterization of the Behavior of Ultrahigh Performance Concrete. Ph.D. Thesis, University of Maryland, College Park, MD, USA, 2005.

31. Yoo, D.Y.; Kim, S.; Kim, M.J. Comparative shrinkage behavior of ultra-high-performance fiber-reinforced concrete under ambient and heat curing conditions. *Const. Build. Mater.* **2018**, *162*, 406–419. [CrossRef]

32. Pan, Z.; Li, B.; Lu, Z. Re-evaluation of CEB-FIP 90 prediction models for creep and shrinkage with experimental database. *Constr. Build. Mater.* **2013**, *38*, 1022–1030. [CrossRef]

33. Jiang, L.Z.; Yan, B.L.; Xiao, Z.M.; Wang, W.Y.; Zhang, D.T. General purposed portland cement. *Cement* **2008**, *4*, 1–2. (In Chinese)

34. Madandoust, R.; Ranjbar, M.M.; Ghavidel, R.; Shahabi, S.F. Assessment of factors influencing mechanical properties of steel fiber reinforced self-compacting concrete. *Mater. Des.* **2015**, *83*, 284–294. [CrossRef]

35. Zheng, W.Z.; Luo, B.F.; Wang, Y. Compressive and tensile properties of reactive powder concrete with steel fibres at elevated temperatures. *Constr. Build. Mater.* **2013**, *41*, 844–851. [CrossRef]

36. GB 50010-2002. *Ministry of Construction of the People's Republic of China, Code for Design of Concrete Structures*; Chinese Architecture and Building Press: Beijing, China, 2002. (In Chinese)

37. GB/T 50081-2009. *Ministry of Housing and Urban-Rural Development of the People's Republic of China. Standard for Test Methods of Long-Term Performance and Durability of Ordinary Concrete*; China Architecture and Building Press: Beijing, China, 2009. (In Chinese)

38. Wang, C.; Liu, F.; Wan, C.; Pu, X. Preparation of Ultra-High Performance Concrete with common technology and materials. *Cem. Concr. Comp.* **2012**, *34*, 538–544. [CrossRef]

39. Gao, D.Y.; Cheng, H.Q.; Zhu, H.T. Splitting Tensile Bonding Strength of Steel Fiber Reinforced Concrete to Old Concrete. *J. Build. Mater.* **2007**, *10*, 505–509. (In Chinese)

40. Ren, G.M.; Wu, H. Effects of steel fiber content and type on static mechanical properties of UHPCC. *Constr. Build. Mater.* **2017**, *163*, 826–839. [CrossRef]

41. Huang, K.; Deng, M.; Mo, L.; Wang, Y. Early age stability of concrete pavement by using hybrid fiber together with MgO expansion agent in high altitude locality. *Constr. Build. Mate.* **2013**, *48*, 685–690. [CrossRef]

42. Miao, B.Q. Influences of fibre content on properties of self-compacting steel fibre reinforced concrete. *J. Chin. Inst. Eng.* **2003**, *26*, 523–530. (In Chinese) [CrossRef]

43. Yu, J.C.; Zhao, Q.X. Effect of steel fibre on creep behaviour of concrete. *J. Chin. Ceram. Soc.* **2013**, *8*, 1087–1093. (In Chinese)

44. Noushini, A.; Vessalas, K.; Arabian, G.; Samali, B. Drying Shrinkage Behaviour of Fibre Reinforced Concrete Incorporating Polyvinyl Alcohol Fibres and Fly Ash. *Adv. Civil. Eng.* **2014**, 356–365. [CrossRef]

45. Mo, J.; Ou, Z.; Wang, Y. Influence of MgO and Hybrid Fiber on the Bonding Strength between Reactive Powder Concrete and Old Concrete. *Adv. Mater. Sci. Eng.* **2016**, *5*, 1–13.

46. ACI Committee209. *Prediction of Creep, Shrinkage, and Temperature Effects in Concrete Structures*; American Concrete Institute: Farmington Hills, MI, USA, 1992.

47. Clark, L.A. CEB-FIP Model Code 1990. *Progr. Usenix Unix Suppl. Doc.* **2008**, *40*, 233–235.

48. Gardner, N.J.; Lockman, M.J. Design Provisions for Drying Shrinkage and Creep of Normal-Strength Concrete. *Aci. Mater. J.* **2001**, *98*, 159–167.

![applied sciences logo] *applied sciences*

MDPI

Article

Residual Properties Analysis of Steel Reinforced Recycled Aggregate Concrete Components after Exposure to Elevated Temperature

Zongping Chen [1,2,*], Rusheng Yao [1], Chenggui Jing [1] and Fan Ning [1]

[1] College of Civil Engineering and Architecture, Guangxi University, Nanning 530004, China; rsy1710@163.com (R.Y.); cgj171040@163.com (C.J.); fn171040@163.com (F.N.)
[2] Key Laboratory of Disaster Prevention and Structure Safety of Chinese Ministry of Education, Nanning 530004, China
* Correspondence: zpchen@gxu.edu.cn; Tel.: +86-138-7880-6048

Received: 6 October 2018; Accepted: 19 November 2018; Published: 24 November 2018

Abstract: The application of recycled aggregate concrete (RAC) has developed rapidly in recent years. But how to evaluate the residual properties of RAC after the fires is more beneficial to the further popularization and application of RAC. This paper presents the residual properties of RAC and steel reinforced recycled aggregate concrete (SRRAC) components after exposure to elevated temperature. A total of 176 specimens (120 rectangular prisms specimens, 24 SRRAC short columns and 32 SRRAC beams) were designed and tested after exposure to elevated temperature. The parameters were considered in the test, including replacement percentage of recycled coarse aggregate (0%, 30%, 50%, 70% and 100%) and exposure to different temperatures (20, 200, 400, 600 and 800 degrees centigrade). According to the test results, heat damage and residual properties of specimens were analyzed in detail, such as surface change, mass loss, bearing capacity degradation, stiffness degradation, ductility and energy dissipation of specimens under the elevated temperature. The results showed that a series of significant physical phenomena occurred on the surface of RAC and SRRAC components after exposure to elevated temperature, such as the color changed from green-grey to gray-white, chapped on the concrete surface after 400 degrees centigrade and the mass loss of concrete is less than 10%. The degradation of mechanical properties degenerated significantly with the increase of temperature, such as the strength of RAC, and compressive capacity, bending capacity, shear capacity and stiffness of SRRAC components, among that, the degradation of the strength of RAC was most obvious, up to 26%. The ductility and energy dissipation of SRRAC components were insignificant affected by the elevated temperature. Mass loss ratio, peak deformation and bearing capacity showed a slight increase trend with the increase of replacement percentage. But the stiffness showed significant fluctuation when replacement percentage was 70% to 100%. And the ductility and energy dissipation showed significant fluctuation when replacement percentage was 30% to 70%.

Keywords: recycled aggregate concrete (RAC); steel reinforced recycled aggregate concrete (SRRAC); elevated temperature; residual properties

---

1. Introduction

Demolished concrete is used to make RAC, which has the advantages of being energy-saving and environment-friendly. Moreover, the steel reinforced concrete (SRC) structure has the advantages of high bearing load capacity and good seismic performance. And SRC can combined with the RAC [1–3] to form the SRRAC structure. SRRAC meets the development direction of modern architecture, which has many advantages, such as being energy-saving and environment-friendly, sustainable development, having good mechanical performance. Therefore, SRRAC has broad application prospects. However, the porosity of recycled aggregate is higher than that of natural aggregate [4]. And the coarse aggregate surface in RAC attaches the cement-based or mortar, and its initial defects are more than natural aggregate, thus, those will affect the mechanical properties of RAC [5,6]. Therefore, many studies were carried out by scholars for RAC. The results show that inferior waste concrete reduces the quality of recycled coarse aggregate and consequently reduces the strength of RAC [7]. The mechanical properties of RAC can be improved by Polyvinyl alcohol (PVA) after soaking in polyvinyl alcohol solution, which the optimum concentration of PVA solution is 10% [8]. The compressive strength of RAC increases when cement is replaced by fly ash [9]. Recycled fine aggregate with demolished concrete has little effect on the mechanical properties of RAC, when the replacement percentage is less than 30% [10]. Meanwhile, Choi, Won Chang et al. [11] conclude that the compressive strength of RAC columns meets the American Certification Institute (ACI) design criteria. Butler, L. et al. [12] consider that the bond strength of RAC is 9 to 19% lower than natural aggregate concrete. Xiao, J. et al. [13] conclude that the seismic performance of recycled concrete frame structure is reduced with the increase of replacement percentage.

Reis, Nuno et al. [14] suggest that the stiffness and cracking load of RAC slabs are slightly lower than those of natural aggregate concrete. According to the literature [15–17], the content of recycled coarse aggregate has an insignificant effect on the flexural or shear properties of RAC beams. Secondly, in order to improve the utilization ratio of RAC in high-rise and super-high-rise buildings, a large number of scholars pay attention to the structure of SRRAC. The results show that the flexural strength [18,19] and shear strength [20] of SRRAC beams are similar to those of SRC beams. SRRAC frame has good seismic performance [21], and the seismic performance of SRRAC components decreases in varying degrees with the replacement percentage of recycled aggregate increases [22–24]. The seismic performance of three beam–column joints with ordinary concrete is similar to the RAC, for which the replacement percentage of the specimens is 30% [25]. The SRRAC structure can be used in high-rise and super-high-rise buildings after reasonable preparation and design.

Fire is one of the major hazards that can affect engineering structures. It is necessary to study the fire behavior of RAC structures. The results show that the residual performance of RAC after exposure to elevated temperature is optimal level when replacement percentage is 50% [26]. The performance of RAC is similar to ordinary concrete after exposure to elevated temperature [27,28]. With the increase of elevated temperature, the elastic modulus of RAC decreases [29]. And its compressive strength increases slightly at first under the temperature of 400 degrees centigrade and then decreases with the temperature increases [30]. The ductility and cracking performance of RAC are improved after exposure to elevated temperature when the steel fibers are added [31]. Meanwhile, the energy dissipation and anti-spalling properties of RAC can improve by the rubber powder [32]. The increase of RAC strength can improve its impact behaviors when the temperature is lower than 500 degrees centigrade. RAC strength has an insignificant effect on the impact behaviors when the temperature is greater than 500 degrees centigrade [33].

In conclusion, research on the residual properties of SRRAC components after exposure to elevated temperature is limited. So that it needs to be further studied. A total of 176 specimens (120 rectangular prisms specimens, 24 SRRAC short columns and 32 SRRAC beams) were designed and tested after exposure to elevated temperature. The parameters were considered in the test, like replacement percentage of recycled coarse aggregate (0%, 30%, 50%, 70% and 100%) and exposure to different temperatures (20, 200, 400, 600 and 800 degrees centigrade). Residual properties of RAC and SRRAC were analyzed to provide reference for further research and the engineering application of SRRAC.

2. Experimental Work

A total of 176 recycled concrete specimens were designed and fabricated, including 120 rectangular prisms specimens (40 groups, three in each group), 24 SRRAC short columns and 32 SRRAC beams. The beam specimens were divided into the bending beam and the shear beam, and the corresponding shear span-to-depth ratios were 2 and 1.2, respectively. The dimensions of rectangular prisms specimens were 150 mm × 150 mm × 300 mm. Section size of SRRAC beams and columns was shown in Figure 1.

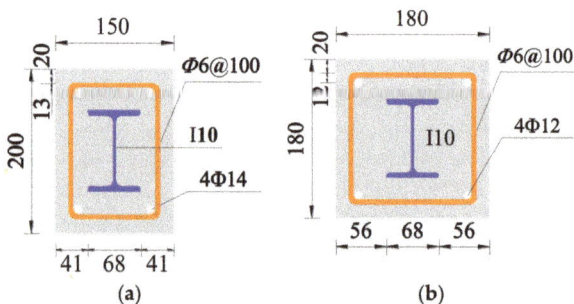

Figure 1. The section of steel reinforced recycled aggregate concrete (SRRAC) beam and column. (**a**) Beam section; (**b**) Column section.

2.1. Test Materials

All of the test materials are from Nanning, Guangxi, China. The materials were used in the test, such as 32.5R Portland cement of Guangxi conch brand, natural river sand, urban water, recycled coarse aggregate and natural gravel. Recycled coarse aggregate is derived from waste concrete specimens in laboratory, and it can be obtained after mechanical crushing, cleaning and sieving of waste concrete. Recycled and natural coarse aggregate were screened under the same conditions. Particle sizes range from 5 mm to 20 mm and continuously graded. The bulk density and water absorption ratio are 1432 kg/m^3 and 3.27%, respectively. The reference value of mix proportion design is to be $r = 0\%$. Total mass of coarse aggregate remains constant, and only changes the ratio of recycled and natural coarse aggregate. Meanwhile, other materials remain constant. The mix proportion of concrete is shown in Table 1. Section steel is I10 of Q235. And the diameter of the stirrup bar of the columns and beams is 6 mm, and the steel bars is HPB300. The diameter of longitudinal bar (HRB335) of the columns and beams is 12 mm and 14 mm, respectively. And the detail of SRRAC beam section and column section was showed in Figure 1.

Table 1. Mix proportions of recycled concrete.

$r/\%$	Cement/kg	Water/kg	Sand/kg	Natural Gravel/kg	Recycled Coarse Aggregate/kg
0	500	215	532	1129	0
30	500	215	532	790.3	338.7
50	500	215	532	564.5	564.5
70	500	215	532	338.7	790.3
100	500	215	532	0	1129

2.2. High-Temperature Installation and Loading System

As showed in Figure 2, RX-45-9 industrial box-type resistance furnace (Detianli Electric Furnace Manufacturer Co., Ltd., Jinan, China) was used for the high-temperature installation. And its maximum temperature can reach 950 degrees centigrade. The target temperature was set separately for batch heating based on the design of the specimen. In order to uniformly distribute the temperature inside the specimen, the temperature in high-temperature installation was constant for one hour when the furnace temperature raised to the target temperature. And then, open the furnace door and let the specimens fall to room temperature under natural conditions. The temperature in the test furnace was recorded during the test. The heating curves was showed in Figure 3. After the specimens fall to room temperature, the test block and column specimens were loaded by RMT-201 test machine

(Institute of Rock and Soil Mechanics, Chinese Academy of Sciences, Wuhan, China) which is showed in Figure 4a. Displacement control loading was adopted in the test. And the loading ratio was about 0.01 mm/s. The load-deformation curve was obtained by the acquisition system of installation. On the other hand, beam specimens were loaded by two points symmetrically. And the specimens were tested under the displacement–force mixed control. Before the pre-estimation of the ultimate load, force control was used in the test, and its step length and holding time for each level are 10 kN and 5 min, respectively. Then, when the specimen was approaching the ultimate load, displacement control loading was applied until the specimen was destroyed. And its displacement gradient was 0.5 mm. The loading installation was showed in Figure 4b.

Figure 2. High-temperature installation.

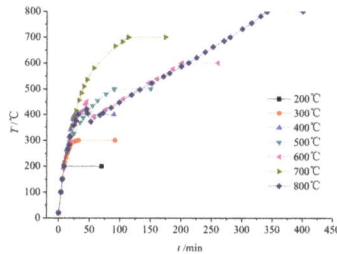

Figure 3. Heating curves.

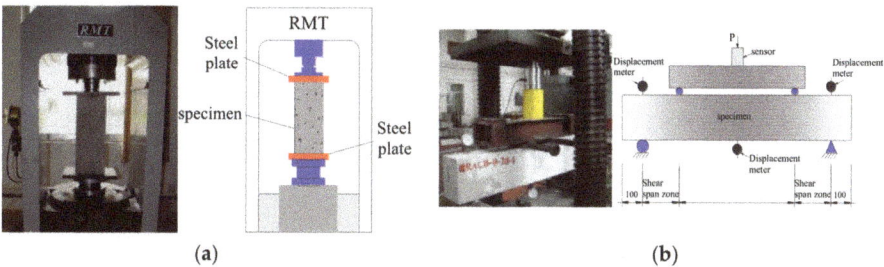

Figure 4. Loading installation. (**a**) Loading installation for block and column; (**b**) Loading installation for beam.

3. Test Results

3.1. Surface Change

A series of physical and chemical reactions had occurred in recycled concrete under elevated temperature. The phenomenon (color change, cracking and spalling on concrete surface) appeared in rectangular prisms specimens, SRRAC columns and SRRAC beams. The physical phenomena of different specimens are basically similar. The color on the concrete surface changes from shallow to deeper as the temperature rises. When T is at 200 degrees centigrade to 400 degrees centigrade, the color on the concrete surface is green-grey and there are no visible cracks. And when T is at 600 degrees centigrade, the color on the concrete surface is brown-grey, and irregular micro-cracks can be found on the concrete surface. When T is at 800 degrees centigrade, the color on the concrete surface is gray-white, and the spalling phenomenon can be seen on the concrete surface. Concrete surface is green-grey due to the hydration reaction to form a little $Ca(OH)_2$. And then, the hydration products of concrete ($C_3S_2H_3$) decompose into CaO as the temperature rises, and CaO is gray-white when it contains impurities [34]. Figure 5 shows apparent morphology of rectangular prisms specimens after exposure to elevated temperature.

Figure 5. Apparent morphology of specimen after elevated temperature.

3.2. Mass Loss

The mass loss of concrete can be found at specimens after exposure to elevated temperature. Mass loss ratio (β_m) is defined to reflect this physical change. Mass loss ratio is to be $\beta_m = (M - M_T)/M \times 100\%$. And M is the mass of specimens before exposure to elevated temperature; M_T is the mass of specimens after exposure to elevated temperature.

Mass loss ratio of specimens after exposure to different elevated temperature is shown in Figure 6. As showed in Figure 6, the β_m increase as the elevated temperature increases. The β_m increases fastest when T is at 200 degrees centigrade to 400 degrees centigrade. Because the amount of water in the concrete evaporates and most of the combustible are burned at the ignition point, when the temperature reaches 200 degrees centigrade. Meanwhile, a large amount of white fog leaks out of the resistance furnace when the temperature reaches at 200 degrees centigrade to 400 degrees centigrade. This phenomenon shows that the moisture in the concrete evaporates most significant at 200 degrees centigrade to 400 degrees centigrade. Mass loss ratio of specimens grows slowly when the temperature is greater than 600 degrees centigrade. Because the moisture and combustible are completely evaporated or burned when the temperature rises to a certain extent. Therefore, mass loss ratio of specimens tends to be stable.

Figure 6d shows that the mean of β_m in each group at the same elevated temperature. According to the Figure 6d, mass loss ratio of rectangular prisms specimens increases the most significant. Mass loss ratio of SRRAC beams increase the least. Figure 6e shows that the mean of mass loss ratio of similar specimens at the same replacement percentage. According to the Figure 6e, mass loss ratio increases with replacement percentage increases. Meanwhile, mass loss ratio of replacement percentage at 100% is about 1.4 times that of the replacement percentage at 0%. Because a lot of old cement paste attach on the surface of recycled coarse aggregate. And then, more water is absorbed by the old cement paste during the stirring process. Therefore, the more recycled coarse aggregate has, the more the moisture of recycled concrete absorbs. Thus, water evaporation increased in recycled concrete after exposure to elevated temperature, and its mass loss ratio increases.

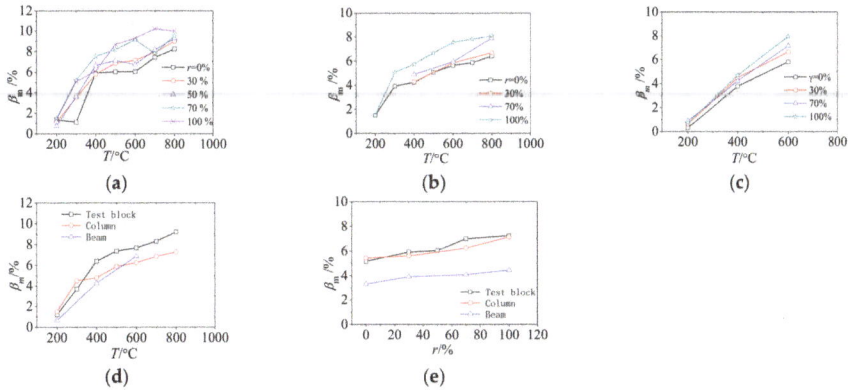

Figure 6. Mass loss ratio of specimens. (**a**) rectangular prisms specimens; (**b**) columns; (**c**) beams; (**d**) The influence of temperature on different components; (**e**) The influence of replacement percentage on different components

3.3. Failure Mode

3.3.1. Failure Mode of Rectangular Prisms Specimens

The interior of RAC has changed after exposure to elevated temperature, so that the failure process and mode of RAC are also different. The difference is more obvious with the temperature increases. The failure modes of the specimens before 400 degrees centigrade are similar to those at room temperature. These show that a few short vertical micro-cracks can be found in the middle of the specimen. As the development of stress, the cracks extend gradually toward the slope. Finally, one or two obvious oblique edge crack or crack zone is formed when the oblique edge crack penetrated specimen. After 400 degrees centigrade, the initial damage of concrete is serious due to the effect of high temperature. After the action of external force, many cracks appeared and accompanied debris falling, and a wider cracks zone is clearly visible. Typical failure modes of the specimen are shown in Figure 7 after exposure to different temperatures.

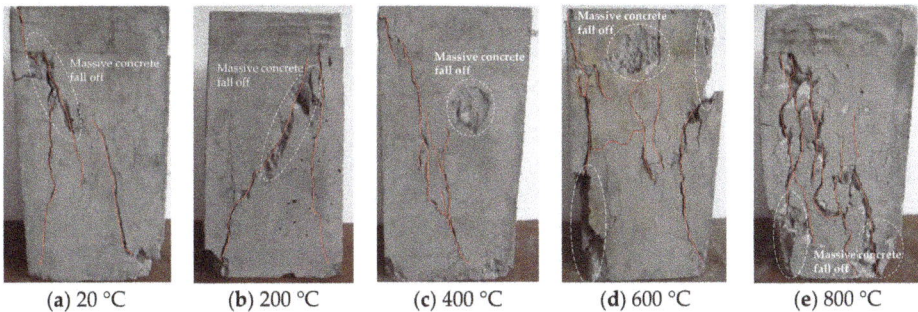

(**a**) 20 °C   (**b**) 200 °C   (**c**) 400 °C   (**d**) 600 °C   (**e**) 800 °C

Figure 7. Typical failure mode of recycled aggregate concrete (RAC).

3.3.2. Failure Mode of SRRAC Columns

It is found that the failure process and failure mode of SRRAC columns are mainly related to the highest temperature after observing the loading process of SRRAC columns under axial compression.

The failure process and mode of specimens at different replacement percentage are similar to each other at the same temperature. The failure process of specimens presents elastic stage, crack stage and failure stage. The mode and failure process of specimens after exposure to elevated temperature are as follows: The higher the temperature is, the earlier the crack appears and penetrates. The failure mechanism of the specimen is different from that of the room temperature specimen when the temperature is higher than 500 degrees centigrade. It is not the crushing failure of edge concrete. But the concrete cover falls off, then SRRAC columns damage in advance before reaching the peak load. The higher the temperature is, the earlier the concrete falls off. The failure modes of the specimen are shown in Figure 8.

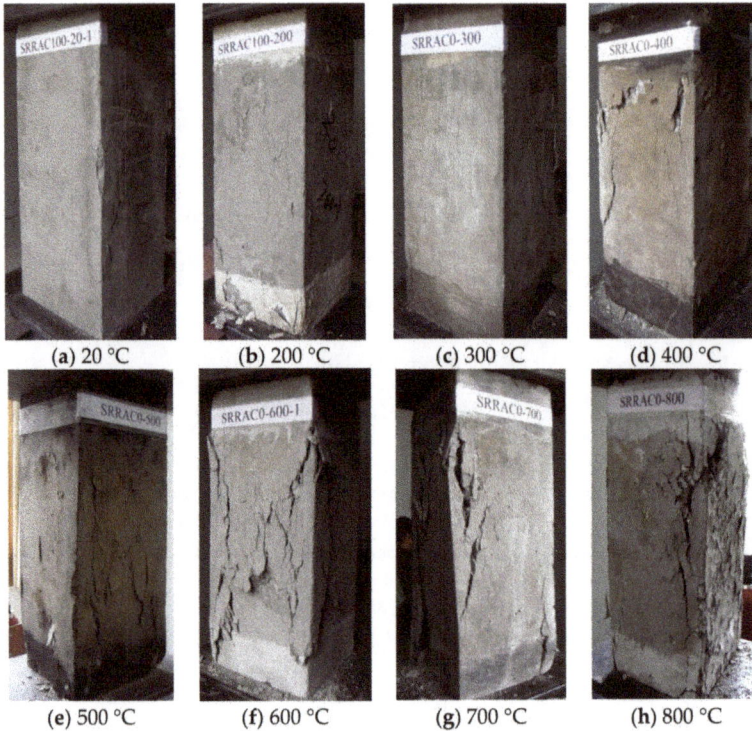

Figure 8. The failure modes of column specimen.

3.3.3. Failure Mode of SRRAC Beams

Failure modes of SRRAC beams after exposure to elevated temperature is the same as those at room temperature, which reflects shear-baroclinic failure, bond failure and bending failure. SRRAC beams with shear span-to-depth ratios of 1.2 show shear-baroclinic failure. SRRAC beams with shear span-to-depth ratios of 2.0 show shear-baroclinic, bond failure and bending failure.

Shear-baroclinic failure is as follows: Vertical cracks can be seen in the mid span and develop slowly upward when the external load is between 0.13 Pu and 0.35 Pu. The oblique crack appears between the loading point and the support when the external load is between 0.3 Pu and 0.6 Pu. And then, oblique cracks increase and widen, the oblique concrete zones crushed, then the specimens were destroyed.

The bending failure process is as follows: Vertical cracks can be seen in the mid span when the external load is between 0.23 Pu and 0.3 Pu. And vertical cracks increase and developed when the external load is between 0.45 Pu and 0.5 Pu. The oblique crack appears between the loading point and the support when the external load is between 0.5 Pu and 0.6 Pu. Vertical cracks develop faster when the external load is at 0.75 Pu. Finally, the concrete is crushed, then the specimen is destroyed.

Bond failure process is as follows: Longitudinal bonding crack was found near the compressive zone of section steel on the outside flange when the external load is at 0.65 Pu. Finally, the concrete to the outside of the bonding crack is split, then the specimen is destroyed. The failure modes of some specimens are shown in Figure 9. The BBi-j, SAi-j and SBi-j are used to represent corresponding pictures. And SA is shearing beam, and its span-to-depth ratio is 2.0; BB mean bending beam, and its span-to-depth ratio is 1.2. And *i* and *j* are replacement percentage and temperature, respectively.

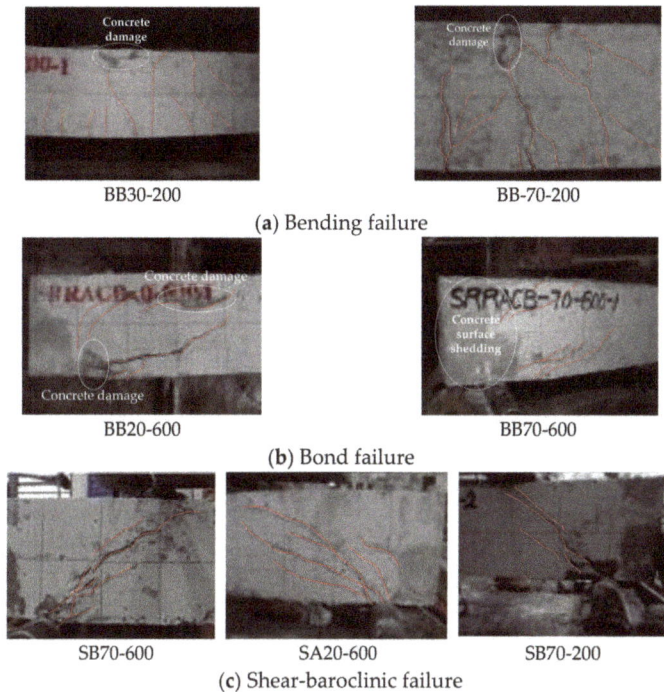

BB30-200   BB-70-200

(**a**) Bending failure

BB20-600   BB70-600

(**b**) Bond failure

SB70-600   SA20-600   SB70-200

(**c**) Shear-baroclinic failure

Figure 9. The failure modes of beam specimens.

3.4. Load-Displacement Curve Analysis

3.4.1. Load-Displacement Curve of Rectangular Prisms Specimens

Load-displacement curves of RAC test block was showed in Figure 10. The data in Figure 10 are the mean values of three identical test blocks in each group. According to Figure 10, the shape of the load-displacement curve of RAC exposure to elevated temperature is similar to that of ordinary concrete. And with the increase of temperature, the peak load of the curve decreases gradually, the peak displacement increases gradually, the descending stage of the curve becomes gentle and the whole curve becomes more and more flat. The effect of temperature on peak load is more and more significant when replacement percentage increases. Specimens in the elastic stage are no crack in the concrete surface, then, the specimens were brittle failure after the load exceeds the elastic limit. The peak displacement gradually increases as the temperature rises. The maximum peak displacement of RAC reaches 3.5 mm when the temperature is 800 degrees centigrade.

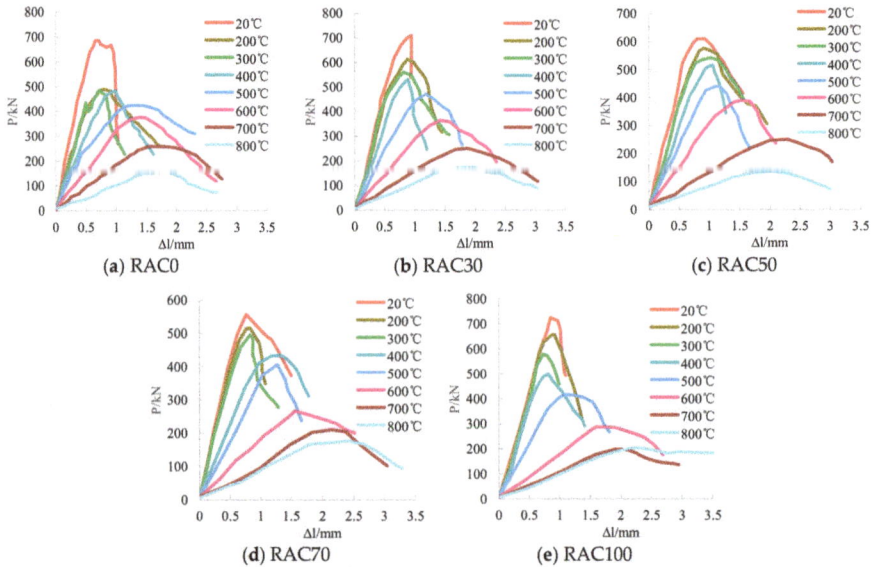

Figure 10. Load-displacement curve of rectangular prisms specimens.

3.4.2. Load-Displacement Curve of SRRAC Columns

According to Figure 11, as the temperature increase, the axial load-displacement curve of the SRRAC columns tends to be flat, and the value of the peak point gradually decreases. Loading process of SRRAC columns mainly includes elastic stage, elastic–plastic stage, stiffness strengthening stage, descending stage and residual stage. The characteristics of each stage are as follows:

During the elastic stage, the elastic deformation of the section steel and concrete is coordinated, and there is no crack in the concrete surface.

The elastic–plastic stage of specimens is as follows: The cracks on the concrete surface emerge and develop continuously after the load exceeds the elastic limit. And then, section steel and longitudinal reinforcement yield gradually. The load-displacement curve at this stage is nonlinear. And bond cracks appear at this stage, therefore, the bond slip between section steel and concrete is more obvious.

The stiffness strengthening stage of specimens is as follows: The load-displacement curve of the specimen appears obvious curvature rising stage before reaching the elastic limit when the exposure temperature of specimen is greater than 600 degrees centigrade. Because the evaporation of free water in the internal void of concrete makes the concrete loose. The loose concrete becomes dense with the increase of compressive load during the loading process. Therefore, the stiffness of the specimen has been improved.

During the descending stage of specimens, the bearing capacity of the specimen decreases obviously after exceeding the limit load.

The residual stage of specimens is as follows: The load decreases gently with the increase of displacement. The longitudinal reinforcement and concrete cover have basically not provided bearing capacity at all. Residual strength is provided by section steel and core-concrete.

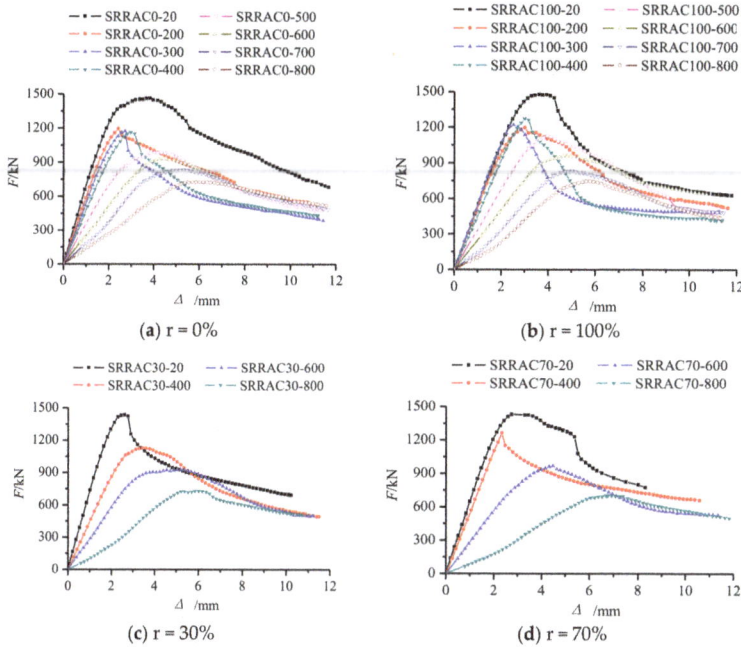

Figure 11. Load-displacement curve of SRRAC columns.

3.4.3. Load-Deflection Curve of SRRAC Beams

The measured load-deflection curve of SRRAC beams is showed in Figure 12. According to the Figure 12, SRRAC beams have undergone three stages: elastic stage, elastic–plastic stage and failure stage. Early loading of SRRAC beams is elastic stage. And it is elastic–plastic stage after cracking. Then, the specimen experienced the failure stage after the peak load.

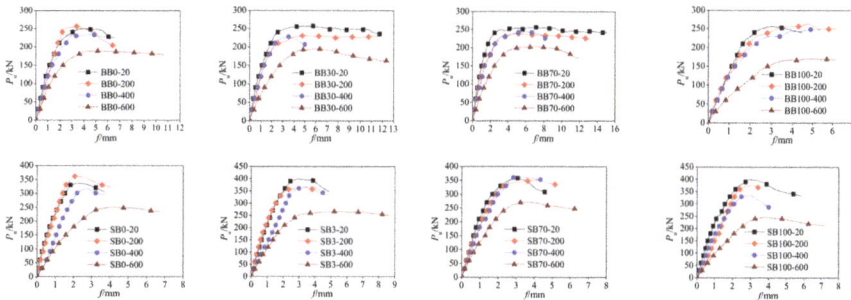

Figure 12. Load–deflection curves.

4. Residual Properties Analysis

In order to analyze the residual properties of RAC and SRRAC components after exposure to elevated temperature, many mechanical performance parameters can be obtained by the load-displacement curve of RAC and SRRAC components, such as bearing capacity, peak deformation, secant stiffness, displacement ductility factors and energy dissipation factor.

The mean performance index of all specimens is obtained at the same temperature and different replacement percentage for comparative analysis the residual properties of different components. And residual properties coefficient (β_N, β_Δ, β_K, β_μ and β_η) can be calculated by normalization based on a performance index at room temperature. Similarly, replacement percentage coefficient (α_N, α_Δ, α_K, α_μ and α_η) can be calculated by normalization based on replacement percentage at 0%.

4.1. Bearing Capacity Degradation

Figure 13 shows the degradation on bearing capacity of specimens after exposure to different elevated temperature. As seen from Figure 13a–d, compressive strength of test blocks after high temperature is deeply downtrend as the temperature ascend. The bearing capacity decreases by about 18% in the scope from 200 degrees centigrade to 400 degrees centigrade. And bearing capacity decreases by 48% when exposure to elevated temperature is at 600 degrees centigrade. Meanwhile, bearing capacity decreased by 74% when exposure to elevated temperature is at 800 degrees centigrade.

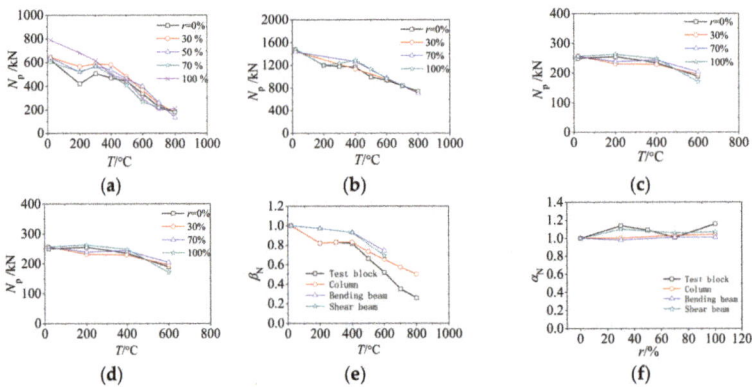

Figure 13. Bearing capacity degradation. (**a**) rectangular prisms specimens; (**b**) columns; (**c**) bending beams; (**d**) shear beams; (**e**) The influence of temperature; (**f**) The influence of replacement percentage.

After exposure to elevated temperature, the evaporation of free water and bound water in concrete leads to internal cracks. And then, the structure of hardened cement paste is crisp pine. The thermal performance of the coarse aggregate and concrete is inconsistent. Therefore, internal cracks continue to develop due to uncoordinated thermal expansion and syneresis micro-deformation. And then, the interface between aggregate and cement paste is further loosened when the temperature is above 700 degrees centigrade. Then, the pore size of cement slurry increases further. The crack between aggregate and slurry expands rapidly, the crack width increases. Therefore, the strength of concrete decreases. It can be proved by the scanning electron microscope (SEM) test in the literature [35].

As seen from Figure 13e, the bearing capacity of SRRAC components decreases significantly, especially with higher exposure to elevated temperature. However, the amplitude of reduction is slower than the test block. The law of bearing capacity degradation of the bending beam and the shear beam is basically the same. The bearing capacity is degraded slowly when T is less than 400 degrees centigrade. The degradation rate speeds up when T is greater than 400 degrees centigrade. Bearing capacity degradation rate is about 30% when T is 600 degrees centigrade. Its value is equaled to 50% of the rectangular prism specimens. The law of bearing capacity degradation of SRRAC columns under compression is similar to the prism block. And the degradation rate of SRRAC column is slower than that of rectangular prisms specimens, when T is greater than 400 degrees centigrade. The bearing capacity degradation of compression components is more serious than that of flexural components after exposure to the same elevated temperature. The β_N of compressive components is about 11%

to 22% smaller than that of flexural components in the scope from 200 degrees centigrade to 600 degrees centigrade.

As seen from Figure 13f, the replacement percentage of recycled coarse aggregate has an insignificant effect on RAC test blocks and SRRAC components after exposure to elevated temperature. As the increase of replacement percentage, the variation ranges of the bearing capacity of prism test block, compression column, bending beam and shear beam are 1~16%, 1~4%, 2~1% and 6~10%, respectively. Generally speaking, the bearing capacity of specimens increases slightly with the increase in the replacement percentage. Free water in concrete evaporates at elevated temperature action to form pore. And then, the bearing capacity of concrete falls. However, the surface of recycled coarse aggregate is rough and porous, which can hold water firmly and reduce the evaporation of free water. So bearing capacity of RAC degenerates slower than ordinary concrete. In conclusion, after exposure to elevated temperature, the bearing capacity of SRRAC components is slightly better than that of ordinary steel reinforced concrete components.

4.2. Stiffness Degradation

The law of secant stiffness degradation of specimens is shown in Figure 14. And secant stiffness is the secant modulus at 0.4 Np. As seen from Figure 14a–e, the laws of stiffness degradation of the specimens are similar to that of bearing capacity. But the degradation of stiffness is larger than that of bearing capacity. The bending stiffness of the beam deteriorated most slowly. Bending stiffness degradation rate is about 49%, when T is 600 degrees centigrade. The axial compressive stiffness of the prism test block deteriorated most rapidly. The axial compressive stiffness degradation rate is about 70% when T is 600 degrees centigrade. And the axial compressive stiffness degradation rate is about 88% when T is 800 degrees centigrade. According to the degradation trend, the stiffness degradation rate of SRRAC components under axial compression is larger than that of the flexural components after exposure to elevated temperature. Stiffness degradation coefficient (β_K) of compression components is about 5~21% less than that of the flexural member when T is at 200 degrees centigrade to 600 degrees centigrade.

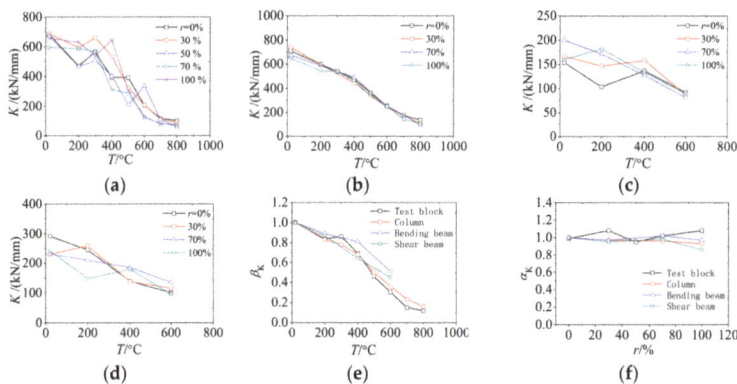

Figure 14. Stiffness degradation. (**a**) rectangular prisms specimens; (**b**) columns; (**c**) bending beams; (**d**) shear beams; (**e**) The influence of temperature; (**f**) The influence of replacement percentage.

As seen in Figure 14f, the stiffness of each component after exposure to elevated temperature is an insignificantly affected by the replacement percentage. As replacement percentage increases, the variation ranges of the axial compression stiffness of prism test block and column, flexural rigidity and shear rigidity of beams are −5~8%, −7~−3%, −3~2% and −14~−2%, respectively. In general, stiffness of a component has a slight descent trend after exposure to elevated temperature with replacement percentage increases. The cause of stiffness degradation of RAC and SRRAC is similar to that of bearing capacity degradation. Both are caused by the evaporation of free water in RAC and bond failure between the cement slurry and aggregate. And the failure mechanism of the RAC after exposure to elevated temperature is basically similar to that of ordinary concrete.

4.3. Peak Deformation

The law of peak deformation of specimens is shown in Figure 15. The peak deformation is the axial deformation corresponding to the peak load for test blocks and columns or the mid span deflection corresponding to the peak load for beams. As seen from Figure 15, peak deformation is insignificant affected by high temperature when T is less than or equal to 400 degrees centigrade. The peak deformation of the test block, column and shear beam increases rapidly with temperature increases when T is higher than 400 degrees centigrade. The peak deformation of the test block is about 2.26 times that of the room temperature. But the peak deformation of bending beams is less affected by temperature. Its peak deformation is approximately 1.24 times that of the room temperature when T is equal to 600 degrees centigrade. The increase of peak deformation is related to the increase of porosity inside concrete after exposure to elevated temperature.

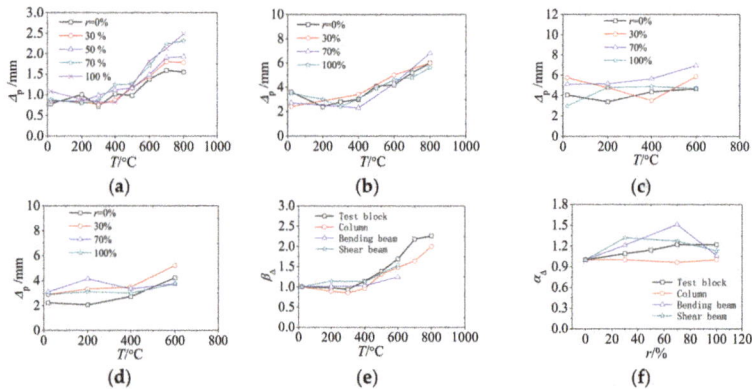

Figure 15. Deformation corresponding to the peak load. (**a**) rectangular prisms specimens; (**b**) columns; (**c**) bending beams; (**d**) shear beams; (**e**) The influence of temperature; (**f**) The influence of replacement percentage.

According to Figure 15f, the effect of replacement percentage on peak deformation of different components is different. The peak deformation linearly increases as the replacement percentage increases for a test block. And when the replacement percentage is 100%, the peak deformation is about 1.22 times that of replacement percentage at 0%. The peak deformation of the bending beams and shear beams increases first and then decreases. And those maximum increment is 39% and 32%, respectively. However, the replacement percentage has an insignificant effect on the peak deformation of axial compression column. And its peak deformation only decreases by 4% when the replacement percentage is 70%.

4.4. Ductility

Ductility ($\mu = \Delta_u / \Delta_y$) is calculated according to the load-deformation curve. And Δu is the value of deflection at 0.85 times Np, and the maximum deformation is taken to be Δu when the load falls below 0.85 Np. Δ_y is the value of initial yield deformation and calculated by the equivalent energy method. As shown in Figure 16, the dimension of OAB is equal to the dimension of YUB.

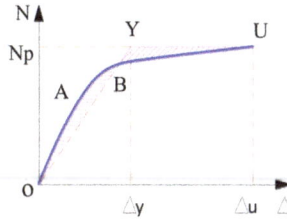

Figure 16. The sketch of energy equivalent method.

Figure 17 shows the displacement ductility factor of each component after exposure to different elevated temperature. As seen from Figure 17a–e, the displacement ductility coefficient of each component fluctuated slightly up or down as exposure to elevated temperature rises. Generally speaking, the ductility factor of the four types specimens decreases with increasing temperature when *T* is between 200 degrees centigrade and 400 degrees centigrade. When *T* is equal to 400 degrees centigrade, the displacement ductility factor of test block, column, bending beam and shear beam decrease by 10%, 28%, 44% and 31%, respectively. The displacement ductility factor of each component varies from −25% to 4% when *T* is higher than 400 degrees centigrade.

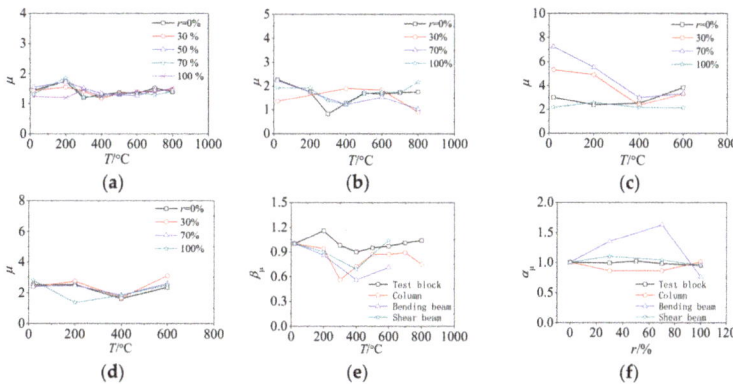

Figure 17. Ductility. (**a**) rectangular prisms specimens; (**b**) columns; (**c**) bending beams; (**d**) shear beams; (**e**) The influence of temperature; (**f**) The influence of replacement percentage.

As seen from Figure 17f, the displacement ductility factor of the bending beam increases first and then decreases. And the displacement ductility factor of bending beam increases 63% when the replacement percentage is 70%. And the displacement ductility factor of bending beam decreases 24% when replacement percentage is 100%. Then, the displacement ductility factor of the other three types of components varies slightly with replacement percentage varies.

4.5. Energy Dissipation

Energy dissipation factors ($\eta = S_{OUYC}/S_{OABC}$) are calculated according to the load-deformation curve. As showed in the Figure 18, the S_{OUYC} is equal to the shadow area surrounded by the load deflection curve. The S_{OABC} is equal to the rectangle area, which is passes through the peak point (U) and the limit deformation point (Y).

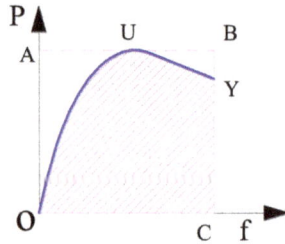

Figure 18. The sketch of energy consumption coefficient.

Figure 19 shows the energy dissipation factors of each component after exposure to different elevated temperature. As seen from Figure 19a–e, the law of energy dissipation factors of each component is similar to that of ductility factor. Energy dissipation factor decreases with the increase of temperature when T is lower than or equal to 400 degrees centigrade. When T is equal to 400 degrees centigrade, the energy dissipation factor of test block, column, bending beam and shear beam decrease by 13%, 11%, 7% and 10%, respectively. The energy dissipation factor increases slightly when T is from 400 degrees centigrade to 600 degrees centigrade. The energy dissipation factor decreases steeply when T is above 600 degrees centigrade. When T is equal to 800 degrees centigrade, the energy dissipation factor of test block and column decreases 36% and 10%, respectively. As seen from Figure 19f, the energy dissipation factor of each component is not obviously affected by the replacement percentage. The energy dissipation factor varies from -10% to 16% with replacement percentage increases.

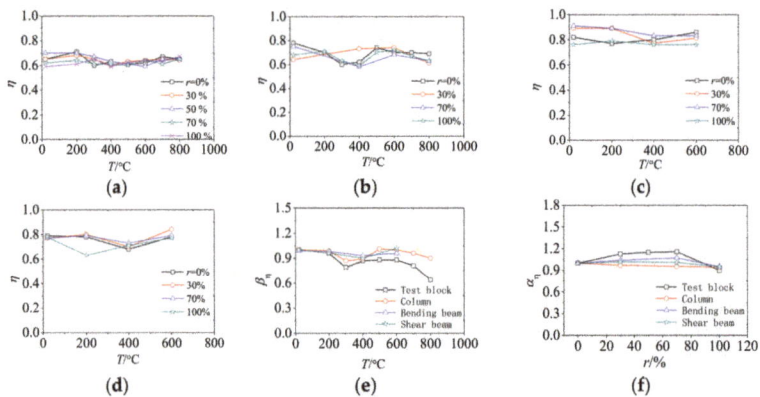

Figure 19. Energy dissipation. (**a**) rectangular prisms specimens; (**b**) columns; (**c**) bending beams; (**d**) shear beams; (**e**) The influence of temperature; (**f**) The influence of replacement percentage.

5. Conclusions

(1) Significant physical changes occurred on RAC surface after exposure to elevated temperature. Firstly, color on RAC surface changed from green-grey to gray-white. And then, chapped phenomenon occurs on the RAC surface when the temperature reached 600 degrees centigrade. Finally, the spalling phenomenon occurs on RAC surface when the temperature reaches 800 degrees centigrade. The phenomenon of mass loss can be found in the RAC after exposure to elevated temperature, and the phenomenon is more significant when the temperature and the replacement percentage increase.

(2) Mechanical properties of RAC test blocks and SRRAC components are significantly degraded after exposure to elevated temperature. And its bearing capacity and stiffness degrade most

obviously. And the performance of the prism test block degraded faster than SRRAC components. Performance degradation of SRRAC beams is slower than RAC test blocks and SRRAC columns.

(3) The bearing capacity, stiffness and peak deformation vary slightly with temperature changes when the temperature is below 400 degrees centigrade. When the temperature exceeds 400 degrees centigrade, the bearing capacity and stiffness decrease steeply, and the peak deformation increases steeply. And ductility and energy dissipation are insignificant affected by elevated temperature.

(4) Loss on ignition of specimen mass, the peak deformation and bearing capacity increase slightly with replacement percentage increases. The stiffness was significant fluctuation when replacement percentage was 70% to 100%. The ductility and energy dissipation were significant fluctuation when replacement percentage was 30% to 70%.

6. Future Research

At present, research on residual properties of SRRAC after exposure to elevated temperature is still in the initial stage. In order to further promote the application of SRRAC components, the following focuses should be further studied or discussed.

(1) This research focuses on the influence of the parameters (replacement percentage of recycled coarse aggregate and exposure to different temperatures) on the residual properties of SRRAC components. However, the coverage of steel bars or steel profiles can be influence on the mechanical properties of SRRAC components according to the regulations. Therefore, the influence of concrete cover and the coverage of steel bars or steel profiles on the residual properties of SRRAC components should be further discussed.

(2) In this research, in order to ensure the same water–cement ratio of natural aggregate concrete and recycled aggregate concrete, the same amounts of water were used in the mix proportion. However, the porosity of recycled aggregate is higher than that of natural aggregate. So that the water absorption of recycled aggregate is higher than that of natural aggregate. Therefore, in order to get more accurate test results, it is necessary to further consider the influence of the water absorption of recycled coarse aggregate on effective water cement ratio.

(3) In order to obtain more comprehensive laws of residual properties of SRRAC components after exposure to elevated temperature, the X-ray diffraction (XRD), thermogravimetric analysis (TGA) and SEM tests should be carried out for further research. Because XRD and TGA tests can reveal the chemical changes produced by the high temperatures in the concrete. And SEM tests and image analysis can identify on a microscopic scale the original, difference and failure procedures presented by the different samples.

Author Contributions: Z.C. conceived the experiments, C.J. and R.Y. designed the experiments and wrote the initial draft of the manuscript. Z.C., R.Y., C.J. and F.N. analyzed the date and wrote the final manuscript.

Funding: This research was funded by the National Natural Science Foundation of China (grant 51578163), and the project of Natural Science Foundation of Guangxi Province (2016GXNSFDA380032). And High-Level Innovation Group and Outstanding Scholar Program Project of Guangxi High Education (No. [2017]38). The authors are grateful for the above supporters.

Acknowledgments: Thanks for Yi. Li's help in the experiment.

Conflicts of Interest: The authors declare no conflict of interest.

References

1. Chen, Z.; Zhang, Y.; Chen, J.; Fan, J. Sensitivity Factors Analysis on the Compressive Strength and Flexural Strength of Recycled Aggregate Infill Wall Materials. *Appl. Sci.* **2018**, *8*, 1090. [CrossRef]
2. Oikonomou, N. Recycled concrete aggregates. *Cem. Concr. Compos.* **2005**, *27*, 315–318. [CrossRef]
3. Dhir, R.K.; Henderson, N.A.; Limbachiya, M.C. Sustainable Construction: Use of recycled concrete aggregate. *Adv. Struct. Eng.* **1998**, *11*, 383–396.

4. Gómez-Soberón, J.M.V. Porosity of recycled concrete with substitution of recycled concrete aggregate: An experimental study. *Cem. Concr. Res.* **2002**, *32*, 1301–1311. [CrossRef]

5. Loo, Y.H.; Tam, C.T.; Ravindrarajah, R.S. Recycled concrete as fine and coarse aggregate in concrete. *Mag. Concr. Res.* **1987**, *39*, 214–220.

6. Abbas, A.; Fathifazl, G.; Fournier, B.; Isgor, O.B.; Zavadil, R.; Razaqpur, A.G.; Foo, S. Quantification of the residual mortar content in recycled concrete aggregates by image analysis. *Mater. Charact.* **2009**, *60*, 716–728. [CrossRef]

7. Tam, V.W.; Wang, K.; Tam, C.M. Assessing relationships among properties of demolished concrete, recycled aggregate and recycled aggregate concrete using regression analysis. *J. Hazard. Mater.* **2008**, *152*, 703–714. [CrossRef] [PubMed]

8. Kou, S.C.; Poon, C.S. Properties of concrete prepared with PVA-impregnated recycled concrete aggregates. *Cem. Concr. Compos.* **2010**, *32*, 649–654. [CrossRef]

9. Hansen, T.C. Recycled concrete aggregate and fly ash produce concrete without portland cement. *Cem. Concr. Res.* **1990**, *20*, 355–356. [CrossRef]

10. Evangelista, L.; Brito, J. Mechanical behaviour of concrete made with fine recycled concrete aggregates. *Cem. Concr. Compos.* **2007**, *29*, 397–401. [CrossRef]

11. Choi, W.C.; Yun, H.D. Compressive behavior of reinforced concrete columns with recycled aggregate under uniaxial loading. *Eng. Struct.* **2012**, *41*, 285–293. [CrossRef]

12. Butler, L.; West, J.S.; Tighe, S.L. The effect of recycled concrete aggregate properties on the bond strength between RCA concrete and steel reinforcement. *Cem. Concr. Res.* **2011**, *41*, 1037–1049. [CrossRef]

13. Xiao, J.; Sun, Y.; Falkner, H. Seismic performance of frame structures with recycled aggregate concrete. *Eng. Struct.* **2006**, *28*, 1–8. [CrossRef]

14. Reis, N.; Brito, J.D.; Correia, J.R.; Arruda, M.R.T. Punching behaviour of concrete slabs incorporating coarse recycled concrete aggregates. *Eng. Struct.* **2015**, *100*, 238–248. [CrossRef]

15. Knaack, A.M.; Kurama, Y.C. Behavior of Reinforced Concrete Beams with Recycled Concrete Coarse Aggregates. *J. Struct. Eng.* **2015**, *141*, B4014009. [CrossRef]

16. Ev, M.M.; Radonjanin, V.; Marinkovi, S.A. Recycled Concrete as Aggregate for Structural Concrete Production. *Sustainability* **2010**, *2*, 1204–1225.

17. Fathifazl, G.; Razaqpur, A.G.; Isgor, O.B.; Abbas, A.; Fournier, B.; Foo, S. Shear capacity evaluation of steel reinforced recycled concrete (RRC) beams. *Eng. Struct.* **2011**, *33*, 1025–1033. [CrossRef]

18. Jia, Y.D.; Guo, Y.K.; Sun, Z.P.; Zhao, X. Experimental Research on Behavior of Composite Beams of Steel-Reinforced Recycled Concrete. *Adv. Mater. Res.* **2013**, *639–640*, 145–148. [CrossRef]

19. Fathifazl, G.; Razaqpur, A.G.; Isgor, O.B.; Abbas, A.; Fournier, B.; Foo, S. Flexural Performance of Steel-Reinforced Recycled Concrete Beams. *ACI Struct. J.* **2009**, *106*, 858–867.

20. Fathifazl, G. *Structural Performance of Steel Reinforced Recycled Concrete Components*; Dissertation Abstracts International; Carleton University: Ottawa, ON, Canada, 2008; Volume 69, p. 1175.

21. Xue, J.; Zhang, X.; Ren, R.; Zhai, L.; Ma, L. Experimental and numerical study on seismic performance of steel reinforced recycled concrete frame structure under low-cyclic reversed loading. *Adv. Struct. Eng.* **2018**, *21*, 1895–1910. [CrossRef]

22. Ma, H.; Xue, J.; Liu, Y.; Zhang, X. Cyclic loading tests and shear strength of steel reinforced recycled concrete short columns. *Eng. Struct.* **2015**, *92*, 55–68. [CrossRef]

23. Xue, J.; Zhai, L.; Bao, Y.; Ren, R.; Zhang, X. Seismic behavior of steel-reinforced recycled concrete inner-beam-column connection under low cyclic loads. *Adv. Struct. Eng.* **2017**, *21*, 631–642. [CrossRef]

24. Ma, H.; Xue, J.; Zhang, X.; Luo, D. Seismic performance of steel-reinforced recycled concrete columns under low cyclic loads. *Constr. Build. Mater.* **2013**, *48*, 229–237. [CrossRef]

25. Gonzalez, V.C.L.; Moriconi, G. The influence of recycled concrete aggregates on the behavior of beam–column joints under cyclic loading. *Eng. Struct.* **2014**, *60*, 148–154. [CrossRef]

26. Govinda, G.G.; Rao, B.S.; Naik, S.M. Behaviour of Recycled Aggregate Concrete on exposed to Elevated Temperature. *Int. J. Civ. Eng.* **2017**, *4*, 5–13.

27. Salau, M.A.; Oseafiana, O.J.; Oyegoke, T.O. Effects of Elevated Temperature on Concrete with Recycled Coarse Aggregates. In *IOP Conference Series: Materials Science and Engineering*; IOP Publishing: Bristol, UK, 2015; Volume 96, p. 012078.

28. Laneyrie, C.; Beaucour, A.L.; Green, M.F.; Hebert, R.L.; Ledesert, B.; Noumowe, A. Influence of recycled coarse aggregates on normal and high performance concrete subjected to elevated temperatures. *Constr. Build. Mater.* **2016**, *111*, 368–378. [CrossRef]

29. Gupta, A.; Mandal, S.; Ghosh, S. Recycled aggregate concrete exposed to elevated temperature. *J. Eng. Appl. Sci.* **2012**, *7*, 100–107.

30. Liu, Y.; Ji, H.; Zhang, J.; Wang, W.; Chen, Y.F. Mechanical properties of thermal insulation concrete with recycled coarse aggregates after elevated temperature exposure. *Mater. Test.* **2016**, *58*, 669–677. [CrossRef]

31. Chen, G.M.; He, Y.H.; Yang, H.; Chen, J.F.; Guo, Y.C. Compressive behavior of steel fiber reinforced recycled aggregate concrete after exposure to elevated temperatures. *Constr. Build. Mater.* **2016**, *128*, 272–286. [CrossRef]

32. Guo, Y.C.; Zhang, J.H.; Chen, G.M.; Xie, Z.H. Compressive behaviour of concrete structures incorporating recycled concrete aggregates, rubber crumb and reinforced with steel fibre, subjected to elevated temperatures. *J. Clean. Prod.* **2014**, *72*, 193–203. [CrossRef]

33. Li, W.; Luo, Z.; Wu, C.; Tam, V.W.Y.; Duan, W.H.; Shah, S.P. Experimental and numerical studies on impact behaviors of recycled aggregate concrete-filled steel tube after exposure to elevated temperature. *Mater. Des.* **2017**, *136*, 103–118. [CrossRef]

34. Gu, L.; Ling, F.; Sheng, Z. Research on properties of concrete and its compositions after high temperature. *Sichuan Build. Sci.* **1991**, *2*, 1–5. (In Chinese)

35. Zai, Q.; Hui, W. Research on Scanning Electron Microscopic of Concrete after High Temperature. *J. Fuzhou Univ.* **1996**, *S1*, 36–40. (In Chinese)

applied
sciences

MDPI

Article

Using Neural Networks to Determine the Significance of Aggregate Characteristics Affecting the Mechanical Properties of Recycled Aggregate Concrete

Zhenhua Duan [1,2], Shaodan Hou [1], Chi-Sun Poon [2,*], Jianzhuang Xiao [1] and Yun Liu [1,*]

[1] Department of Structural Engineering, Tongji University, Shanghai 200092, China;
     zhduan@tongji.edu.cn (Z.D.); hsd2017@tongji.edu.cn (S.H.); jzx@tongji.edu.cn (J.X.)
[2] Department of Civil and Environmental Engineering, The Hong Kong Polytechnic University, Hung Hom,
     Kowloon, Hong Kong, China
*   Correspondence: cecspoon@polyu.edu.hk (C.-S.P.); liuyun@tongji.edu.cn (Y.L.);
     Tel.: +852-2766-6024 (C.-S.P.); +86-21-6598-3320 (Y.L.)

Received: 29 September 2018; Accepted: 1 November 2018; Published: 6 November 2018

Abstract: It has been proved that artificial neural networks (ANN) can be used to predict the compressive strength and elastic modulus of recycled aggregate concrete (RAC) made with recycled aggregates from different sources. This paper is a further study of the use of ANN to analyze the significance of each aggregate characteristic and determine the best combinations of factors that would affect the compressive strength and elastic modulus of RAC. The experiments were carried out with 46 mixes with several types of recycled aggregates. The experimental results were used to build ANN models for compressive strength and elastic modulus, respectively. Different combinations of factors were selected as input variables until the minimum error was reached. The results show that water absorption has the most important effect on aggregate characteristics, further affecting the compressive strength of RAC, and that combined factors including concrete mixes, curing age, specific gravity, water absorption and impurity content can reduce the prediction error of ANN to 5.43%. Moreover, for elastic modulus, water absorption and specific gravity are the most influential, and the network error with a combination of mixes, curing age, specific gravity and water absorption is only 3.89%.

Keywords: recycled aggregate; recycled aggregate concrete; artificial neural networks; aggregate characteristic; input variable

---

1. Introduction

There is no doubt that the utilization of recycled aggregate concrete (RAC) has been the best way to resolve the problem of the increasing amount of construction and demolition (C&D) waste and further attain sustainable development. The improved environmental performance of recycled aggregate concrete (RAC) [1–3] has led to research on recycled aggregate (RA) and RAC, a popular topic in the last decades [4–7]. Recycled coarse aggregate and recycled fine aggregate were both used in concrete to make full use of C&D wastes; meanwhile, the cementitious materials supplied, such as fly ash and silica fume, were used together with RA for high-performance RAC [8–11]. Though there has been a large amount of research on the properties of RAC, the results have been varied because the properties of RAs from different sources (such as the demolition of bridges, buildings and airport pavements) and produced using different recycling methods (e.g., the type and effort of the crushers used) vary greatly [12]. It is generally accepted that the properties of RA and the hardened properties of RAC made with such RA are both largely affected by the nature of the attached old mortar [13–17]. However, it is difficult to establish an accurate relationship between the two, since at present there is no established method for accurately measuring the quantity and quality of the attached mortar in RA. On the other hand, RAs may also contain impurities, such as bricks, glass, tiles, asphalt, plastics, gypsum, wood and clay, etc. In small amounts, however, their presence may seriously deteriorate the quality of RA. The presence of other impurities makes it more complicated to predict the properties of RAC.

Therefore, there are at least two major difficulties in building a model that can predict the performance of hardened RAC made with RAs from different sources: (1) the model should act as an expert system covering the factors that may affect the properties of RAC, such as cement content, water to cement ratio, aggregate to cement ratio, cement type and particle size of aggregates, etc.; (2) an optimal combination of RA characteristics should be included in the model so that it can be applicable to the majority of RAs from different sources. A previous study [18] used regression analysis to propose a number of equations relating the hardened properties (compressive strength) of RAC with the water absorption or density of different types and combinations of aggregates obtained from different sources. However, the accuracy of the prediction is limited since the properties of RA cannot be completely represented by the density or water absorption values of RA. Tam and Tam [19] suggested that there were six main factors that characterize the properties of RA: (1) particle size distribution; (2) particle density; (3) porosity and absorption; (4) particle shape; (5) strength and toughness; and (6) chloride and sulphate contents. Through a comparison and analysis of ten sources of RAs and one type of natural aggregate (NA), they constructed relationships among these factors and indicated that the RA properties could be assessed by only measuring three of the six parameters mentioned. However, whether the model is suitable for RAs obtained from other sources has not been verified.

As a modeling tool, artificial neural networks (ANN) have been widely used since the mid-1980s, and have also been demonstrated to have superior capacities in modeling more complex relationships. Among all the ANN structures, the back-propagation network (BPN) is generally regarded as one of the simplest and most applicable networks used in simulating concrete properties. As shown in Figure 1, a typical BPN model consists of an input layer, one or more hidden layers and an output layer, and each layer consists of numerous neurons. During the training set, feed-forward propagation and back-propagation propagation run in turn to reach the required criteria. The former propagation can first transform the input mode onto the hidden layer, and then pass the weighted sum of inputs to the output layer through an activation function, resulting in one output value. In this stage, the sigmoidal function (f (.)) is generally used, and the output can be calculated according to Equation (1). Immediately after that, the back-propagation propagation works by passing the error of network backwards from the output layer to the input layer, with the weights adjusted based on some learning strategies to reduce the network error.

$$f_j = \frac{1}{1 + \exp(-\sum w_{ji} o_i + b)} \tag{1}$$

where w_{ji} is the connection weight from neuron i in the lower layer to neuron j in the upper layer and an initially small random value, o_i is the output of neuron i, and b is the bias value.

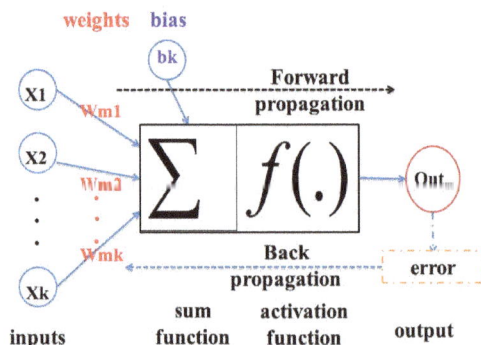

Figure 1. A typical artificial neural networks (ANN) model.

Duan et al. [20] carried out a study on predicting the compressive strength of RAC at the curing time of 28 days using an ANN model. The authors collected a large amount of published data on the 28-day compressive strength of RAC, which came from previous research. The RA used was derived from different countries and sources. The data were used for the construction of the ANN model, and the predicted results of the ANN model were quite accurate. Moreover, the same author also established another ANN model for predicting the elastic modulus of RAC [21]. This model was built based on regression analysis and performed better predictions. Sensitivity analysis, an uncertainty analysis technique in relation to quantitative analysis, is a study to assess the sensitivity of the prediction results of the model to the change in the selected input variables [22]. It also determines the significance of these uncertain factors on the results [23–25]. Therefore, it is of interest to apply the sensitivity analysis to the constructed ANN model to further study the influence of each input variable on the output. By conducting a sensitivity analysis, Jain et al. [26] determined the effect of the constituents of concrete mixes on the desired workability.

To predict the compressive strength of NAC (natural aggregate concrete) using ANN, the concrete mix proportions used [27–30] and the time of testing of the compressive strength [31–33] were generally selected as the input variables. For concrete made with RAs from different sources, the difference between the properties of different RAs should be taken into account. The aggregate characteristics, such as water absorption, specific gravity, and aggregate crush value are closely related to the properties of the old mortar attached, which can affect the properties of RAC by different levels. In theory, the more factors are taken into consideration, the more accurate the model is. However, in practice it is unsuitable to use all the affected factors due to the complicated calculation and the measuring error. Therefore, it is important to determine the significance of each RA characteristic and the optimal combination of factors, which aims to ensure the factors can be applicable to the majority of RAs from different sources. In other words, how to fully represent the aggregate properties in the ANN model is an important issue.

The purpose of this study is to examine the relative importance of the different characteristics of RA in affecting RAC properties. Moreover, it also aims to determine which factor or combination of factors is most suitable for representing RA properties when used in ANN model for compressive strength and elastic modulus prediction. In this study, the following steps were used for this purpose.

2. Methodologies

2.1. Building the ANN Models and the Sensitivity Analysis

First, experiments on the mechanical properties of RAC with different RAs were carried out in the laboratory, which had 46 concrete mixes and were divided into 3 groups. The RAs were categorized into 3 groups according to their sources: (1) RAs derived from 3 different sources and crushed using different methods; (2) RAs derived from concrete cubes made in the laboratory with different compressive strengths (35–85 MPa); (3) RAs contained different amounts of masonry added (clay bricks or tiles). As many sources of natural and recycled aggregates were used in these mixes, 8 aggregate characteristics, including fineness modulus of the fine aggregate (FM), residual mortar content (M_C), 10% fines value (TFV), aggregate crushing value (ACV), water absorption value (W_a), specific gravity (SG_{SSD}), impurity content (δ) and masonry content (m) of the coarse aggregate, were comprehensively measured and quantified. These factors, together with the mix proportions and concrete curing time, were selected as the input variables of the ANN for modeling the compressive strength and elastic modulus. To facilitate the analysis, factors including the mix proportions (5 variables) and curing time (1 variable) were designated as "certainties", while the other factors (8 variables) were named "uncertainties".

The experimental results obtained from the above mixes at different ages were divided into three groups, acting as: (i) the training set; (ii) the validation set; and (iii) the testing set, respectively. The corresponding ANN model could be established using the procedures described previously [20]. For each model, the ANN network parameters were determined when the error values reached the minimum. Based on the comparison of the error of integral testing set after a series of trials, the initial network architecture and parameters used in this study were as follows:

- Number of input layer units = 14
- Number of hidden layers = 1
- Number of hidden layer units = 40
- Number of output layer units = 1
- Momentum rate = 0.9
- Learning rate = 0.01
- Learning cycle = 15,000

In this study, the mean absolute percentage error (*MAPE*), root-mean-squared error (*RMS*) and absolute fraction of variance (R^2) computed using Equations (2)–(4) were used to access the accuracy of the ANN model developed.

$$MAPE = \left(\frac{o_j - t_j}{o_j} \right) \tag{2}$$

$$RMS = \sqrt{\frac{1}{p} \times \Sigma_j \left| t_j - o_j \right|^2} \tag{3}$$

$$R^2 = 1 - \left(\frac{\Sigma_j \left| t_j - o_j \right|^2}{\Sigma_j (o_j)^2} \right) \tag{4}$$

where *t*: the predicted output of the network; *o*: the actual output of the network; *p*: the total number of training and testing patterns; t_j: the predicted output of *j*th pattern of the network; o_j: the actual output of *j*th pattern of the network.

After the construction of the ANN models, the sensitivity analysis was then conducted according to Figure 2.

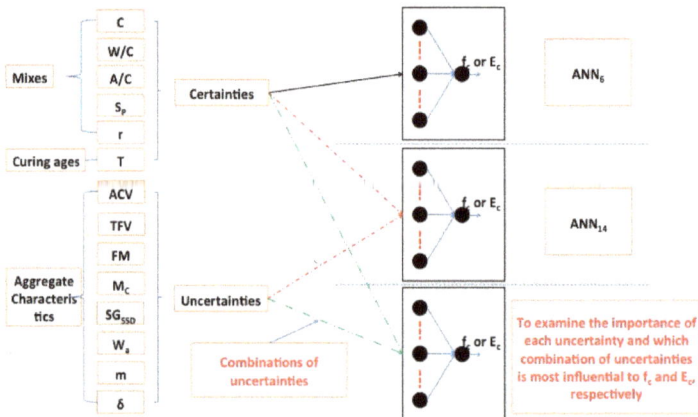

Figure 2. Flow chart of the sensitivity analysis

Step 1: A comparison of the performance between the models (ANN_{14}) with all variables (14 variables) and that (ANN_6) with certainties (only 6 variables) was first made, while keeping the other networks parameters constant.

Step 2: Various combinations of the uncertainties (aggregate characteristics) together with the "certainties" were used as the inputs of each model to find the best model with the minimum error. However, it would take a huge amount of time if all the combinations were tried out one by one. Considering the interaction and constraints among the aggregate characteristics, a simple method developed to determine the best combination of variables is shown in Figure 3.

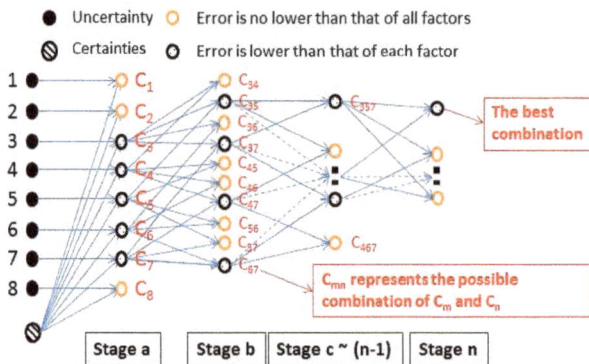

Figure 3. The determination of the best combination of input variables.

- At Stage a, the resulted error of ANN_6 was compared with that of the networks (ANN_7) when each "uncertainty" was sequentially added as an input variable. If the error could not be reduced, then the "uncertainty" added was regarded as negative for the output and would not be further studied in the next stages. The larger the reduction in the error value, the more important the respective "uncertainty", and vice versa.
- At Stage b, the variables that could reduce the error of ANN_6 were retained. Sequentially, each variable was paired with the others as the added inputs of ANN_6 to build a new model (ANN_8), and the resulted error values were compared with those of the networks (ANN_7) when only one variable was added to ANN_6. If the resulted error of the ANN_8 was not less than that of each ANN_7, the pair of variables was not further studied. For example, as shown in Figure 3,

in the further right column, items 3 and 4 represent two aggregate characteristics, respectively. Accordingly, C3 and C4 (in the second right column) represent a combination of "certainty" variables with items 3 and 4, respectively. Assuming that the addition of either item 3 or item 4 to the certainties could reduce the network error of ANN_6, C3 and C4 would be both retained and used to form a new combination C34, which contained 8 input variables (items 3, 4 and certainties). When comparing the *MAPE* values of the networks using C34, C3 and C4 as inputs, respectively, and if the first one was lower than both the latter two values, C34 would be retained for the next stage. Otherwise, it would be discarded.

- This above approach was continued until the networks error could not be further reduced. In this way, after trying out all the possible combinations, the most influential factor or a combination of factors to the compressive strength and elastic modulus of RAC could be identified.

Considering that the predicted results of the networks would change slightly even when using the same model, each of the networks was trained 5 times and the average value of the *MAPEs* of the testing set and validation set was used as the final indicator of the network error.

2.2. Experimental Program

It is not necessary to use ANN to model the effect of RA on the properties of RAC when only one type of RA is used, since in this case the complexity of RA cannot be reflected and the predictive ability of ANN is generally no better than that of traditional methods like regression analysis.

When RAs from different sources were used, ANN models, which are more capable of modeling complex non-linear relationships, may be more suitable for predicting the hardened properties of RAC. The published data of RAs used can be divided into two cases: (1) several types of RAs used by a single researcher; (2) the data of RAs from different literature sources. The factors that influence the properties of RAC and used as the input variables of the networks in the two cases are quite different. For the former case, the types of materials other than aggregate, specimen size and operator error are essentially the same, so only the mix proportions and RA characteristics are chosen as the input variables; for the latter case, on the other hand, in addition to the mix proportions and RA characteristics, more factors such as cement type, specimen size, etc. should be included to establish a generalized model. In this paper, only the first case is considered. The second case will be dealt with in a separate paper.

(1) The source of the data

As introduced above, experiments on the mechanical properties of RAC with different RAs were carried out in the laboratory, which had 46 concrete mixes and were divided into 3 groups. The properties of these aggregates are shown in Table 1. Except for the chosen aggregate characteristics, the particle size was also listed in Table 1 for different aggregate types. The details of the mixes and the corresponding hardened properties of the concrete prepared are shown in Tables 2 and 3, respectively.

Table 1. Properties of aggregates.

Sources	Aggregate Type	Particle Size mm	SG_{SSD} g/cm^3	W_a %	M_c %	ACV (%) 10–14 mm	TFV (KN) 10–14 mm	m %	δ %	FM
Source 1	NA1	20	2.6	1.01	0	21.7	155	0	0	/
	RA1	20	2.48	3.36	21	22.5	143	1.4	0.4	/
	RA2	20	2.36	6.14	35.1	23.4	133	2.9	1.1	/
	RA3	20	2.36	6.44	62	23.9	127	1	0	/
	FNA1	5	2.63	0.94	0	/	/	0	0	2.19

Appl. Sci. **2018**, *8*, 2171

Table 1. *Cont.*

Sources	Aggregate Type	Particle Size mm	SG$_{SSD}$ g/cm^3	W$_a$ %	M$_c$ %	ACV (%) 10–14 mm	TFV (KN) 10–14 mm	m %	δ %	FM
Source 2	NA2	20	2.66	0.71	0	15.8	259	0	0	/
	RA4	20	2.41	6.38	0	19.3	131	0	0	/
	RA5	20	2.42	5.18	0	19.7	154	0	0	/
	RA6	20	2.44	5.36	0	19.5	151	0	0	/
	RA7	20	2.45	5.3	0	20.3	147	0	0	/
	RA8	20	2.46	5.36	0	20.4	155	0	0	/
	FNA2	5	2.62	0.76	0	/	/	0	0	3.28
Source 3	NA1	20	2.6	1.01	0	21.7	155	0	0	/
	NA2	20	2.66	0.71	0	15.8	259	0	0	/
	RA9	20	2.49	3.85	22	21.5	149	2	1.1	/
	brick	20	1.99	21.74	0	27.1	44	100	100	/
	tile	10	2.03	14.82	0	19.1	105	100	100	/
	FNA2	5	2.62	0.76	0	/	/	0	0	3.28
	FNA3	5	2.61	0.44	0	/	/	0	0	2.94

NA1-NA2, RA1-RA9, FNA1-FNA3 represent natural coarse aggregate, recycled coarse aggregate, and natural fine aggregate from different sources or batches, respectively. SGSSD: specific gravity; W$_a$: water absorption value; M$_c$: residual mortar content; ACV: aggregate crushing value; TFV: 10% fines value; FM: fineness modulus.

Table 2. Mix proportions of recycled aggregate concrete (RAC) made with aggregates from different sources (kg/m^3).

Sources	Mixes	W	Cement	Sand	NA	RA	Aggregate Used
Source 1	NA30	205	300	697	1143	0	NA1
	RC30-1	205	300	697	0	1075	RA1
	RC30-2	205	300	697	0	1027	RA2
	RC30-3	205	300	697	0	1027	RA3
	NC45	180	350	706	1158	0	NA1
	RC45-1	180	350	706	0	1089	RA1
	RC45-2	180	350	706	0	1041	RA2
	RC45-3	180	350	706	0	1041	RA3
	NA60	185	425	696	1092	0	NA1
	RC60-1	185	425	696	0	1028	RA1
	RC60-2	185	425	696	0	982	RA2
	RC60-3	185	425	696	0	982	RA3
	NA80	165	485	685	1089	0	NA1
	RC80-1	165	485	685	0	1039	RA1
	RC80-2	165	485	685	0	979	RA2
	RC80-3	165	485	685	0	982	RA3
	MC45-2	180	350	675	0	1089	RA2
	MC45-3	180	350	654	0	1041	RA3
	MC60-2	185	425	637	0	1028	RA2
	MC60-3	185	425	618	0	982	RA3
Source 2	NAC	155	440	666	1166	0	NA2
	R30	155	440	666	0	1070	RA4
	R45	155	440	666	0	1077	RA5
	R60	155	440	666	0	1083	RA6
	R80	155	440	666	0	1090	RA7
	R100	155	440	666	0	1094	RA8
Source 3	Control	190	380	710	1110	0	NA2 + FNA3
	T5	190	380	710	1055	44	NA2 + FNA3 + tile
	T10	190	380	710	999	88	NA2 + FNA3 + tile
	T15	190	380	710	944	132	NA2 + FNA3 + tile
	b5	190	380	710	1055	43	NA2 + FNA3 + brick
	b10	190	380	710	999	86	NA2 + FNA3 + brick
	b15	190	380	710	944	129	NA2 + FNA3 + brick
	b5r50	185	370	732	545	481	NA1 + RA9 + FNA2 + brick
	b5r100	185	370	732	0	961	RA9 + FNA2 + brick
	b10r50	185	370	732	545	475	NA1 + RA9 + FNA2 + brick
	b10r100	185	370	732	0	948	RA9 + FNA2 + brick

Table 2. *Cont.*

Sources	Mixes	W	Cement	Sand	NA	RA	Aggregate Used
	b15r50	185	370	732	545	526	NA1 + RA9 + FNA2 + brick
	b15r100	185	370	732	0	1049	RA9 + FNA2 + brick
	T5r50	185	370	732	545	486	NA1 + RA9 + FNA2 + tile
	T5r100	185	370	732	0	970	RA9 + FNA2 + tile
	T10r50	185	370	732	545	511	NA1 + RA9 + FNA2 + tile
	T10r100	185	370	732	0	1018	RA9 + FNA2 + tile
	ro	185	370	732	1090	0	NA1 + FNA2
	r50	185	370	732	545	463	NA1 + RA9 + FNA2
	r100	185	370	732	0	924	RA9 + FNA2

NA: natural aggregate; RA: recycled aggregate.

Table 3. Mechanical properties of RAC.

Sources	Mixes	f_c (MPa)					E_c (GPa)	
		1 Day	4 Days	7 Days	28 Days	90 Days	28 Days	90 Days
Source 1	NA30				34.5	39.4	25.1	26.6
	RC30-1				35	39.8	20.85	25.18
	RC30-2				29.2	34	21.9	22.83
	RC30-3				27.7	28.4	20.49	21.5
	NC45				48.3	53	30.68	31.1
	RC45-1				47.6	51.3	28.86	30.68
	RC45-2				42	47	24.46	25.91
	RC45-3				42.9	46.3	26.55	27.22
	NA60				61.6	69.6	32.36	34.5
	RC60-1				60	67.7	29.42	33.42
	RC60-2				53.7	55.5	24.61	26.3
	RC60-3				53.2	58.6	28.5	27.94
	NA80				80.5	88.3	35.43	36.88
	RC80-1				78.2	84.1	34.76	35.49
	RC80-2				71.2	74.3	29.52	29.92
	RC80-3				65.4	73.3	30.62	30.74
	RC45-1				49.2	51.5	29.5	31.2
	RC45-2				43.6	50.1	25.48	26.35
	RC60-1				60.4	68	30.7	33.6
	RC60-2				57.3	62.7	26.99	27.3
Source 2	NAC	29.3	54.8	59.7	69.6	75.3	32.3	36.1
	R30	24	50.7	54.1	59.4	63	27.43	28.67
	R45	31	56	60.2	69.8	76.3	27.26	30.9
	R60	22.9	50.1	57.6	67.8	74.8	27.02	30.98
	R80	24.8	52.6	59.4	68.7	72.7	26.85	30
	R100	20.1	46.5	55.2	62.1	66.3	26.79	28.48
Source 3	Control				54.4	60.5	29.85	31.51
	T5				54.4	59.9	28.42	30.94
	T10				54.9	60	27.44	29.14
	T15				52.5	57.6	27.09	28.08
	b5				54.2	59.4	27.49	30.27
	b10				52.3	57.6	25.46	28.05
	b15				46.9	54.8	23.18	24.24
	b5r50	18.6	38.2 [a]	41.7	48.4	54.1	29.03	30
	b5r100	15.9	34.7 [a]	35	44	45.9	27.1	27.9
	b10r50	25.3 [b]		39.1	47.5	54	27	28.26
	b10r100	23.9 [b]		34.6	42.4	45.4	26.69	27.85
	b15r50	21.6	37.5	38.8	46.7	50.5	24.42	26.14
	b15r100	17.5	31.5	33.9	41.1	42.1	24.15	25.58

Table 3. *Cont.*

Sources	Mixes	f_c (MPa)					E_c (GPa)	
		1 Day	4 Days	7 Days	28 Days	90 Days	28 Days	90 Days
	T5r50	20	34.9 [c]	41.4	49.1	54	27.39	30.13
	T5r100	18.5		36.5	44.7	47.4	25.69	26.15
	T10r50	19.2	37.6	42.5	50.7	52.8	26.72	28.87
	T10r100	14.4	28.6	34.4	39.9	42	24.55	25.55
	r0	21.3		42.1	48.2	51.3	30.45	33.86
	r50	20	40.2	44.1	50.3	53.6	29.58	30.35
	r100	18	40.1	43	49.2	51.3	26.78	27.86

[a], [b], [c] measured at the age of 5 days, 2 days and 3 days, respectively.

(2) Construction of the ANN models

As shown in Table 3, the experiment had a total of 145 and 92 results for compressive strength and elastic modulus, respectively, which were divided randomly into 3 groups used to construct the ANN models. The 3 groups were used as the training, testing and validation sets, respectively. The testing and validation sets were intended to establish the model with the generalization ability. After training, the optimal models for simulating the compressive strength (ANN_{14}-f_c) and elastic modulus (ANN_{14}-E_c) using all 14 variables were constructed (Figure 4), and the network architecture and parameters selected were as follows, in line with the similar procedure previously established [15].

- Number of input layer units = 16
- Number of hidden layers = 1
- Number of hidden layer units = 40
- Number of output layer units = 1
- Momentum rate = 0.9
- Learning rate = 0.01
- Learning cycle = 10,000

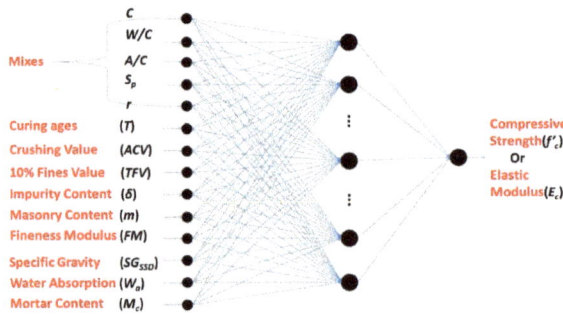

Figure 4. ANN model constructed for compressive strength or elastic modulus.

3. Results and Discussion

The performance of the constructed ANN models (ANN_{14}-f_c, ANN_{14}-E_c) in predicting the compressive strength and elastic modulus of RAC with all 14 variables and compared to the models (ANN_6-f_c, ANN_6-E_c) using only the "certainties" as input variables is shown in Table 4 and Figure 5.

Table 4. Performance of ANN models. *MAPE*: the mean absolute percentage error; RMS: root-mean-squared error.

Sets	Model	R^2	RMS	MAPE (%)	Model	R^2	RMS	MAPE (%)
Training		0.9984	2.067	3.531		0.9999	0.2825	0.73
Testing	ANN_{14}-f_c	0.9952	3.445	5.859	ANN_{14}-E_c	0.9965	1.6986	4.72
Validation		0.9949	3.562	6.032		0.9968	1.563	4.399
Training		0.997	2.764	4.743		0.9941	0.9558	2.641
Testing	ANN_6-f_c	0.987	5.234	8.67	ANN_6-E_c	0.9931	2.3066	6.437
Validation		0.992	4.41	7.557		0.9913	2.5234	7.264

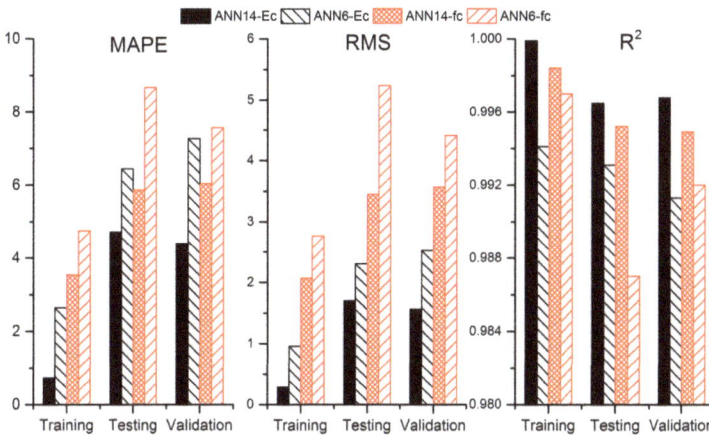

Figure 5. Performance of the ANN models constructed. *MAPE*: the mean absolute percentage error; RMS: root-mean-squared error.

The correlation coefficient R^2 of the networks in modeling the compressive strength and elastic modulus reached 0.9984 and 0.9999, respectively, indicating that the correlations between the predictions and the true results were very good. The R^2 values of both models were all above 0.994 in the testing and validation sets and these further proved that the constructed models, ANN_{14}-f_c and ANN_{14}-E_c, had not only good simulating abilities, but also good generalization capabilities.

When only the mix proportions and the curing ages were used as the inputs of the networks, the R^2 values of ANN_6-f_c and ANN_6-E_c in the training sets were still up to 0.997 and 0.9941. However, the generalization performance (testing and validation sets) of both networks were significantly poorer, with the R^2 values reduced to the range of 0.987–0.9931 and the predicted errors *MAPE* increased by about 50% in both the validation and testing sets. This might explain why many established formulae (based on regression analysis) could not be used for practical applications although they had good correlation coefficients.

For compressive strength, Figure 6 shows that the predicted error of the networks (ANN_6-f_c) using only "certainties" as inputs was about 8.11%, and the performance of the networks could be enhanced with the addition of each aggregate characteristic to the inputs. It can be found that the predicted error of compressive strength was lower when the water absorption or masonry content of the coarse aggregate were taken into consideration, which was almost close to the model (ANN_{14}-f_c) with all variables as inputs.

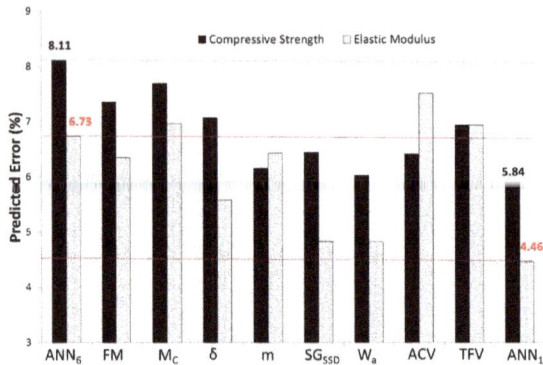

Figure 6. Influence of each uncertainty on the properties of RAC relative to models (ANN_6) with only "certainties" as inputs—Stage a.

As shown in Table 5, the performance of the networks (ANN_8) with the combinations of two aggregate characteristics added as inputs of ANN_6-f_c was not necessarily better than those with only one aggregate characteristic added to the inputs. This was mainly due to the fact that the degrees of influence related to these aggregate characteristics were inconsistent, which may have misled the correlation of the inputs and outputs. However, the results demonstrated that the use of seven combinations (italics) of aggregate characteristics as inputs to ANN_6-f_c could improve the predicted capability of the networks; these combinations were SG_{SSD} and W_a, SG_{SSD} and TFV, W_a and δ, W_a and TFV, FM and TFV, m and δ, and m and M_c.

Table 5. The errors of networks for compressive strength with different input variables—Stage b (%).

Factors	SG_{SSD}	W_a	FM	M_c	δ	m	TFV	ACV
SG_{SSD}	**6.44**	<u>5.94</u>	6.95	7.12	6.8	7.63	<u>6.11</u>	6.54
W_a		**6.03**	6.6	6.99	<u>5.76</u>	6.89	<u>5.92</u>	6.55
FM			**7.36**	7.34	7.48	6.91	<u>6.55</u>	8.02
M_c				**7.7**	<u>5.84</u>	<u>5.99</u>	6.99	7.25
δ					**7.07**	6.88	8.09	7.23
m						**6.16**	8.32	6.64
TFV							**6.94**	6.87
ACV								**6.42**

For Tables 5–8, each figure represents the mean absolute percentage error (*MAPE*) value of the networks with "certainties" and the uncertain factors indicated in the 1st row and 1st column as inputs; the bold figures represent the *MAPE* value of the networks with "certainties" and the uncertain factors indicated either in the 1st row or 1st column as inputs; the underlined figures are *MAPE* values lower than those of the corresponding bold letters.

Then, these combinations of two characteristics were used to examine whether they could form new combinations of three or four characteristics that could further improve the prediction of the networks. The results listed in Table 6 show that the error of networks could be reduced to 5.43–5.91 when the following combinations of aggregate characteristics, together with "certainties", were adopted as the inputs of networks: SG_{SSD} + W_a + δ, SG_{SSD} + m + TFV + M_c, SG_{SSD} + W_a + TFV and FM + m + TFV + M_c. Moreover, the further combinations of these factors was no longer useful to reduce the error of prediction of the networks (Table 7).

Table 6. The errors of networks for compressive strength with different input variables—Stage c (%).

Factors	$SG_{SSD} + W_a$	$SG_{SSD} + TFV$	$W_a + \delta$	$W_a + TFV$	$FM + TFV$	$m + \delta$	$m + M_c$
$SG_{SSD} + W_a$	5.94	5.61	5.43	5.61	6.26	6.93	6.55
$SG_{SSD} + TFV$		6.11	6.4	5.61	6.67	6.85	5.53
$W_a + \delta$			5.76	5.81	6.33	6.52	6.16
$W_a + TFV$				5.92	5.96	5.96	7.87
$FM + TFV$					6.55	6.8	5.91
$m + \delta$						5.84	6.95
$m + M_c$							5.99

Table 7. The errors of networks for compressive strength with different input variables—Stage d (%).

Factors	$SG_{SSD} + W_a + TFV$	$SG_{SSD} + W_a + \delta$	$SG_{SSD} + m + TFV + M_c$	$FM + m + TFV + M_c$
$SG_{SSD} + W_a + TFV$	5.61	5.91	5.93	6.07
$SG_{SSD} + W_a + \delta$		5.43	5.94	6.28
$SG_{SSD} + m + TFV + M_c$			5.53	6.19
$FM + m + TFV + M_c$				5.91

To sum up, the addition of any one of the eight aggregate characteristics to ANN_6-f_c could help achieve a better prediction of the compressive strength of RAC. When these characteristics were added to the input variables of ANN_6-f_c alone, water absorption contributed to the largest reduction in the error of networks, from about 8.11% to only 6.03%. The use of some combinations of these eight characteristics could further decrease the error of networks, even lower than that of ANN_{14}-f_c. The network error was only 5.43% when a combination of SG_{SSD}, W_a, δ, and the "certainties" (mix proportions and curing ages) were used as the input variables.

The case was slightly different for the elastic modulus. As shown in Figure 5, the error of the networks (ANN_6-E_c) with only the mix proportions and the curing ages as inputs was about 6.73%. When each one of the eight aggregate characteristics was added to the input variables alone, the results showed that three characteristics (viz mortar content, aggregate crushing value and 10% fines value) could not improve the prediction, while the other five characteristics could help to optimize the model; among the eight aggregate characteristics, the SG_{SSD} and W_a played the most significant influence, being capable of reducing the error to about 4.84% and 4.83%, respectively.

However, only the combination of SG_{SSD} and W_a could further decrease the network error to about 3.89%, as shown in Table 8. Therefore, in this study the best combination of parameters for modeling the elastic modulus of RAC was mix proportions, curing ages, and the specific gravity and water absorption values of the RA.

Table 8. The errors of networks for elastic modulus with different input variables—Stage b (%).

Factors	M_c	FM	δ	SG_{SSD}	W_a
M_c	6.43	6.93	6.76	5.5	5.15
FM		6.35	6.98	5.01	4.94
δ			5.58	5.76	5.52
SG_{SSD}				4.84	3.89
W_a					4.83

4. Conclusions

The purpose of this paper was to analyze the significance of each aggregate characteristic and determine the best combinations of factors which further influence the compressive strength and elastic modulus of RAC using the ANN model. The ANN model was trained and built on the basis of a series of experimental results including 46 concrete mixes. The research took eight factors into consideration as the inputs of the ANN model. The results are as follows.

(1)   The predicted results of RAC prepared with different sources of RAs were not satisfactory using the ANN models (ANN_6-f_c and ANN_6-E_c), although the learning abilities of these models were still good. This was because these ANN models only adopted the mix proportions and curing age as the input variables, without considering the aggregate characteristics of RA, which were quite different from the natural aggregate.

(2)   The water absorption of RA played a most important role in affecting the compressive strength of RAC, the addition of which could reduce the error of ANN_6-f_c from 8.11% to 6.03%. The combination of specific gravity, water absorption and impurity content could further decrease the error to only about 5.43%.

(3)   As regards elastic modulus, characteristics like mortar content, aggregate crushing value and 10% fines value were proved to be not important in affecting the prediction. In addition to the mix proportions and curing ages, water absorption and specific gravity were the most significant aggregate characteristics. The addition of each of them as networks inputs could decrease the error to less than 4.85%, and the network error could even be reduced to only about 3.89% when the inputs were a combination of mix proportions, curing time, specific gravity and water absorption of the coarse aggregate.

Author Contributions: Funding acquisition, C.-S.P. and Y.L.; Investigation, S.H. and J.X.; Methodology, Z.D. and C.-S.P.; Software, Y.L.; Writing—original draft, Z.D.

Funding: This research was funded by the National Natural Science Foundation of China (grant number 51708419) and Shanghai Pujiang Talent Fund (grant number 17PJ1409500).

Acknowledgments: The authors wish to acknowledge the financial support of the Hong Kong Polytechnic University and Sun Hung Kai Properties Ltd.

Conflicts of Interest: The authors declare no conflict of interest.

References

1.   Kurda, R.; Silvestre, J.D.; Brito, J.D. Life cycle assessment of concrete made with high volume of recycled concrete aggregates and fly ash. *Resour. Conserv. Recycl.* **2018**, *139*, 407–417. [CrossRef]
2.   Colangelo, F.; Petrillo, A.; Cioffi, R.; Borrelli, C.; Frcina, A. Life cycle assessment of recycled concretes: A case study in southern Italy. *Sci. Total Environ.* **2018**, *615*, 1506–1517. [CrossRef] [PubMed]
3.   Colangelo, F.; Forcina, A.; Farina, I.; Petrillo, A. Life cycle assessment (LCA) of different kinds of concrete containing waste for sustainable construction. *Buildings* **2018**, *8*, 70. [CrossRef]
4.   Silva, R.V.; Brito, J.D.; Dhir, R.K. Properties and composition of recycled aggregates from construction and demolition waste suitable for concrete production. *Constr. Build. Mater.* **2014**, *65*, 201–217. [CrossRef]
5.   Poon, C.S.; Shui, Z.H.; Lam, L. Effect of microstructure of ITZ on compressive strength of concrete prepared with recycled aggregates. *Constr. Build. Mater.* **2004**, *18*, 461–468. [CrossRef]
6.   Guo, Z.G.; Tu, A.; Chen, C.; Lehman, D.E. Mechanical properties, durability, and life-cycle assessment of concrete building blocks incorporating recycled concrete aggregates. *J. Clean. Prod.* **2018**, *199*, 136–149. [CrossRef]
7.   Xiao, J.Z. *Recycled Aggregate Concrete Structures*; Spring: Berlin, Germany, 2018; pp. 65–97. ISBN 978-3-662-53985-9.
8.   Pedro, D.; Brito, J.D.; Evangelista, L. Durability performance of high-performance concrete made with recycled aggregates, fly ash and densified silica fume. *Cem. Concr. Compos.* **2018**, *93*, 63–74. [CrossRef]
9.   Colangelo, F.; Cioffi, R. Mechanical properties and durability of mortar containing fine fraction of demolition wastes produced by selective demolition in south Italy. *Compos. Part B* **2017**, *115*, 43–50. [CrossRef]
10. Evangelista, L.; Brito, J.D. Durability of crushed fine recycled aggregate concrete assessed by permeability-related properties. *Mag. Concr. Res.* **2018**. [CrossRef]
11. Rodríguez, C.; Miñano, I.; Aguilar, M.Á.; Ortega, J.M.; Parra, C.; Sánchez, I. Properties of Concrete Paving Blocks and Hollow Tiles with Recycled Aggregate from Construction and Demolition Wastes. *Materials* **2017**, *10*, 1374. [CrossRef] [PubMed]

12. Kou, S.C.; Poon, C.S. Mechanical properties of 5-year-old concrete prepared with recycled aggregates obtained from three different sources. *Mag. Concr. Res.* **2008**, *60*, 57–64. [CrossRef]

13. Gonçalves, P.; Brito, J.D. Recycled aggregate concrete (RAC)-comparative analysis of existing specifications. *Mag. Concr. Res.* **2010**, *62*, 339–346. [CrossRef]

14. Jayasuriya, A.; Adams, M.P.; Bandelt, M.J. Understanding variability in recycled aggregate concrete mechanical properties through numerical simulation and statistical evaluation. *Constr. Build. Mater.* **2018**, *178*, 301–312. [CrossRef]

15. Hu, Z.; Mao, L.X.; Xia, J.; Liu, J.B.; Gao, J.; Yang, J.; Liu, Q.F. Five-phase modelling for effective diffusion coefficient of chlorides in recycled concrete. *Mag. Concr. Res.* **2017**, *70*, 583–594. [CrossRef]

16. Sco, D.S.; Choi, H.B. Effects of the old cement mortar attached to the recycled aggregate surface on the bond characteristics between aggregate and cement mortar. *Constr. Build. Mater.* **2014**, *59*, 72–77. [CrossRef]

17. Etxeberria, M.; Vazquez, E.; Mari, A.; Barra, M. Influence of amount of recycled coarse aggregates and production process on properties of recycled aggregate concrete. *Cem. Concr. Res.* **2007**, *37*, 735–742. [CrossRef]

18. Brito, J.D.; Robles, R. Recycled aggregate concrete (RAC) methodology for estimating its long-term properties. *Indian J. Eng. Mater. Sci.* **2010**, *17*, 449–462.

19. Tam, V.W.Y.; Tam, C.M. Parameters for assessing recycled aggregate and their correlation. *Waste Manag. Res.* **2009**, *27*, 52–58. [CrossRef] [PubMed]

20. Duan, Z.H.; Kou, S.C.; Poon, C.S. Prediction of compressive strength of recycled aggregate concrete using artificial neural networks. *Constr. Build. Mater.* **2013**, *40*, 1200–1206. [CrossRef]

21. Duan, Z.H.; Kou, S.C.; Poon, C.S. Using artificial neural networks for predicting the elastic modulus of recycled aggregate concrete. *Constr. Build. Mater.* **2013**, *44*, 524–532. [CrossRef]

22. Sensitivity Analysis. Available online: http://en.wikipedia.org/wiki/Sensitivity_analysis (accessed on 14 December 2017).

23. Coronado, C.A.; Lopez, M. Sensitivity analysis of reinforced concrete beams strengthened with FRP laminates. *Cem. Concr. Compos.* **2006**, *28*, 102–114. [CrossRef]

24. El-Dash, K.M.; Ramadan, M.O. Effect of aggregate on the performance of confined concrete. *Cem. Concr. Res.* **2006**, *36*, 599–605. [CrossRef]

25. Khatri, R.P.; Sirivivatnanon, V. Characteristic service life for concrete exposed to marine environments. *Cem. Concr. Res.* **2004**, *34*, 745–752. [CrossRef]

26. Jain, A.; Jha, S.K.; Misra, S. Modeling and analysis of concrete slump using artificial neural networks. *J. Mater. Civ. Eng.* **2008**, *20*, 628–633. [CrossRef]

27. Nikoo, M.; Sadowski, L.; Moghadam, F.T. Prediction of Concrete Compressive Strength by Evolutionary Artificial Neural Networks. *Adv. Mater. Sci. Eng.* **2015**, *2015*, 849126. [CrossRef]

28. Liang, C.; Qian, C.; Chen, H.; Kang, W. Prediction of compressive strength of concrete in wet-dry environment by bp artificial neural networks. *Adv. Mater. Sci. Eng.* **2018**, *2018*, 6204942. [CrossRef]

29. Parichatprecha, R.; Nimityongskul, P. Analysis of durability of high performance concrete using artificial neural networks. *Constr. Build. Mater.* **2009**, *23*, 910–917. [CrossRef]

30. Siddique, R.; Aggarwal, P.; Aggarwal, Y. Prediction of compressive strength of self-compacting concrete containing bottom ash using artificial neural networks. *Adv. Eng. Softw.* **2011**, *42*, 780–786. [CrossRef]

31. Chithra, S.; Kumar, S.R.R.S.; Chinnaraju, K.; Ashmita, F.A. A comparative study on the compressive strength prediction models for High Performance Concrete containing nano silica and copper slag using regression analysis and Artificial Neural Networks. *Constr. Build. Mater.* **2016**, *114*, 528–535. [CrossRef]

32. Sarıdemir, M.; Topçu, I.B.; Ozcan, F.; Severcan, M.H. Prediction of long-term effects of GGBFS on compressive strength of concrete by artificial neural networks and fuzzy logic. *Constr. Build. Mater.* **2009**, *23*, 1279–1286. [CrossRef]

33. Hodhod, O.A.; Ahmed, H.I. Developing an artificial neural network model to evaluate chloride diffusivity in high performance concrete. *HBRC J.* **2013**, *9*, 15–21. [CrossRef]

applied
sciences

MDPI

Article

Analysis of Life Cycle Environmental Impact of Recycled Aggregate

Won-Jun Park [1], Taehyoung Kim [2],*, Seungjun Roh [3] and Rakhyun Kim [4]

1   Department of Architectural Engineering, Kangwon National University, Sancheck, Kangwon-Do 25913, Korea; wjpark@kangwon.ac.kr
2   Department of Living and Built Environment Research, Korea Institute of Civil Engineering and Building Technology, Goyang-Si, Gyeonggi-Do 10223, Korea
3   School of Architecture, Kumoh National Institute of Technology, Gumi 39177, Korea; roh@kumoh.ac.kr
4   Sustainable Building Research Center, Hanyang University, Ansan 15588, Korea; redwow6@hanyang.ac.kr
*   Correspondence: kimtaehyoung@kict.re.kr; Tel.: +82-31-995-0838

Received: 31 December 2018; Accepted: 22 February 2019; Published: 12 March 2019

Abstract: This study assessed the influence of matter discharged during the production (dry/wet) of recycled aggregate on global warming potential (GWP) and acidification potential (AP), eutrophication potential (EP), ozone depletion potential (ODP), biotic resource depletion potential (ADP), photochemical ozone creation potential (POCP) using the ISO 14044 (LCA) standard. The LCIA of dry recycled aggregate was 2.94×10^{-2} kg-CO_{2eq}/kg, 2.93×10^{-5} kg-SO_{2eq}/kg, 5.44×10^{-6} kg-$PO_4^{3}{}_{eq}$/kg, 4.70×10^{-10} kg-CFC_{11eq}/kg, 1.25×10^{-5} kg-C_2H_{4eq}/kg, and 1.60×10^{-5} kg-$Antimony_{eq}$/kg, respectively. The environmental impact of recycled aggregate (wet) was up to 16~40% higher compared with recycled aggregate (dry); the amount of energy used by impact crushers while producing wet recycled aggregate was the main cause for this result. The environmental impact of using recycled aggregate was found to be up to twice as high as that of using natural aggregate, largely due to the greater simplicity of production of natural aggregate requiring less energy. However, ADP was approximately 20 times higher in the use of natural aggregate because doing so depletes natural resources, whereas recycled aggregate is recycled from existing construction waste. Among the life cycle impacts assessment of recycled aggregate, GWP was lower than for artificial light-weight aggregate but greater than for slag aggregate.

Keywords: recycled aggregate; concrete; life cycle assessment; environmental impact

---

1. Introduction

In South Korea, redevelopment and reconstruction are being actively performed due to the economic and functional service-life expiry of older constructed structures. Such development inevitably generates a rapidly increasing quantity of construction waste. Waste concrete accounts for approximately 60% of the waste produced in redevelopment and reconstruction projects, and is not being effectively recycled despite attempts underway in countries such as the United States and Japan. Technology is currently under development to utilize recycled aggregate, obtained by crushing waste concrete, as a high-value-add material with applications such as aggregate for road construction and concrete. The field application of concrete using recycled aggregate is being performed with slowly increasing frequency in a variety of areas, including road construction.

The construction industry has been making efforts to become an economically and environmentally sustainable industry, while continuing to face social problems such as an increase in industrial and construction waste and the measures devised to process the waste cause further environmental problems, such as pollution and resource depletion [1].

Moreover, the use of natural aggregate is gradually decreasing due to changes in environmental awareness and the depletion of resources. The use of crushed stone aggregate and marine sand is increasingly replacing natural aggregate, but these materials also cause various problems, such as ecosystem destruction and transportation-distance challenges [2].

Recycled aggregates are alternative resources to river or forest aggregates. Their supply proportion is continuously increasing and will continue to play an important role as a recycled resource. If natural aggregates are consumed at the current level without promoting the usage of recycled aggregates at a rate of 30 million tons per year, all natural aggregates are expected to be exhausted in the next 20 years [3].

However, the process of producing recycled aggregate generates large environmental loads because it requires more processing than the collection of natural aggregate. In addition, the reliability of the quality of recycled aggregate is low, making it most suitable for utilization in landfill or embankments. Studies have been conducted on the quality of recycled aggregate, but there remains a paucity of research on the reduction of the environmental load generated by the production of recycled aggregate. Although the use of recycled aggregate that utilizes construction waste is being recognized as a positive phenomenon in terms of the efficient utilization of resources and the positive environmental aspects, minimizing the environmental loads inevitably generated by aggregate production remains an important issue.

Accordingly, a number of studies on the emissions and the reduction in the environmental impact of concretes that are mixed with recycled aggregates have been conducted overseas. A quantitative evaluation of concretes mixed with recycled and general aggregates is required to assess the various environmental impact categories (global warming potential (GWP), ozone depletion potential (ODP), photochemical ozone creation potential (POCP), eutrophication potential (EP), acidification potential (AP), and biotic resource depletion potential (ADP)).

This study assessed the life cycle environmental impact of recycled aggregate using life cycle assessment (LCA). It divided the process into raw materials, transport, and manufacturing of recycled aggregate, and identified the materials used in each stage and energy consumption amount in order to assess environmental impact. The study also compared and analyzed the environmental impact of recycled aggregate with artificial light-weight and slag aggregate.

2. Literature Review

Stefania Butera et al. [4] demonstrated that the utilization of construction and demolition waste in road construction as a replacement material for natural gravel was preferable to landfilling for most environmental-impact categories.

Laís Peixoto Rosado et al. [5] compared the natural and mixed recycled aggregate production for use as road base through a life cycle assessment. The primary data have been collected in a natural aggregate production facility and in a recycling facility of mixed aggregate in Southeast Brazil.

Vivian W. Y. Tam et al. [6] reviewed the literature on the production and utilization of recycled aggregate in concrete, concrete pavements, roadway construction, and other civil engineering works and some discussion on the savings on CO_2 emissions have been included.

Md. Uzzal Hossain et al. [7] assessed the environmental impacts of aggregate production from these waste materials, and compared them with the aggregate production from virgin materials that can be utilized for the production of lower grade concrete products. Ardavan Yazdanbakhsh et al. [8] studied the influence of the choice of geographic boundaries on the results of regional LCA studies in a large and dense metropolitan area. Specifically, the study incorporates an LCA on the construction and demolition waste (CDW) produced, processed, and used in concrete.

Patrizia Ghisellini et al. [9] reviewed the recent literature within the framework of the circular economy to explore how its key principles (reduce, reuse, and recycle) apply to the management of construction and demolition waste.

A. Julliena et al. [10] have assessed the energy consumption and impacts due to aggregate production within the LCA framework. Towards this end, a methodology has been presented that is specific to the impacts of aggregate production for various quarries.

Rawaz Kurda et al. [11] compared the environmental impacts of concrete mixes, which contain different incorporation ratios of fly ash and recycled concrete aggregates, with and without Superplasticizer. Mayuri Wijayasundara et al. [12] attempted to simulate the manufacturing set up to produce RAC by integrating processes involved in concrete waste recycling and concrete production environments.

S. Marinkovic et al. [13] determined the potentials of recycled aggregate concrete for structural applications and compared the environmental impact of the production of ready-mixed concrete: natural aggregate concrete made entirely with river aggregate and recycled aggregate concrete.

Rawaz Kurda et al. [14] mainly focused on the effect of high incorporation ratios of fly ash and recycled concrete aggregates on the carbonation resistance of concrete. Francesco Colangelo et al. [15] applied the standard protocol of LCA to three different concrete mixtures composed of wastes from construction and demolition, marble sludge, and cement kiln dust in order to compare the environmental and energy impacts.

Hossain, MU et al. [16] developed a social sustainability assessment tool based on the established standards and guidelines. The case study showed that four subcategories are crucial social concerns for construction materials. S.B.Marinković et al. [17] focused on the LCA of aggregates obtained by recycling of demolished concrete–recycled concrete aggregates, and concrete made with such aggregates–recycled aggregate concrete.

Desirée Rodríguez-Robles et al. [18] presented a literature review on expected environmental impacts inherent to the production and use of recycled aggregates and other common concrete constituents, as well as a LCA concerning some key issues when dealing with recycled aggregate concrete. Nikola Tošić et al. [19] determined the optimal choice of aggregate type and transport scenario in concrete production, employing a multicriteria optimization method taking into account technical, economic, and environmental limits and constraints.

Mayuri W. et al. [20] evaluated "cradle-to-gate" embodied energy of recycled concrete aggregate received at a construction site, in comparison to natural aggregate concrete. G.M. Cuenca-Moyano et al. [21] developed the life cycle inventory of masonry mortars made of natural fine aggregate and recycled fine aggregate. In order to create the inventory, the data used were those provided by producers [22].

Nicolas Serres et al. [23] evaluated environmental impacts associated with mixing compositions of concrete made of waste materials by using LCA. Environmental performances of natural formulation with the same mechanical strength regarding the functional unit, were evaluated.

Xin Shan et al. [24] presented LCA of a base case building in Singapore as well as material LCA with customized life cycle inventory datasets by considering importation and transportation of material in particular, and by considering the effects of adopting locally recycled aggregates.

F. Colangelo et al. investigate the physical and mechanical characteristics of different kind of construction and demolition waste obtained from selective and traditional demolition techniques [25].

3. Analysis of Life Cycle Impact Assessment (LCIA)

3.1. Method

This study assessed the environmental impact of recycled aggregate using LCA as defined by ISO standards. As environmental impact categories for life cycle impact assessment (LCIA), global warming, acidification, eutrophication, abiotic resource depletion, ozone layer destruction, and the production of photochemical oxides were chosen [26].

The functional unit for the life cycle impact assessment of recycled aggregate was 1 kg, and the system boundary was product stage of concrete (Cradle to Gate) [27].

The production stage of recycled aggregate was divided into the raw material, transport, and manufacturing stage, and the environmental impact of all input and output matter on atmospheric and water systems was assessed for each stage [28]. The Life Cycle Index Database (LCI DB) of Korea was examined and applied as shown in Table 1 [29].

Table 1. The Life Cycle Index (LCI) Database.

Division		Reference	Country
Raw material	Waste concrete	National LCI	Korea
Energy	Electric	National LCI	Korea
	Diesel	National LCI	Korea
	Kerosene	National LCI	Korea
Transportation	Truck	National LCI	Korea

3.2. Process of LCIA

Life cycle impact assessment is divided into the following [30,31]: classification, which collects substances drawn from inventory analysis into their relevant environmental impact categories; characterization, which quantifies the impact of items classified as environmental impact categories; normalization, which divides the impact on environmental impact categories by the total environmental impact of a certain region or period; and weighting which determines the relative importance of impact categories [32]. This study assessed all stages up to characterization since the normalization and weighting factors of recycled aggregates that fit the current situation of Korea have not been developed [33,34].

The substances discharged from recycled aggregate production impact the atmosphere and water quality, and the resulting environmental issues include global warming potential (GWP), ozone depletion potential (ODP), photo-chemical oxidant creation potential (POCP), abiotic depletion potential (ADP), eutrophication potential (EP), and acidification potential (AP) [35].

$$CI_i = \sum CI_{i,j} = \sum (Load_j \times eqv_{i,j}) \tag{1}$$

Here, CI_i is the value of the impact by all inventory items (j) included in impact category i in the belonging impact categories; $CI_{i,j}$ is the impact size by inventory item j on impact category i; $Load_j$ is the environmental load of inventory item j; and $eqv_{i,j}$ is the characterization coefficient value of inventory item j in impact category i.

3.2.1. Raw Material Stage

This study assessed the environmental impact according to the production process of waste concrete, a raw material for recycled aggregate [36]. Recycling breaks and separates waste concrete into small pieces and recycled aggregate are produced for use. Waste concrete is generated during the demolition and deconstruction of structures and is classified as construction waste if it is not used as another product. During the deconstruction of a building, a large amount of energy is used and environmental impact substances are discharged; however, since waste concrete is a by-product, its previous industrial process is not included in the environmental impact assessment category. Thus, the process of waste concrete discharge during the demolition and deconstruction of a structure was excluded from the categories of environmental impact assessment.

$$Raw\ material_{E.I} = \sum [(M(i) / Wc) \times t \times D]\ (i = Waste\ concrete) \tag{2}$$

Here, raw material E.I is the environmental impact of raw materials, (kg-unit/kg), M(i) is waste concrete (ton), Wc is (i) the volume (ton) of waste concrete collecting equipment, T is work hours (h), and D is characterization value (kg-unit/L) of diesel by environmental impact category.

3.2.2. Transportation Stage

The waste concrete collected from the building demolition/deconstruction sites was transported to a recycled aggregate production factory by truck, and the transport distance was 30 km on average. From the loading capacity of transportation means and the amount of waste concrete, the number of transport equipment units was calculated and transport was applied to assess environmental impact.

$$\text{Transportation}_{E.I} = \sum [(M(i) / Wt) \times d \times D] \ (i= \text{Waste concrete}) \tag{3}$$

Here, Transportation E.I is transport stage environmental impact (kg-unit/kg), $M_{(i)}$ is waste concrete (ton), Wt is the loading amount (ton) of (i) transport equipment, d is transport distance (km), and D is the characterization value of crude oil by environmental impact categories (kg-unit/L).

3.2.3. Manufacturing Stage

In this stage, waste concrete is selected and crushed to manufacture recycled aggregate [37]. The manufacturing method can be divided into the dry method and the wet method. To assess the environmental impact of each manufacturing method, the amount of electric power used in the manufacturing facility was examined and the environmental impact according to the usage of the energy source was assessed.

$$\text{Manufacture}_{E.I} = \sum [(E_{(i)} / W_a) \times C] \ (i = 1: \text{Electric}) \tag{4}$$

Here, Manufacture E.I is environmental impact (kg-unit/kg) in the manufacturing stage, W_a is the amount of recycled aggregate production (kg), $E_{(i)}$ is the amount of energy source use (kwh), and C is characterization value (kg-unit/kwh) of electric power by environmental impact categories.

Dry Method

The dry production method uses a technology that eliminates mortar on an aggregate surface by crushing waste concrete and easily separates a fair amount of fine powder generated to produce high-quality recycled aggregate as shown in Table 2. The number of crushing and differentiation work taken during the production process was divided for each equipment and shown in Figure 1. The environmental impact emission amount by electric power usage for each equipment unit was calculated for the process of producing recycled aggregate through the dry production method.

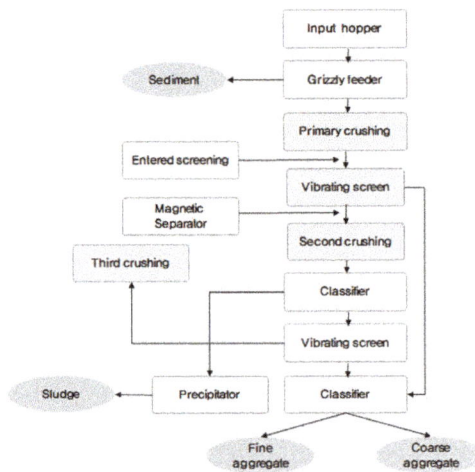

Figure 1. Dry production method.

Table 2. Manufacturing facility of dry recycled aggregate.

Dry System		Operation Count	Total Electricity Usage
Feeders	Grizzly feeder	1 time	183.33 kW
Screen	Vibrating screen	4 times	169.01 kW
Crusher	Jaw Crusher	1 time	593.75 kW
	Cone Crusher	3 times	529.41 kW
Magnetic Separator		1 time	250.00 kW
Classifier		3 times	289.47 kW
Precipitator		1 time	750.00 kW

This study assumed that the production efficiency of aggregate was 70% and set that cone and jaw crushers were used twice to produce high-quality recycled aggregate.

Wet Method

The wet production method includes a selection process that is not included in the dry production method. It has an air blower as a selection device in the middle, and, unlike the dry production method in which two types of crushers are applied, the wet production method uses only an impact crusher. The wet production system needs significant initial cost and vast land since facilities such as cleaning, precipitation, and filter presses are needed. Due to the issue of reclaiming sludge cake generated from production instead of recycling it, this method also has many difficulties in recycled sand production.

However, the wet system can more easily produce high-quality recycled sand compared with the dry system; aggregates have good particle size and form, and the process is favorable for high volume production, so the wet system using water has been applied in Korea. The number of crushing and differentiation work taken in the production process by equipment was classified by equipment type in Figure 2. This study assumed the production efficiency of aggregate was 70% and assumed that crushing was carried out by using a cone crusher twice and a jaw crusher once to produce quality aggregate as shown in Table 3.

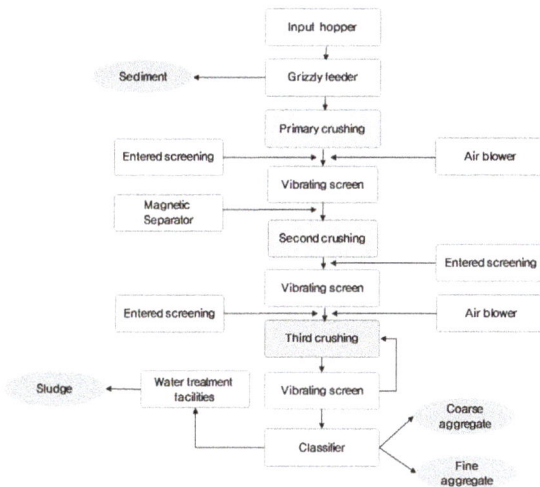

Figure 2. Wet production method process.

Table 3. Manufacturing facility for wet recycled aggregate.

Dry System		Operation Count	Total Electricity Usage
Feeders	Grizzly feeder	1 time	315.79 kW
Screen	Vibrating screen	3 times	240.00 kW
Crusher	Jaw crusher	1 time	600.00 kW
	Cone crusher	1 time	450.00 kW
	Impact crusher	1 time	950.00 kW
Air blower		2 times	840.00 kW
Magnetic Separator		3 times	120.00 kW
Classifier		1 time	1050.00 kW

3.3. Result of LCIA

The life cycle impact assessment (LCIA) of recycled aggregate (dry) was 2.94×10^{-2} kg-CO_{2eq}/kg, 2.93×10^{-5} kg-SO_{2eq}/kg, 5.44×10^{-6} kg-$PO_4^{3}{}_{eq}$/kg, 4.70×10^{-10} kg-CFC_{11eq}/kg, 1.25×10^{-5} kg-C_2H_{4eq}/kg, and 1.60×10^{-5} kg-$Antimony_{eq}$/kg for GWP, AP, EP, ODP, POCP, and ADP, respectively, as shown in Table 4.

Table 4. Life cycle impact assessment of recycled aggregate; GWP: Global Warming Potential; ADP: Abiotic Depletion Potential; ODP: Ozone Depletion Potential; AP: Acidification Potential; EP: Eutrophication Potential; POCP: Photochemical Ozone Creation Potential.

Division		GWP (kg-CO_{2eq}/kg)	AP (kg-SO_{2eq}/kg)	EP (kg-$PO_4^{3-}{}_{eq}$/kg)	ODP (kg-$CFC11_{eq}$/kg)	POCP (kg-$C_2H_{4\,eq}$/kg)	ADP (kg-$Antimony_{eq}$/kg)
Raw material stage		3.10×10^{-3}	3.90×10^{-6}	7.90×10^{-7}	3.00×10^{-11}	8.30×10^{-6}	1.00×10^{-6}
Transportation stage		4.00×10^{-3}	7.50×10^{-6}	1.08×10^{-6}	7.00×10^{-11}	1.10×10^{-6}	1.30×10^{-5}
Manufacture stage	Dry	2.23×10^{-2}	1.79×10^{-5}	3.57×10^{-6}	3.70×10^{-10}	3.10×10^{-6}	2.00×10^{-6}
	Wet	3.10×10^{-2}	2.99×10^{-5}	4.46×10^{-6}	4.80×10^{-10}	5.90×10^{-6}	7.00×10^{-6}
TOTAL	Dry	2.94×10^{-2}	2.93×10^{-5}	5.44×10^{-6}	4.70×10^{-10}	1.25×10^{-5}	1.60×10^{-5}
	Wet	3.81×10^{-2}	4.13×10^{-5}	6.33×10^{-6}	5.80×10^{-10}	1.53×10^{-5}	2.10×10^{-5}

LCIA of recycled aggregate (wet) was 3.81×10^{-2} kg-CO_{2eq}/kg, 4.13×10^{-5} kg-SO_{2eq}/kg, 6.33×10^{-6} kg-$PO_4^{3}{}_{eq}$/kg, 5.80×10^{-10} kg-CFC_{11eq}/kg, 1.53×10^{-5} kg-C_2H_{4eq}/kg, and 2.10×10^{-5} kg-$Antimony_{eq}$/kg for GWP, AP, EP, ODP, POCP, and ADP, respectively.

This shows that the LCIA of recycled aggregate (wet) was greater than that of recycled aggregate (dry), and that GWP and ADP of recycled aggregate (wet) were 30% higher than for recycled aggregate (dry). ODP and POCP were also greater by 20% and AP was greater by 40%. EP showed a difference of about 16%, because it was determined according to the electric power of the additional facility (impact crusher) used for producing recycled aggregate (wet). This was because more carbon dioxide (CO_2), trichlorofluoromethane (CFC-11), chlorodifluoromethane (HCFC-22), phosphate (PO_4^{3-}), crude oil, ammonia (NH_3), nitrogen oxides (NOx), and ethylene (C_2H_{4eq}), which impact the environment as more electric power was consumed, were directly and indirectly released.

Among environmental impact categories, the manufacturing stage of the life cycle constituted approximately 81%, 72%, 70%, and 83% of GWP, AP, EP, and ODP, respectively.

However, the raw material stage was responsible for the highest proportion of POCP at 66% and the transportation stage for 81% of ADP.

4. Comparison of Life Cycle Impact Assessment

4.1. Method

The LCIA results for recycled aggregate were compared with those for natural aggregate and artificial light-weight aggregate, slag aggregate. Previous studies were referred to for the LCIA results of natural aggregate and artificial light-weight aggregate, slag aggregate.

4.2. Information on Other Aggregates

4.2.1. Natural Aggregate

Natural aggregate is made by natural actions, and includes both sand and gravel sourced from sea, land, and mountains. This study focused on gravel. Gravel is collected and classified in plants by size (25 m, 45 m, or 75 m) for later application, and is mainly used as aggregate for concrete or filler for roads and septic tanks.

4.2.2. Artificial Light-Weight Aggregate

First, an artificial light-weight aggregate is manufactured by processing coal material (bottom ash), shale, and clay generated from coal powerplants and through plastic bulging (firing), to obtain required physical characteristics [38].

Artificial light-weight aggregate is produced through the mixing, crushing, molding, and plastic process of bottom ash for dry coal material, and the material satisfies wide-ranging functions required in the complex construction field such as light-weight characteristics, insulation, sound absorption, and thermal resistance.

Dredged soil and bottom ash are the main ingredients of artificial light-weight aggregate; energy such as electricity, LPG, and coal are used for manufacturing.

4.2.3. Slag Aggregate

Steel slag is largely divided into shaft slag and hard steel slag depending on the production process, and it is 100% recycled according to the current statistics [39].

Shaft slag, which is recycled more than 75% as the raw material for cement, is used as a relatively high-value addition. However, since most hard steel slag is used as a low-value addition for roads and fill-up, development of value-added technology is needed.

Slag aggregate is produced by watering and cooling the shaft of hard steel slag in an open-air storage yard and through quarrying and crushing processes. During quarrying and crushing, energy such as electricity, oil, and water are used.

Slag aggregate, which uses slag as the main raw ingredient, is used for various purposes such as roads and ceramics and is mostly used for civil construction.

4.3. Comparison of Environmental Impact by Aggregate Type

As shown in Table 5, the environmental impact of recycled aggregate was compared with that of natural aggregate and artificial light-weight aggregate, slag aggregate.

Table 5. Environmental impact comparison by aggregate type.

Division		GWP (kg-CO_{2eq}/m^3)	AP (kg-SO_{2eq}/m^3)	EP (kg-$PO_4{}^{3-}{}_{eq}$/m^3)	ODP (kg-$CFC11_{eq}$/m^3)	POCP (kg-C_2H_{4eq}/m^3)	ADP (kg-Antimony$_{eq}$/m^3)
Recycled aggregate	Dry	2.94×10^{-2}	2.93×10^{-5}	5.44×10^{-6}	4.68×10^{-10}	1.25×10^{-5}	1.60×10^{-5}
	Wet	3.81×10^{-2}	4.13×10^{-5}	6.33×10^{-6}	5.80×10^{-10}	1.53×10^{-5}	2.10×10^{-5}
Artificial Lightweight aggregate		5.16×10^{-2}	1.73×10^{-4}	2.80×10^{-5}	4.01×10^{-9}	5.28×10^{-5}	1.94×10^{-4}
Slag aggregate		2.26×10^{-2}	1.61×10^{-4}	1.61×10^{-4}	6.49×10^{-9}	2.43×10^{-5}	9.08×10^{-5}
Natural aggregate (Gravel)		1.43×10^{-2}	1.98×10^{-5}	3.67×10^{-6}	3.06×10^{-10}	1.41×10^{-5}	3.82×10^{-4}

According to Figure 3, the environmental impact of recycled aggregate was found to be up to twice as high as that of natural aggregate. Compared to natural aggregate, the global warming potential (GWP) was approximately 106% (dry) and 166% (wet) higher, AP was 48% (dry) and 109% (wet) higher, EP was 48% (dry) and 72% (wet) higher, and ODP was 53% (dry) and 90% (wet) higher for recycled aggregate. The manufacturing process for natural aggregate consumes very little energy because it involves simple processes, such as collection, transportation, and (minimal) processing.

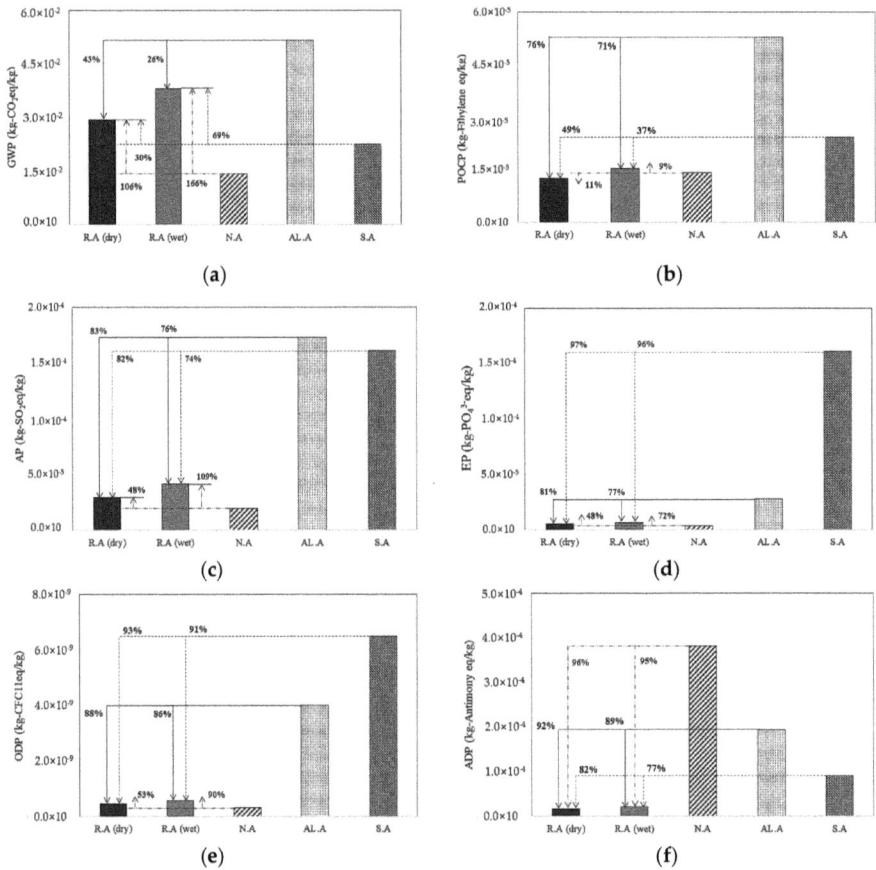

Figure 3. Environmental impact comparison by aggregate. (**a**) GWP: Global Warming Potential; (**b**) POCP: Photochemical Ozone Creation Potential; (**c**) AP: Acidification Potential; (**d**) EP: Eutrophication Potential; (**e**) ODP: Ozone Depletion Potential; (**f**) ADP: Abiotic Depletion Potential

Recycled aggregate, on the other hand, requires a lot of energy for physical and chemical processes, such as crushing, sorting, transportation, processing, and particle adjustment. However, the abiotic depletion potential (ADP), an important environmental impact measure, was found to be approximately 20 times higher for natural aggregate than for recycled aggregate. This is because the recycled aggregate is simply recycled construction waste, but natural aggregate depletes natural resources. Among environmental impacts emitted during recycled aggregate production GWP was 43% (dry) and 26% (wet) lower than for artificial light-weight aggregate, but was 30% (dry) and 69% (wet) greater than for slag aggregate.

This is because the amount of energy used, such as electricity, LNG (liquid natural gas), and coal in the mixing, molding, and plasticity processes, during artificial light-weight aggregate production was higher than the amount of energy used by recycled aggregate. This was due to the high emissions of CO_2, CH_4, and N_2O, which are the main compounds impacting GWP. At the same time, the amount of energy used in the differentiation and crushing processes during recycled aggregate production was higher than the amount of energy used by slag aggregate. AP, EP, ODP, POCP, and ADP also impact according to the emissions during recycled aggregate production were assessed to be lower by 37% to 93% compared with artificial light-weight aggregate and slag aggregate.

This is because NOx, NH_3, SO_2, NH_4, Halon, CFCs (Chlorofluorocarbons), soft coal, hard coal, and crude oil that impact AP, EP, ODP, POCP, and ADP during recycled aggregate production were emitted less than during artificial light-weight aggregate and slag aggregate production.

Due to the use of LNG and coal energy during the manufacturing of lightweight aggregate and slag aggregate, matters that impact acidification, eutrophication, and ozone depletion, such as sulfur dioxide (SO_2), sulfuric acid (H_2SO_4), and nitrate (NO_3^-) are emitted. Ammonia (NH_3), ammonium (NH_4^+), phosphate (PO_4^{3-}), nitrogen oxide (NOx), etc., according to electric energy usage are emitted as well. This is because matters impacting the environment are emitted when coal input in thermoelectric powerplant is combusted to produce electric energy.

Especially, the ADP of recycled aggregate was lower by 77% to 92% compared with artificial light-weight aggregate and slag aggregate. This is because recycled aggregate uses industrial by-products as raw materials, a small amount of ADP's main impact matters such as iron (Fe), natural gas, hard coal, lead (Pb), and uranium (U) were emitted.

5. Discussion

Recently, the application of recycled aggregate has been expanded to major concrete structure elements, such as the columns and beams of construction structures. For concrete, the major material used in the construction industry, environmental impact analysis is required because it discharges many substances with a high environmental load over its life cycle from production to construction, maintenance, and destruction/disposal. Concrete is a mixture of cement, aggregate, and admixture. Cement production consumes a large amount of energy during the processes of extracting limestone and clay and manufacturing clinker. Moreover, soil erosion or ecosystem destruction may occur in the process of collecting the necessary aggregate. Energy is also consumed in the course of transporting materials, such as cement and aggregate, to concrete manufacturers, and when producing concrete in batch plants, various high-environmental-load substances are discharged into the air, water, and soil.

As the concrete production process impacts the environment in a variety of ways, it was necessary to assess various environmental impact categories as shown Figure 4. First, for concrete with a compressive strength of 24 MPa, a life cycle assessment was performed according to the volume fraction (substitution rate) of recycled aggregate (R.G) used instead of natural aggregate. As the volume fraction increased, GWP, one of the environmental impact categories, increased by 11–34% compared to when only using natural aggregate. This is because CO_2, CH_4, and N_2O emissions, which all have a major effect on GWP, are larger during the production of recycled aggregate than during the production of natural aggregate. As the volume fraction of recycled aggregate increased to 10%, 20%, and 30%, the ADP decreased by 9–29% compared to the ADP of the concrete with natural aggregate only. This is because antimony (Sb), hard coal, and crude oil emissions, which affect ADP, were smaller during the production of recycled aggregate than during the production of natural aggregate.

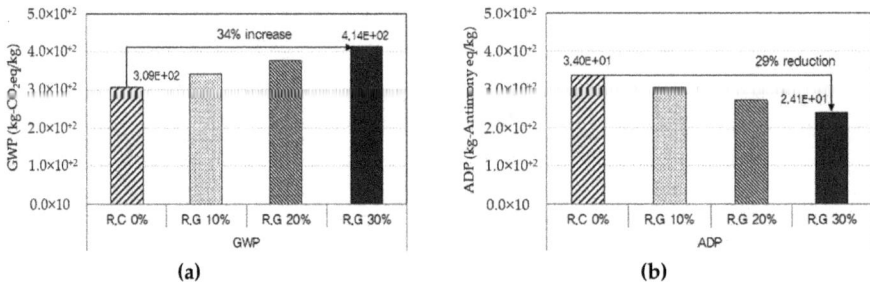

Figure 4. Comparison of environmental impact by recycled aggregate mixing ratio. (**a**) GWP: Global Warming Potential; (**b**) ADP: Abiotic Depletion Potential

6. Conclusions

By observing the ISO 14044 (LCA) standard, this study assessed the life cycle environmental impact of matter discharged in the production (dry/wet) of recycled aggregate on GWP, AP, EP, ODP, ADP, and POCP.

The environmental impact of recycled aggregate (wet) was up to 16%(EP)~40%(AP) higher compared with recycled aggregate (dry); the amount of energy used by impact crushers while producing wet recycled aggregate was the main cause for this result.

Comparing the methods of production, while the environmental impact of the recycled aggregate produced by the wet method is somewhat high, the wet method remains the most practical method, and is beneficial in terms of the aggregate quality attained due to the high fine-powder-removal effect achieved by washing with water. The recycled aggregate produced by the dry method has a simpler production process and a lower production cost, but its quality requires improvement because the impurity and adhesive mortar removal efficiency is low.

The environmental impact of using recycled aggregate was found to be up to twice as high as that of using natural aggregate, largely due to the greater simplicity of production of natural aggregate requiring less energy. However, ADP was approximately 20 times higher in the use of natural aggregate because doing so depletes natural resources, whereas recycled aggregate is recycled from existing construction waste.

Among the life cycle impacts assessment of recycled aggregate, GWP was lower than for artificial light-weight aggregate but greater than for slag aggregate. This is because energy use, such as electricity, LNG, and coal in the mixing, molding, and frit sealing process during the production of artificial light-weight aggregate was higher than the amount of energy used in the production of recycled aggregate. However, in the environmental impact of recycled aggregate, AP, EP, ODP, POCP, and ADP were lower by 37% to 93% than that of artificial light-weight aggregate and slag aggregate. Although the concrete into which recycled aggregate was substituted exhibited a somewhat higher environmental impact than natural aggregate in terms of GWP, it was found to be more environmentally friendly in terms of ADP.

The results drawn through this study do not represent the environmental impact index of all recycled aggregates, and the range of environmental impact indices must be assessed through more analyses in the future.

Author Contributions: Author contributions for this research article are as follows; conceptualization, W.-J.P.; methodology and investigation, S.-J.R.; data curation, R.-H.K.; writing—original draft preparation, T.-H.K.; writing and editing, W.-J.P.; review and editing, T.-H.K.

Funding: This work was supported by both a grant (19CTAP-C141186-02) from Technology Advancement Research Program (TARP) funded by Ministry of Land, Infrastructure and Transport of Korean Government and the National Research Foundation of Korea (NRF) grant (No. NRF-2018R1D1A3B07045700) funded by the Korea government.

Conflicts of Interest: The authors declare no conflict of interest.

References

1. Ministry of the Interior. *Law of Low-Carbon on Green Growth*; Ministry of Government Legislation: Sejong-si, Korea, 2013.

2. Department of Climate Policy. *Climate Change Handbook, Korea Meteorological Administration*; Korea Meteorological Administration: Seoul, Korea, 2009.

3. Ministry of Environment. *Waste Recycling: Extended Producer Responsibility*; Ministry of Environment: Sejong-si, Korea, 2012.

4. Butera, S.; Christensen, T.H.; Astrup, T.F. Life cycle assessment of construction and demolition waste management. *Waste Manag.* **2015**, *44*, 196–205. [CrossRef] [PubMed]

5. Rosado, L.P.; Vitale Penteado, C.S.G.; Arena, U. Life cycle assessment of natural and mixed recycled aggregate production in Brazil. *J. Clean. Prod.* **2017**, *151*, 634–642. [CrossRef]

6. Tam, V.W.Y.; Soomro, M.; Evangelista, A.C.J. A review of recycled aggregate in concrete applications (2000–2017). *Constr. Build. Mater.* **2018**, *172*, 272–292. [CrossRef]

7. Hossain, M.U.; Poon, C.S.; Lo, I.M.C.; Cheng, J.C.P. Comparative environmental evaluation of aggregate production from recycled waste materials and virgin sources by LCA. *Resour. Conserv. Recycl.* **2016**, *109*, 67–77. [CrossRef]

8. Yazdanbakhsh, A.; Lagouin, M. The effect of geographic boundaries on the results of a regional life cycle assessment of using recycled aggregate in concrete. *Resour. Conserv. Recycl.* **2019**, *143*, 201–209. [CrossRef]

9. Ghisellini, P.; Ripa, M.; Ulgiati, S. Exploring environmental and economic costs and benefits of a circular economy approach to the construction and demolition sector. *J. Clean. Prod.* **2018**, *178*, 618–643. [CrossRef]

10. Julliena, A.; Proust, C.; Martaud, T.; Rayssac, E.; Ropert, C. Variability in the environmental impacts of aggregate production. *Resour. Conserv. Recycl.* **2012**, *62*, 1–13. [CrossRef]

11. Kurda, R.; Silvestre, J.D.; de Brito, J. Life cycle assessment of concrete made with high volume of recycled concrete aggregates and fly ash. *Resour. Conserv. Recycl.* **2018**, *139*, 407–417. [CrossRef]

12. Wijayasundara, M.; Mendis, P.; Zhang, L.; Sofi, M. Financial assessment of manufacturing recycled aggregate concrete in ready-mix concrete plants. *Resour. Conserv. Recycl.* **2016**, *109*, 187–201. [CrossRef]

13. Marinkovic, S.; Radonjanin, V.; Malešev, M.; Ignjatovic, I. Comparative environmental assessment of natural and recycled aggregate concrete. *Waste Manag.* **2010**, *30*, 2255–2264. [CrossRef]

14. Kurda, R.; de Brito, J.; Silvestre, J.D. Carbonation of concrete made with high amount of fly ash and recycled concrete aggregates for utilization of CO_2. *J. CO2 Util.* **2019**, *29*, 12–19. [CrossRef]

15. Colangelo, F.; Petrillo, A.; Cioffi, R.; Borrelli, C.; Forcina, A. Life cycle assessment of recycled concretes: A case study in southern Italy. *Sci. Total Environ.* **2018**, *615*, 1506–1517. [CrossRef] [PubMed]

16. Hossain, M.U.; Poon, C.S.; Dong, Y.H.; Lo, I.M.C.; Cheng, J.C.P. Development of social sustainability assessment method and a comparative case study on assessing recycled construction materials. *Int. J. Life Cycle Assess.* **2018**, *23*, 1654–1674. [CrossRef]

17. Marinković, S.B.; Ignjatović, I.; Radonjanin, V. *Handbook of Recycled Concrete and Demolition Waste (Life-Cycle Assessment of Concrete with Recycled Aggregates)*; Woodhead Publishing Series in Civil and Structural Engineering; Woodhead Publishing: Cambridge, UK, 2013; pp. 569–604.

18. Rodríguez-Robles, D.; Van Den Heede, P.; De Belie, N. *New Trends in Eco-Efficient and Recycled Concrete (Life Cycle Assessment Applied to Recycled Aggregate Concrete)*; Woodhead Publishing Series in Civil and Structural Engineering; Woodhead Publishing: Cambridge, UK, 2019; pp. 207–256.

19. Tošić, N.; Marinković, S.; Dašić, T.; Stanić, M. Multicriteria optimization of natural and recycled aggregate concrete for structural use. *J. Clean. Prod.* **2015**, *87*, 766–776. [CrossRef]

20. Wijayasundara, M.; Crawford, R.H.; Mendis, P. Comparative assessment of embodied energy of recycled aggregate concrete. *J. Clean. Prod.* **2017**, *152*, 406–419. [CrossRef]

21. Cuenca-Moyano, G.M.; Zanni, S.; Bonoli, A.; Valverde-Palacios, I. Development of the life cycle inventory of masonry mortar made of natural and recycled aggregates. *J. Clean. Prod.* **2017**, *140*, 1272–1286. [CrossRef]
22. Taehyoung, K.; Sungho, T.; Seongjun, R. Assessment of the CO_2 emission and cost reduction performance of a low-carbon-emission concrete mix design using an optimal mix design system. *Renew. Sustain. Energy Rev.* **2013**, *25*, 729–741.
23. Serres, N.; Braymand, S.; Feugeas, F. Environmental evaluation of concrete made from recycled concrete aggregate implementing life cycle assessment. *J. Build. Eng.* **2016**, *5*, 24–33. [CrossRef]
24. Shan, X.; Zhou, J.; Chang, V.; Yang, E.-H. Life cycle assessment of adoption of local recycled aggregates and green concrete in Singapore perspective. *J. Clean. Prod.* **2017**, *164*, 918–926. [CrossRef]
25. Colangelo, F.; Cioff, R. Mechanical properties and durability of mortar containing fine fraction of demolition wastes produced by selective demolition in South Italy. *Compos. Part B* **2017**, *115*, 43–50. [CrossRef]
26. Kim, T.H.; Tae, S.H. Proposal of Environmental Impact Assessment Method for Concrete in South Korea: An Application in LCA (Life Cycle Assessment). *Int. J. Environ. Res. Public Health* **2016**, *13*, 1074. [CrossRef] [PubMed]
27. Kim, T.H.; Tae, S.H.; Suk, S.J.; George, F.; Yang, K.H. An Optimization System for Concrete Life Cycle Cost and Related CO_2 Emissions. *Sustainability* **2016**, *8*, 361. [CrossRef]
28. Ministry of Land, Transport and Maritime Affairs of the Korean government. *National D/B for Environmental Information of Building Products*; Ministry of Land, Transport and Maritime Affairs of the Korean Government: Sejong, Korea, 2008.
29. Korea Environmental Industry and Technology Institute. National Life Cycle Index Database Information Network. Available online: http://www.edp.or.kr (accessed on 31 October 2018).
30. IPCC Guidelines for National Greenhouse Gas Inventories. Available online: http://www.ipccnggip.iges.or.jp/public/2006gl/ (accessed on 31 October 2018).
31. Guinee, J.B. Development of a Methodology for the Environmental Life Cycle Assessment of Products: with a Case Study on Margarines. Ph.D. Thesis, Leiden University, Leiden, The Netherlands, 1995.
32. Heijungs, R.; Guinée, J.B.; Huppes, G.; Lamkreijer, R.M.; Udo de Haes, H.A.; Wegener Sleeswijk, A.; Ansems, A.M.M.; Eggels, P.G.; van Duin, R.; de Goede, H.P. *Environmental Life Cycle Assessment of Products. Guide (Part1) and Background (Part 2)*; CML Leiden University: Leiden, The Netherlands, 1992.
33. World Metrological Organization (WMO). *Scientific Assessment of Ozone Depletion: Global Ozone Research and Monitoring Project*; WHO: Geneva, Switzerland, 1991; p. 25.
34. Derwent, R.G.; Jenkin, M.E.; Saunders, S.M.; Piling, M.J. Photochemical ozone creation potentials for organic compounds in Northwest Europe calculated with a master chemical mechanism. *Atmos. Environ.* **1998**, *32*, 2429–2441. [CrossRef]
35. Jenkin, M.; Hayman, G. Photochemical Ozone Creation Potentials for oxygenated volatile organic compounds: Sensitivity to variation is in kinetic and mechanistic parameters. *Atmos. Environ.* **1999**, *33*, 1275–1293. [CrossRef]
36. Jung, J.S.; Lee, J.S.; An, Y.J.; Lee, K.H.; Bae, K.S.; Jun, M.H. Analysis of Emission of Carbon Dioxide from Recycling of Waste Concrete. *Archit. Inst. Korea* **2008**, *24*, 109–116.
37. Kim, T.H.; Tae, S.H.; Chae, C.U.; Choi, W.Y. The Environmental Impact and Cost Analysis of Concrete Mixing Blast Furnace Slag Containing Titanium Gypsum and Sludge in South Korea. *Sustainability* **2016**, *8*, 502. [CrossRef]
38. Lee, D.H.; Jun, M.H.; Bae, K.S. Correlation between density and absorption of domestic recycled aggregate. *Korean Recycl. Constr. Res. Inst.* **2011**, *11*, 95–96.
39. Kim, T.H.; Tae, S.H.; Chae, C.U. Analysis of Environmental Impact for Concrete Using LCA by Varying the Recycling Components, the Compressive Strength and the Admixture. *Sustainability* **2016**, *8*, 389. [CrossRef]

applied
sciences

MDPI

Article

Mechanical Characteristics and Water Absorption Properties of Blast-Furnace Slag Concretes with Fly Ashes or Microsilica Additions

Dora Foti [1,*]**, Michela Lerna** [1]**, Maria Francesca Sabbà** [1] **and Vitantonio Vacca** [2]

[1]   Polytechnic University of Bari, Department of Civil Engineering Sciences and Architecture (DICAR),
     Via Orabona 4, 70125 Bari, Italy; michela.lerna@poliba.it (M.L.); mf.sabba@libero.it (M.F.S.)
[2]   National Research Council (CNR), Institute of Environmental Geology and Geo-Engineering,
     Via Salaria km 29,300, 00015 Montelibretti (Roma), Italy; vitantonio.vacca@hotmail.it
*   Correspondence: dora.foti@poliba.it; Tel.: +39-320-171-0524

Received: 20 February 2019; Accepted: 18 March 2019; Published: 27 March 2019

Abstract: The paper shows the results of an experimental tests campaign carried out on concretes with recycled aggregates added in substitution of sand. Sand, in fact, has been totally replaced once by blast-furnace slag and fly ashes, once by blast-furnace slag and microsilica. The aim is both to utilize industrial by-products and to reduce the use of artificial aggregates, which impose the opening of pits with high environmental damage. The results show that in the concretes so made the water absorption capacity has reduced and durability has improved. The test campaign and the results described in the present article are certainly useful and can be especially utilized for research on a larger scale in this field.

Keywords: concrete; aggregates; fly-ash; silica fume; blast-furnace slag; mechanical properties; water absorption

1. Introduction

In recent times, the problem of preservation of reinforced and pre-stressed concrete structures has been developing rapidly. In fact, nowadays, as a consequence of the use of a bad mix design and inaccurate casting and execution of structures that occurred around the fifties, many structures present serious degradation phenomena. Moreover, there is a higher concentration of industrial and natural aggressive factors in the environment that can cause material degradation when compared to about forty years ago [1].

With a correct mix design, the appropriate use of super-plasticizers, low water/cement ratios, and, especially, the addition of materials constituted by very small granules obtained from the pozzolanic activity, it is possible to reduce the empty spaces in the particles of the concrete mix and get waterproof concretes with good mechanical strength characteristics.

The field of concrete technologies and its production is a current research topic. In recent years, recycling and re-use have become critical issues to develop concrete technologies and build reinforced concrete structures with a long service life and that, at the same time, are able to satisfy economic and environmental issues [2,3]. As a consequence, the use of industrial by-products or solid wastes in concrete production, such as fly ashes (FA), blast-furnace slag, and silica fume (SF) is increasing. Quite good mechanical properties, in fact, are obtained from concretes that use as aggregates calcareous rubble and, in total substitution of sand, blast-furnace slag [4].

Slag is constituted by scoriae at the liquid state; they remain in the furnace and are eliminated during the cast iron production. Approximately 300 kg of slag is generated per ton of pig iron. In particular, the slag cools in extremely quickly and prevents the crystallization process. As a consequence, it solidifies in glass granules mixed with foamy fragments [5,6]. From this process it is possible to obtain the so-called granulated blast slag, an industrial by-product, which can be classified as waste [7], thus increasing problems associated with environmental pollution. Therefore, the aim of this study is to try to increase its possible uses, especially in the field of construction [8,9]. Due to its low iron content it can be safely used in the manufacturing of cement. Two types of blast furnace slag, air-cooled slag and granulated slag, are being generated from the steel plants. The color of granulated slag is whitish. The air-cooled slag is used as aggregate in road making, while the granulated slag is used for cement manufacturing. Although it has no behavioral problems, its use as aggregate in concrete is not very common and there has been little research work done on the subject.

The optimum cement replacement level with granulated blast-furnace slag is often quoted to be about 50% and sometimes as high as 70% and 80% to get an improvement of the mechanical and durability properties of concrete and to generate less heat of hydration.

On the other hand, concrete based on Portland cement is the most widely used material in the world. Compared to other materials like steel, aluminum and plastics, it is the most viable option for the construction industry considering economic and environmental costs. Nonetheless, it is estimated that 7% of anthropogenic CO_2 corresponds to the ordinary Portland cement production; considering the current global environmental situation, it is obvious that cement and concrete specialists must search for ways to reduce that figure, or at least to avoid its growth. One of the possibilities is the massive usage of industrial wastes like blast-furnace slag, to turn them into useful environmentally friendly and technologically advantageous cementitious materials [10]. In a previous study, the slag partially replaced 30%, 50% and 70% of Portland cement, and the cement's strength reduced as the amount of slag increased [11].

The use of blast-furnace slag on a large scale as a material to replace natural aggregates is a most promising concept because its impact strength is higher than that of natural aggregate. It is important in those areas where artificial aggregates are almost exclusively used. In this way, the crushing process of rocks, for example, which is commonly practiced in Apulia, Italy, would be reduced and consequently, the opening of new quarries—unfavorable in terms of environmental impact—would be reduced too [12].

To have the lowest possible environmental impact it is necessary to consider many aspects in the field of concrete technologies and in concrete production, such as the cost of construction and materials, its durability, and especially its compatibility with the protection of the environment. A solution to protect the environment is the use and/or re-use of industrial by-products or solid wastes to add to the concrete mix, such as fly-ash, blast-furnace slag, and silica fume [13].

In the literature, mortars and concretes added with different materials in the mix design have been studied and tested with the aim to improve their mechanical properties [14–16] and, at the same time, to reduce its environmental impact and the amount of waste sent to landfills [17–20]. The latter is especially due to their environmental friendliness and high durability properties. In Foti and Paparella [21], Foti [22], and Metha [23–25], alternative technologies for concrete production were considered by analyzing the cost of materials and construction, durability, and especially environmental friendliness.

Hefni et al. [26] used a high-volume of fly-ashes as a partial replacement of cement in concrete with the addition of different chemical activators and glass fibers and by investigating the concrete strength at room temperature and after exposure to elevated temperature. Usually, the slag in its original granulometric composition has very few fine granules [27]; hence, it is not possible to use slag in place of sand: the concrete so obtained would not be very compact and, consequently, very porous and with a very low strength. However, it is possible to fill the empty spaces with sieved slag made of very small granules; in this way the aggregates are rubble, original slag and sieved slag.

To obtain better performance it is possible to use fly-ashes instead of sieved slag. Fly-ashes, which are by-products of an industrial process (more precisely, of carbon combustion in power-stations), are constituted of very small granules with properties similar to those of the natural pozzolan. For this reason, fly-ashes can be used in partial replacement of cement or as aggregates of mortar, similar to sand, that is, the very thin fraction of the mix design [28,29].

However, the above considerations lead us to believe that an improvement is possible by adding to the slag, in its original composition, other materials made up of very fine particles that replace the sieved fraction. The so-called "fly-ashes" or "light ashes" are particularly suitable for this purpose. With the imported coals and the most commonly used boilers today, the residual ash as a whole is about 13% by weight of the starting fuel. Of this, a small part falls to the bottom of the combustion chamber in a special tank containing cooling water and forms the so-called "heavy" ashes. The remaining part, equal to about 85% of the total ash, follows the path of the gaseous combustion products that, by law, must be dedusted before being dispersed into the atmosphere. Without dwelling on the manner in which this process is implemented, it is sufficient to note that the so-called light ash obtained in this way is a very good quality fine material, with a composition and binding properties not very different from those of natural pozzolans. They are suitable to be used in partial substitution of cement, and as aggregates in concrete with sands poor in thin elements that have already been successfully tested.

An alternative is to replace fly-ashes with microsilica (silica fume), another artificial product obtained by the reduction process of quartzite during the production of alloys like iron-silicon or metallic silicon. Microsilica is constituted of granules smaller than fly-ashes. Their origin is from oxidation, hardening and condensation of silica fume at 2000 °C, following a process similar to the natural pozzolan [30]. Microsilica is utilized in many countries to make concretes with high durability [31]. They are composed of spherical granules with a diameter of 0.1 μm; this size allows the granules to fill all of the empty spaces of concrete. Microsilica is mainly comprised of dioxide of silicon; dioxide activates the transformation of hydrates of calcium (dangerous for concrete durability) in an insoluble, stable and effective product to fill all the pores. The inclusion of silica fume at high replacement levels significantly increases the autogenous shrinkage of concrete due to the refinement of pore size distribution that leads to a further increase in capillary stress and more contraction of the cement paste [32]. In fact, microsilica fume addition produces a positive influence on the dynamic and static mechanical properties of concrete, especially on the resonant frequencies, the dynamic and static moduli of elasticity, damping ratio and strength [33,34]. In particular, the results obtained by Giner et al. [28,29] evidence that microsilica additions or replacements reduce both the dynamic modulus of elasticity and damping ratio of concrete.

In other studies [35–37], some fibers (like carbon or steel fibers) were added to silica fume concretes, obtaining an improvement in the tensile and compressive strengths of concrete.

Bhanjaa and Senguptab [38], and Langana et al. [39] determined the influence of silica fume over a wide range of water-binder ratios and cement replacement percentages. In particular, the incorporation of blast-furnace slag into silica fume concrete reduces the water demand and this combination offers increased resistance to the alkali-silica reaction expansion and chloride ingress than the use of one of these materials alone [40].

Charlee et al. [41] and Elahi et al. [42] showed how the increase of fly-ashes replacement in concrete clearly reduces the chloride penetration (CP) and the steel corrosion in concrete. The last characteristic is also explained because these fine by-products make the microstructure of concrete denser, reducing its diffusivity [43,44].

In the present paper, the good results of experimental tests carried out on concretes in which sand is totally replaced by blast-furnace slag and microsilica, or by blast-furnace slag and fly-ashes are shown and discussed. Mechanical characterization tests and water absorption tests are performed. Blast furnace slag was integrated in the finest fraction by means of appropriate percentages of microsilica or fly-ashes. The tests, although preliminary by nature, confirm the possibility to obtain concretes with satisfactory mechanical properties and lower water absorption values using either fly-ashes or microsilica. Therefore, the results and comparisons reported in this paper are important bases towards the aim of carrying out wider research in this field.

2. Materials and Methods

A series of strength and water absorption tests were performed on specimens made with two different mixes:

(1) Concrete + slag + fly-ashes;
(2) concrete + slag + microsilica.

All tests were carried out at the Laboratory for Testings and Materials "M. Salvati" of the Polytechnic University of Bari, Italy. Different shape specimens were manufactured depending on the kind of test and the characteristics to be determined.

2.1. Materials

The specimens were manufactured utilizing two different mixes made of the following materials:

-   High strength Portland cement (following the EC2 prescriptions [45]);
-   Blast-furnace slag produced by the ILVA factory, Taranto, Italy (Figure 1a);
-   Fly ashes coming from ENEL power station of Brindisi, Italy (Figure 1b); it is constituted on average by 90% of granules with dimensions smaller than 0.3 mm, by 60% of granules smaller than 0.04 mm, and by 10% of granules smaller than 0.01 mm;
-   Microsilica of ELKEM MATERIALS a/s Kristiansand, Norway (Figure 1c);
-   Calcareous rubble;
-   Super-plasticizer RHEO-BUILD 1000 from MAC S.p.A.

(a)   (b)   (c)

Figure 1. (a) Blast furnace slag, (b) fly ash and (c) microsilica from Elkem utilized for the tests.

The characteristics of the materials and their chemical compositions were provided by the manufacturers; they are shown in Table 1.

Table 1. Properties and chemical composition of the materials utilized in the concrete mix.

	Cement (daN/m³)	Slag (daN/m³)	Fly Ashes (daN/m³)	Microsilica (daN/m³)
Absolute specific weight	3100	2720	2140	2160
Bulk density	1400	1110	700	737
Chemical Composition %				
SiO_2	17–25	30–40	40–55	88–98
Al_2O_3	2–8	6–18	20–30	0.5–3
CaO	60–67	38–50	3–7	0.1–0.5
MgO	0.1–5	2–6	1–4	0.3–1.5
SO_3	1–4.8	-	0.4–2	-
Na_2O	-	-	-	0.2–1.4
K_2O	0.2–1.5	-	1–5	0.4–1

The granulometric curve of the blast-furnace slag is shown in Figure 2.

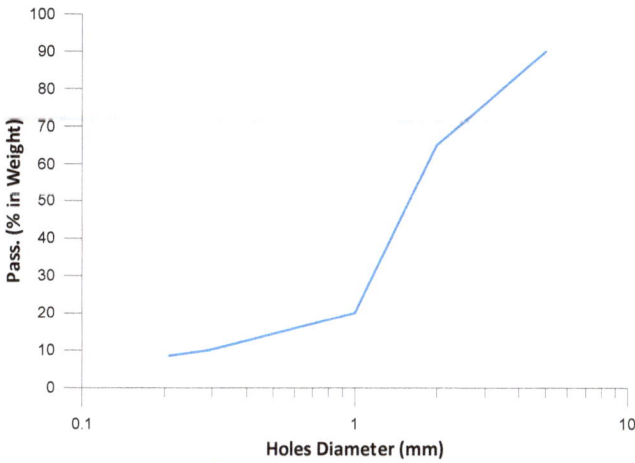

Figure 2. Granulometric curve of blast-furnace slag.

Calcareous rubble was utilized with two different granulometric curves represented in Figure 3. Kind A rubble was utilized in concretes with slag and fly-ashes; kind B rubble has been utilized for concretes with slag and microsilica.

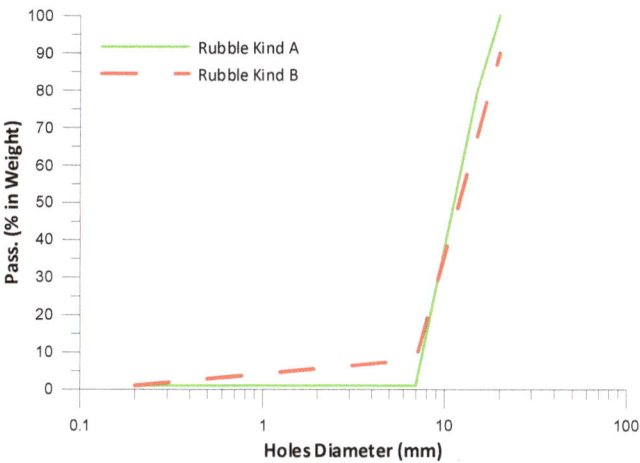

Figure 3. Granulometric curves of rubble.

2.1.1. Concretes with Slag and Fly Ashes (FA)

The concrete with slag and fly-ashes in total substitution of sand and rubble was realized with a calibrated percentage of the aggregates. Therefore, defining the maximum diameter of the aggregate and the consistency of concrete, the mix design was prepared to have a characteristic strength R_{ck} equal to 30 N/mm^2, following A.C.I., 1991 [46] and DOE, 1988 [47] prescriptions. In this case, by setting a maximum diameter of the aggregate equal to 20 mm, a "plastic" consistency and an average strength equal to the value previously assumed, it was possible to define the necessary quantity of water, the volume of blocked air, the water/cement ratio and the quantity of binder required [48].

The residual amounts of aggregates were finally fixed (rubble and slag), taking care that the characteristics of the latter did not allow to refer to a continuous type granulometric curve (Figures 2 and 3).

The composition of the aggregates was studied on the basis of the suggested criteria for discontinuous granulometries [49], which allows a quantity of about 700 N/mm^3 to pass through a 1 mm sieve. The test program takes into account that the slag has only 21% of granules of dimensions smaller than 1 mm and, vice-versa, concrete and fly-ashes have the size of all granules below it. In the end, the known volumes of water, cement, fly-ashes, slag, and blocked air, and the amount of rubble were fixed for difference. Then, the resulting composition was optimized by preliminary tests; the final composition with the ratios of the components that were assumed in the concrete mix is shown in Table 2.

Table 2. Composition of slag and fly ashes (FA) concrete mix per unity of volume and ratios of the components.

Rubble	Fly Ashes	Water	Slag	Cement	Super-Plasticizer
1080 daN/m^3	130 daN/m^3	220 L/m^3	522 daN/m^3	340 daN/m^3	4.7 L/m^3
Components Ratio					
water/cement	water/(ashes + cement)		ashes/cement	super-plasticizer/(ashes + cement)	
0.65	0.47		0.38	0.01	

The granulometric curve of the aggregates is shown in Figure 4.

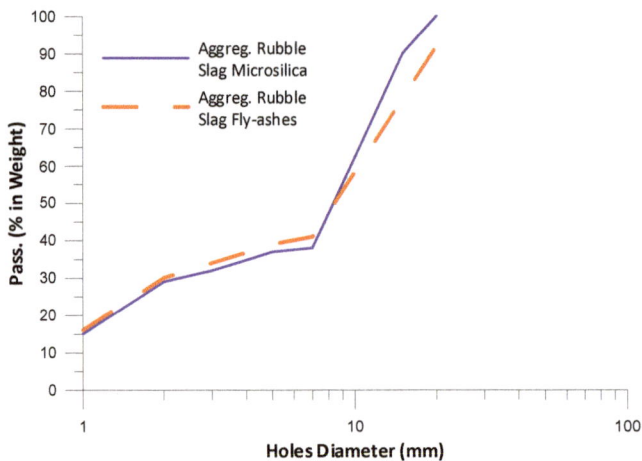

Figure 4. Granulometric curves of the aggregates of slag and FA concrete, and slag and SF concrete.

2.1.2. Concretes with Slag and Microsilica (SF)

In this case, the concrete with slag and microsilica was obtained by totally substituting the fly-ashes added to the previous concrete mix with an equal quantity of microsilica, whose characteristics are shown in Table 1. The preliminary tests also show the necessity to increase water and super-plasticizer to obtain a concrete with the same consistency and workability of the concrete of slag and fly-ashes previously described. The final composition of the mix is shown in Table 3.

The granulometric curve of the aggregates is also shown in Figure 4. Comparing the two concrete mixes, it is possible to notice that the two curves almost coincide for diameters smaller than 3 mm; that is the most significant field. The ratios shown in Table 3 have been assumed in the concrete mix.

Table 3. Composition of slag and microsilica (SF) concrete mix per unit of volume and component ratios utilized.

Rubble	Microsilica	Water	Slag	Cement	Super-Plasticizer
1080 daN/m^3	131 daN/m^3	296 L/m^3	522 daN/m^3	340 daN/m^3	7.22 L/m^3
Components Ratio					
water/cement	water/(microsilica + cement)		microsilica/cement	super-plasticizer/(microsilica + cement)	
0.87	0.63		0.38	0.015	

For comparison purposes, cubic specimens were manufactured with ordinary concrete (ORD); the control mix composition is shown in Table 4. It is a well cured and a high strength concrete (47 N/mm^2 at 28 days of curing). Like for slag and FA and slag and SF concretes, the quantity of water was assumed in order to get the same consistency and workability. In this case the ratio water/cement = 0.60.

Table 4. Composition of ordinary concrete (ORD) mix per unit of volume.

Rubble	Sand	Water	Cement	Super-Plasticizer
1080 daN/m^3	700 daN/m^3	204 L/m^3	340 daN/m^3	4.0 L/m^3

2.2. Test Methods

Strength tests and water absorption tests were performed on the specimens. The tests were carried out at the Laboratory for Testings and Materials "M. Salvati" of the Polytechnic University of Bari, Italy.

Three concrete specimens were manufactured for each strength test. Compressive, bending and traction tests up to failure were performed on the specimens. For the tests, 300 kN Metrocom equipment was utilized; for traction tests, an appropriate grip was manufactured, while the 3-point bending tests were performed by mean of an ad-hoc accessory for the machine. It was also determined that the compressive modulus of elasticity E after 28 days of curing, in the stress field 0–$\sigma_r/3$, where σ_r is the failure stress.

Water absorption tests were carried out on three cylindrical specimens for each type of concrete (with rubble, slag and fly-ashes and with rubble, slag and microsilica) in accordance with UNI 7699, 2005 [50]. In the test procedure, the specimens were dried in an oven at 100–110 °C for 24 h, and then their weights were measured. Next, the specimens were fully immersed in a water tank for 24 h, then they were taken out. Their surfaces were dried with a cloth and finally their weights were measured again. The test was conducted on the specimens with different curing ages (7, 14, 21, and 28 days).

2.3. Results

2.3.1. Strength Tests

Table 5 shows the average values of the failure stresses obtained after 7, 28, and 90 days of curing for slag and fly-ashes concretes. In this case, the compressive elastic modulus E, after 28 days of curing and in the stress field 0–$\sigma_r/3$ is equal to about 25,900 N/mm^2.

Table 5. Mean values and standard deviations of the mechanical characteristics of slag and fly ashes concrete.

Curing Days	Specific Weight (daN/m^3)	Compressive Failure Stress (N/mm^2)		Flexural Failure Stress (N/mm^2)		Tensile Failure Stress (N/mm^2)	
		Value	σ	Value	σ	Value	σ
7	2222	26.0	±3.74	-		-	
28	2185	36.5	±2.12	4.4	± 3.0	2.3	±3.61
90	2185	43.0	±3.74	-		-	

Table 6 shows the average values of the failure stresses after 7, 28 and 60 days of curing for slag and microsilica concretes. In this case, the compressive elastic modulus *E*, after 28 days of curing and in the stress field $0-\sigma_r/3$ is equal to about 25,140 N/mm^2. In Table 7 the mean values for compression strength of the different concretes are reported.

Table 6. Mean values and standard deviations of the mechanical characteristics of slag and microsilica concrete.

Curing Days	Specific Weight (daN/m^3)	Compressive Failure Stress (N/mm^2)		Flexural Failure Stress (N/mm^2)		Tensile Failure Stress (N/mm^2)	
		Value	σ	Value	σ	Value	σ
7	2140	31.2	±3.10	-		-	
28	2140	42.6	±5.724	3.3	±3.74	2.2	±1.73
60	2115	45.3	±4.32	-		-	
90	2113	47.1	±1.34	-		-	

Table 7. Compressive strength for the different concrete types.

Curing Days	Ordinary Concrete (N/mm^2)	Slag and Fly Ashes Concrete (N/mm^2)		Slag and Microsilica Concrete (N/mm^2)	
	Value	Value	σ	Value	σ
28	47	36.5	±2.12	42.6	±5.724

The methods utilized for the mix designs of the concrete utilized in the tests, appropriately refined by the preliminary tests, are shown to be good enough. However, it is evident from the results that the pozzolanic activity, both with fly-ashes, and especially with microsilica, played an important role in getting a good value for compressive strength.

The water/cement ratios are both quite high: if utilized to prepare an ordinary concrete, it would have been difficult to reach strength of the order of those obtained from the tests if there was no pozzolanic activity for fly-ashes and microsilica. Therefore, the ratios water/(fly-ashes + cement) and water/(microsilica + cement) are considered more significant in both cases.

The higher compressive strength obtained utilizing microsilica instead of fly-ashes is justified by the higher pozzolanic activity of silica fume and, especially, by the dimensions of their granules, which were much smaller than fly-ashes and typical of microsilica. The better behavior is obtained even if, when microsilica are used, it is necessary to add more water and super-plasticizers in order to obtain the same consistency as concrete with fly-ashes (see Table 3).

From the tests it seems that higher strengths are obtained for concrete with slag and microsilica, while the specific weight is a little lower than the weight of concretes with slag and fly-ashes. Moreover, for both concretes it seems that while the tensile strengths are comparable with the compressive strengths, the bending strengths are quite low, especially for concretes with slag and microsilica.

A similar result is obtained for the elastic longitudinal moduli, which are not very high if compared to the compressive strength, and lower than an ordinary concrete with the same compressive strength. The latter result is probably due to the discontinuous composition of the aggregates utilized and to the prevailing smaller fractions.

2.3.2. Water Absorption Tests

The percentage of water absorption by the different specimens was calculated from the following Equation (1):

$$\text{Water absorption (\%)} = (Wi - Ws)/Ws \cdot 100 \tag{1}$$

where:

Wi = average weight of dry sample (g),
Ws = the average weight of the wet sample (g).

Table 8 reports the sizes of the specimens, their weights after curing, the weight increments during the phase of immersion in water (Figure 5), and the water absorption percentages (Figure 6). For the purpose of comparison, the results of the tests carried out on cubic specimens of ordinary concrete (well cured and with a high strength (47 N/mm^2)) are reported.

Table 8. Results of the absorption tests.

		Concrete of Slag and Fly Ashes			Concrete of Slag and Microsilica			Ordinary Concrete (Ord)		
	Specimens	1	2	3	1	2	3	1	2	3
	D × H (or L for cubic specimens) [mm]	96 × 96	96 × 96	96 × 96	96 × 96	96 × 99	96 × 92	10 × 10 × 10	10 × 10 × 10	10 × 10 × 10
	Weight after curing = Ms [N]	15.19	15.15	15.34	15.01	15.46	14.09	22.34	22.26	22.78
	After 1 h	15.31	15.20	15.39	15.05	15.46	14.10	22.52	22.44	22.82
	After 3 h	15.74	15.63	15.73	15.25	15.92	14.41	22.81	22.72	23.08
Weight increasing	After 8 h	15.79	15.70	15.87	15.39	15.96	14.49	23.02	22.97	23.31
during the phase	After 24 h	15.80	15.71	15.88	15.45	16.01	14.53	23.40	23.30	23.75
of immersion in	After 48 h	15.82	15.73	15.90	15.49	16.04	14.56	23.41	23.31	23.76
water [N]	After 72 h	15.83	15.74	15.91	15.51	16.05	14.57	23.42	23.31	23.76
	After 96 h	15.83	15.74	15.91	15.52	16.07	14.59	23.42	23.31	23.76
	After 120 h = Mi	15.84	15.74	15.92	15.53	16.08	14.60	23.42	23.31	23.76
	Absorbed Water [%] = W$_{max}$ *	4.20	3.95	3.77	3.45	4.03	3.68	4.87	4.75	4.31

$$* \, W_{max} = \frac{100 \, (M_i - M_s)}{M_s}.$$

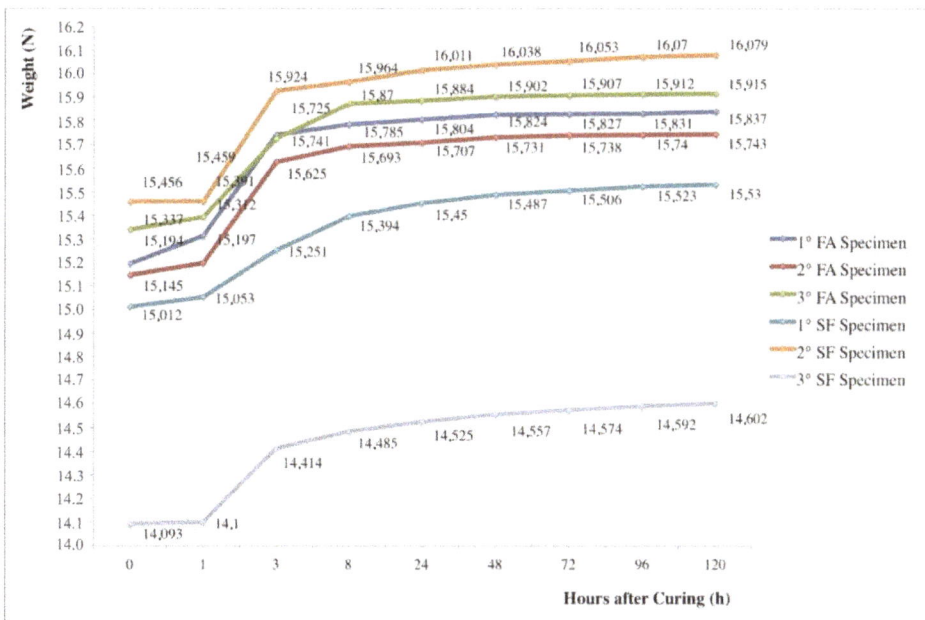

Figure 5. Weight increment during the phase of immersion in water of the specimens (FA = fly ash concrete and SF = silica fume concrete).

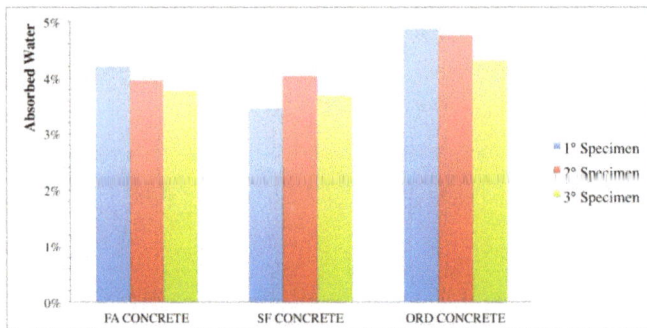

Figure 6. Percentage of water absorbed by the specimens.

For a clearer approach, Figures 5 and 6 show the curves and the histograms relative to the results of the water absorption tests. In Figure 5, h = 0 represents the time at which the immersion of the specimens in the water occurred.

The material porosity decreases both with the addition of fly ashes and microsilica fume. In fact, in the first case the reduction of total pore volume is probably due to the outcome of the continuous generation of pozzolanic reaction products from the hydration of FA that fill the pores. In the second case for microsilica fume concrete, the reduction of total pore volume is caused by the high pozzolanic reactivity, but also by the very small dimensions of SF particles and their pore-filling effect (they subdivide the pore space by amassing themselves between the cement grains).

Figure 6 shows the percentage of water absrbed by the different concrete specimens. It is possible to notice that the higher amount of absorbed water is obtained for the ordinary concrete specimens. In fact, the percentage of absorbed water with respect to the total weight is equal to 4.87% for an ordinary concrete, and 4.20% and 4.03% for FA and SF concretes, respectively.

2.4. Discussion

The overall results are presented in Tables 5–8 and they show that it is possible to obtain concretes with a very satisfactory mechanical strength and a low water absorption capability if sand is completely substituted with slag and fly-ashes, or with slag and microsilica.

The methods used to proportion the mix design, suitably refined through preliminary tests, have proved to be valid. From the results it is evident that the pozzolanic activity of fly-ashes, and especially silica fume, played a key role in reaching a very good compressive strength [51–53]. In fact, water/cement ratios are very high in both cases: these ratios, if utilized for an ordinary concrete mix without the pozzolanic activity of microsilica and fly-ashes, would have hardly given the same high strength values [54,55]. For this reason, it is worth considering the water/(fly-ashes + cement) and water/(microsilica + cement) ratios.

Higher compressive strength is achieved utilizing silica fume instead of fly-ashes [56]. In fact, it is known that the pozzolanic activity of concretes added with SF is higher with respect to other materials, as demonstrated also by Malhotra and Carette [57]. This is due to the higher pozzolanic activity and the smaller sizes of microsilica granules, even if more quantities of water and super-plasticizer are necessary to get the same consistency of fly-ashes concretes. In fact, the water/cement ratio from the value of 0.65 for concretes with slag and fly-ashes, reaches the value of 0.87 for concretes with slag and microsilica; the water/(fly-ashes + cement) ratio is 0.47, while water/(microsilica + cement) ratio is 0.63.

From the results of the tests it has been also possible to notice that microsilica concretes, apart from the higher mechanical strength, present a specific weight lower than for slag and fly-ashes concretes. Moreover, it seems that for both concretes the tensile strength is consistent with the compressive strength, while bending strength is rather low, especially for slag and microsilica concretes. Similarly, the longitudinal elastic moduli are not very high in comparison to the compressive strength and are lower than in an ordinary concrete of the same strength. It is probably due to the discontinuous composition of the aggregates utilized in the mix design and to the higher amount of mortar.

Finally, the water absorption test results showed that the mean values of water absorbed by the different concrete specimens were lower for slag and fly-ashes (or microsilica) concretes in comparison with ordinary concrete, and slag and microsilica concretes with respect to slag and fly-ashes concretes (Table 8, Figures 5 and 6).

In summary, these concretes that use by-products and scoriae of industrial production have mechanical properties comparable with a high strength ordinary concrete. Moreover, they show a lower water absorption capacity and a higher durability with the advantage to improve re-use and recycling of waste materials.

3. Conclusions

On the whole, the results obtained from the present study indicate that the methods utilized to proportion the mixes, appropriately refined through the preliminary tests, are valid. However, it is evident from the analysis of the results that the pozzolanic activity, both of the ashes and even more so of microsilica, played an important role in allowing the achieval of a good compressive strength, with the advantages of using concretes that re-use industrial by-products or solid wastes. Moreover, the two types of concrete, one in which sand is replaced by microsilica and the other in which sand is replaced by fly-ashes, are accurately compared, indicating precisely the pros and cons of each type with respect to their physical and mechanical properties.

The results obtained can be summarized as follows:

- The use of fly-ashes, and especially silica fume, together with slag in concrete enhances the compressive strength of concrete mixes and shows very high water/cement ratios.
- Microsilica concretes present a specific weight lower than slag and fly-ash concretes.
- For both concretes, the tensile strengths are consistent with the compressive ones, while bending strengths are rather low, especially for slag and silica fume concretes.
- Compared to an ordinary concrete, the types of concrete examined in this research have a lower water absorption value, especially the silica fume concrete.

Author Contributions: D.F. carried out the experimental tests and the writing of the text together with M.L., while M.F.S. and V.V. elaborated the results and plotted the graphs.

Funding: This research received no external funding.

Conflicts of Interest: The authors declare no conflict of interest.

References

1. Glasser, F.P.; Marchand, J.; Samson, E. Durability of concrete—Degradation phenomena involving detrimental chemical reactions. *Cem. Concr. Res.* **2008**, *38*, 226–246. [CrossRef]
2. Meyer, C. The greening of the concrete industry. *Cem. Concr. Compos.* **2009**, *31*, 601–605. [CrossRef]
3. Martinez-Abella, F.; Vazquez-Herrero, C.; Perez-Ordonez, J.L. Properties of plain concrete made with mixed recycled coarse aggregate. *Constr. Build. Mater.* **2012**, *37*, 171–176. [CrossRef]
4. Li, G.; Zhao, X. Properties of concrete incorporating fly-ash and ground granulated blast-furnace slag. *Cem. Concr. Compos.* **2003**, *25*, 293–299. [CrossRef]
5. Václavík, V.; Dirner, V.; Dvorský, T.; Daxner, J. The use of blast furnace slag. *Metalurgija* **2012**, *51*, 461–464.
6. Sekhar, D.C.; Nayak, S. Utilization of granulated blast furnace slag and cement in the manufacture of compressed stabilized earth blocks. *Constr. Build. Mater.* **2018**, *166*, 531–536. [CrossRef]
7. Pazhani, K.; Jeyaraj, R. Study on durability of high performance concrete with industrial wastes. *Appl. Technol. Innov.* **2010**, *2*, 19–28. [CrossRef]
8. Hiraskar, K.G.; Patil, C. Use of Blast furnace slag aggregate in concrete. *Int. J. Sci. Eng. Res.* **2013**, *4*, 95–98.

9. Ulubeyli, G.C.; Artir, R. Sustainability for Blast Furnace Slag: Use of Some Construction Wastes. *Procedia Soc. Behav. Sci.* **2015**, *195*, 2191–2198. [CrossRef]

10. Yuksel, I. Blast-furnace slag. In *Waste and Supplementary Cementitious Materials in Concrete*; Siddique, R., Cachim, P., Eds.; Elsevier Science: Amsterdam, The Netherlands, 2018; pp. 361–415.

11. Escalante-Garcia, J.I.; Espinoza-Perez, L.J.; Gorokhovsky, A.; Gomez-Zamorano, L.Y. Coarse blast furnace slag as a cementitious material, comparative study as a partial replacement of Portland cement and as an alkali activated cement. *Constr. Build. Mater.* **2009**, *23*, 2511–2517. [CrossRef]

12. Gaurav, S.; Souvik, D.; Abdulaziz, A.A.; Showmen, S.; Somnath, K. Study of Granulated Blast Furnace Slag as Fine Aggregates in Concrete for Sustainable Infrastructure. *Procedia Soc. Behav. Sci.* **2015**, *195*, 2272–2279.

13. Senani, M.; Ferhoune, N.; Guettala, A. Substitution of the natural sand by crystallized slag of blast furnace in the composition of concrete. *Alex. Eng. J.* **2018**, *57*, 851–857. [CrossRef]

14. Ivorra, S.; Foti, D.; Bru, D.; Baeza, F.J. Dynamic Behavior of a Pedestrian Bridge in Alicante (Spain). *J. Perform. Constr. Facil.* **2015**, *29*, 04014132. [CrossRef]

15. Foti, D.; Vacca, S. Comportamiento mecánico de columnas de hormigón armado reforzadas con mortero reoplástico/Mechanical behavior of concrete columns reinforced with rheoplastic mortar. *Mater. Constr.* **2013**, *63*, 267–282. [CrossRef]

16. Foti, D.; Romanazzi, A. Experimental analysis of fiber-reinforced mortar for walls in rectified brick blocks [Analisi sperimentale di malte fibrorinforzate per pareti in blocchi di laterizio rettificati]. *C Ca* **2011**, *41*, 109–118.

17. Foti, D. Preliminary analysis of concrete reinforced with waste bottles PET fibers. *Constr. Build. Mater.* **2011**, *25*, 1906–1915. [CrossRef]

18. Mehta, P.K. Reducing the environmental impact of concrete. *Concr. Int.* **2001**, *23*, 61–65.

19. Mehta, P.K. Building durable structures in the 21st century. *Concr. Int.* **2001**, *23*, 57–63.

20. Foti, D. Use of recycled waste pet bottles fibers for the reinforcement of concrete. *Compos. Struct.* **2013**, *96*, 396–404. [CrossRef]

21. Foti, D.; Paparella, F. Impact Behavior of Structural Elements in Concrete Reinforced with Pet Fibers. *Mech. Res. Commun.* **2014**, *57*, 57–66. [CrossRef]

22. Foti, D. Innovative techniques for concrete reinforcement with polymers. *Constr. Build. Mater.* **2016**, *112*, 202–209. [CrossRef]

23. Mehta, P.K. Durability critical issues for the future. *Concr. Int.* **1997**, *19*, 69–76.

24. Mehta, P.K. Advancements in concrete technology. *Concr. Int.* **1999**, *21*, 27–33.

25. Mehta, P.K. Greening of the concrete industry for sustainable development. *Concr. Int.* **2002**, *24*, 23–27.

26. Hefni, Y.; El Zaher, Y.A.; Wahab, M.A. Influence of activation of fly ash on the mechanical properties of concrete. *Constr. Build. Mater.* **2018**, *172*, 728–734. [CrossRef]

27. Babu, K.G.; Kumar, V.S.R. Efficiency of GGBS in concrete. *Cem. Concr. Res.* **2000**, *30*, 1031–1036. [CrossRef]

28. Giner, V.T.; Ivorra, S.; Baeza, F.J.; Zornoza, E.; Ferrer, B. Silica fume admixture effect on the dynamic properties of concrete. *Constr. Build. Mater.* **2011**, *25*, 3272–3277. [CrossRef]

29. Giner, V.T.; Baeza, F.J.; Ivorra, S.; Zornoza, E.; Galao, Ó. Effect of steel and carbon fiber additions on the dynamic properties of concrete containing silica fume. *Mater. Des.* **2012**, *34*, 332–339. [CrossRef]

30. Fidjestøl, P.; Lewis, R. Microsilica as an Addition. In *Lea's Chemistry of Cement and Concrete*, 4th ed.; Hewlett, P., Ed.; Elsevier Science: Amsterdam, The Netherlands, 1998; Chapter 12; pp. 679–712.

31. Pedro, D.; De Brito, J.; Evangelista, L. Durability performance of high-performance concrete made with recycled aggregates, fly ash and densified silica fume. *Cem. Concr. Compos.* **2018**, *93*, 63–74. [CrossRef]

32. Siddique, R. Utilization of silica fume in concrete: Review of hardened properties. *Resour. Conserv. Recycl.* **2011**, *55*, 923–932. [CrossRef]

33. Denisiewicz, A.; Kula, K.; Socha, T.; Kwiatkowski, G. Influence of Silica Fume Addition on Selected Properties of Fine-Grained Concrete. *Civ. Environ. Eng. Rep.* **2018**, *28*, 166–176. [CrossRef]

34. Pedro, D.; De Brito, J.; Evangelista, L. Mechanical characterization of high performance concrete prepared with recycled aggregates and silica fume from precast industry. *J. Clean. Prod.* **2017**, *164*, 939–949. [CrossRef]

35. Bhanjaa, S.; Senguptab, B. Modified water-cement ratio law for silica fume concretes. *Cem. Concr Res.* **2003**, *33*, 447–450. [CrossRef]

36. Barbhuiya, S.A.; Gbagbo, J.K.; Russell, M.I.; Basheer, P.A.M. Properties of fly ash concrete modified with hydrated lime and silica fume. *Constr. Build. Mater.* **2009**, *23*, 3233–3239. [CrossRef]

37. Köksal, F.; Altun, F.; Yiğit, I.; Şahin, Y. Combined effect of silica fume and steel fiber on the mechanical properties of high strength concretes. *Constr. Build. Mater.* **2008**, *22*, 1874–1880. [CrossRef]
38. Bhanjaa, S.; Senguptab, B. Influence of silica fume on the tensile strength of concrete. *Cem. Concr. Res.* **2005**, *35*, 743–747. [CrossRef]
39. Langan, B.W.; Weng, K.; Ward, M.A. Effect of silica fume and fly ash on heat of hydration of Portland cement. *Cem. Concr. Res.* **2002**, *32*, 1045–1051. [CrossRef]
40. Bleszynski, R.; Hooton, R.D.; Thomas, M.D.A.; Rogers, C.A. Durability of Ternary Blend Concrete with Silica Fume and Blast-Furnace Slag: Laboratory and Outdoor Exposure Site Studies. *ACI Mater. J.* **2002**, *99*, 499–508.
41. Chalee, W.; Ausapanit, P.; Jaturapitakkul, C. Utilization of fly ash concrete in marine environment for long term design life analysis. *Mater. Des.* **2010**, *31*, 1242–1249. [CrossRef]
42. Elahi, A.; Basheer, P.A.M.; Nanukuttan, S.V.; Khan, Q.U.Z. Mechanical and durability properties of high performance concretes containing supplementary cementitious materials. *Constr. Build. Mater.* **2010**, *24*, 292–299. [CrossRef]
43. Song, H.-W.; Jang, J.-C.; Saraswathy, V.; Byun, K.-J. An estimation of the diffusivity of silica fume concrete. *Build. Environ.* **2007**, *42*, 1358–1367. [CrossRef]
44. Song, H.-W.; Pack, S.-W.; Nam, S.-H.; Jang, J.-C.; Saraswathy, V. Estimation of the permeability of silica fume concrete. *Constr. Build. Mater.* **2010**, *24*, 315–321. [CrossRef]
45. Eurocode, C.E.N. 2: *Design of Concrete Structures–Part 1-1: General Rules and Rules for Buildings: EN 1992-1-1*; European Committee for Standardization: Brussels, Belgium, 2004.
46. A.C.I. Commettee 211. *Standard Practice for Selecting Proportions for Normal, Heavyweight and Mass Concrete*; American Concrete Institute: Detroit, MI, USA, 2002.
47. British DOE (Department of Environment). *Method—Design of Normal Concrete Mixes: BSI*; British DOE (Department of Environment): London, UK, 1988.
48. Puccio, M.; Ferrari, F. L'uso Delle Ceneri Leggere da Carbone nei Conglomerati Cementizi, 1° e 2° Parte. *La Prefabbricazione* **1986**, *5*. (In Italian)
49. Mačiulaitis, R.; Vaičiene, M.; Žurauskiene, R. The effect of concrete composition and aggregates properties on performance of concrete. *J. Civ. Eng. Manag.* **2009**, *15*, 317–324. [CrossRef]
50. EN UNI 7699:2018. *Prova sul Calcestruzzo Indurito—Determinazione Dell'assorbimento di Acqua Alla Pressione Atmosferica*; Ente Nazionale Italiano di Unificazione: Roma, Italy, 2018. (In Italian)
51. Fraay, A.L.A.; Bijen, J.M.; De Haan, Y.M. The reaction of fly ash in concrete a critical examination. *Cem. Concr. Res.* **1989**, *19*, 235–246. [CrossRef]
52. Hanehara, S.; Tomosawa, F.; Kobayakawa, M.; Hwang, K. Effects of water/powder ratio, mixing ratio of fly ash, and curing temperature on pozzolanic reaction of fly ash in cement paste. *Cem. Concr. Res.* **2001**, *31*, 31–39. [CrossRef]
53. Papadakis, V.G. Experimental investigation and theoretical modeling of silica fume activity in concrete. *Cem. Concr. Res.* **1999**, *29*, 79–86. [CrossRef]
54. Bagheri, A.; Zanganeh, H.; Alizadeh, H.; Shakerinia, M.; Marian, M.A.S. Comparing the performance of fine fly ash and silica fume in enhancing the properties of concretes containing fly ash. *Constr. Build. Mater.* **2013**, *47*, 1402–1408. [CrossRef]
55. Elsayed, A.A. Influence of silica fume, fly ash, super pozz and high slag cement on water permeability and strength of concrete. *Jordan J. Civ. Eng.* **2011**, *159*, 1–13.
56. Uzal, B.; Turanlı, L.; Yücel, H.; Göncüoğlu, M.C.; Çulfaz, A. Pozzolanic activity of clinoptilolite: A comparative study with silica fume, fly ash and a non-zeolitic natural pozzolan. *Cem. Concr. Res.* **2010**, *40*, 398–404. [CrossRef]
57. Malhotra, V.M.; Carette, G.G. Silica fume. A pozzolan of new interestfor use in some concretes. *Concr. Constr.* **1982**, *27*, 443–446.

applied sciences

MDPI

Article

Material Characterization for Sustainable Concrete Paving Blocks

Xinyi Wang [1], Chee Seong Chin [1,2,*] and Jun Xia [1,2]

1   Civil Engineering Department, Xi'an Jiaotong-Liverpool University, Suzhou 215123, China;
    Xinyi.wang@xjtlu.edu.cn (X.W.); jun.xia@xjtlu.edu.cn (J.X.)
2   Institute of Sustainable Material and Environment, Xi'an Jiaotong-Liverpool University,
    Suzhou 215123, China
*   Correspondence: Chee.Chin@xjtlu.edu.cn

Received: 13 February 2019; Accepted: 16 March 2019; Published: 21 March 2019

Featured Application: This research provided maximum replacement levels of several recycled materials for concrete paving block, and a reference data for the further investigation about the concrete paving block mixed with multiple types of recycled materials.

Abstract: Recycled aggregates have been widely studied and used in concrete products nowadays. There are still many waste materials that can be used as recycled aggregates other than crushed concrete particles. This paper aims to study the property variations of sustainable concrete paving block incorporating different contents of construction wastes. Five different types of waste materials were used in this project, including: recycled concrete coarse aggregate (RCCA), recycled concrete fine aggregate (RCFA), crushed glass (CG), crumb rubber (CB), and ground granulated blast furnace slag (GGBS). According to the test results of the properties of blocks mixed with different levels of wastes materials, it is concluded that adding both RCCA and RCFA in the block can decrease its strength and increase the water absorption. The suggested replacement levels for RCCA and RCFA are 60% and 20%, respectively. Mixing crushed glass in the concrete paving blocks as a type of coarse aggregates can improve the blocks' strength and decrease the blocks' water absorption. Addition of crumb rubber causes a significant deterioration of blocks' properties except for its slip resistance.

Keywords: recycled concrete aggregate; crumb rubber; crushed glass; compressive strength; tensile splitting strength; water absorption

1. Introduction

In China, the construction industry has developed rapidly since late 1990s [1]. High-rise buildings are established, and the old buildings are demolished at the same time. Due to the irreplaceability of concrete in the construction industry, large quantities of concrete were used in the past two decades, and this situation leads to the generation of large volumes of construction waste [2]. Using recycled demolished concrete as the aggregate in the new construction products can reduce the utilization of the limited nature resource and reduce the environmental burden at the same time.

Relevant studies about using construction wastes in concrete products were investigated by many researchers. Concrete products that incorporates recycled materials can help minimize the CO_2 footprint and potable water usage via the avoidance of using raw materials. Therefore, concrete containing waste material is generally regarded as a type of "sustainable concrete". In this case, studying on the maximum possible amount of recycled materials in concrete, and extending the types of waste materials that can be used in concrete are important in the sustainable concrete research area.

In early 1985, Ravindrarajah and Tam [3] studied the effects of using recycled-concrete aggregate instead of natural aggregate. In recent years, many researchers have studied the impact of using different types of recycled aggregates on the properties of concrete products. Not just crushed concrete, some other common construction wastes, such as crushed clay bricks, are used as recycled-aggregate [4]. In general, using recycled aggregates instead of natural aggregates in concrete will reduce the strength of concrete products [3–5]. As one type of concrete product, blocks are usually composed of 85%–90% fine coarse aggregates and fine aggregates, mainly small limestone particles and sand [6].

Construction wastes not only refer to recycled concrete aggregates, which contain crushed concrete, crushed bricks, and granites, but also contain wasted glass, wood, plastic, and rubber [7]. In order to utilize the crushed glass as aggregate, and produce the concrete incorporating recycled glass, the experiments to incorporate glass waste in concrete have been carried out since the 1960s [8]. Rubber is considered as a type of recycled aggregate that can be used in concrete [9–13]. In general, the research studies suggested that the amount of rubber shall not exceed 25% of the total aggregate volume [11–13]. Furthermore, as a commonly found industrial waste material, ground granulated blast furnace slag (GGBS) is also investigated as a material to replace some of the cement during the concrete products' manufacturing [14].

There are no strict requirements on strength and durability for concrete paving blocks, and such products are easy to manufacture and to cure [15]. Besides compressive strength and tensile splitting strength, the water absorption, abrasion resistance and slip resistance of blocks are also important properties for concrete paving blocks [15,16]. The performance requirements for concrete paving blocks are summarized and shown in Table 1, according to British standard BS EN-1338: 2003 [15] and Chinese standard GB 28635-2012 [16]. The common size of concrete paving blocks is 200 mm × 100 mm × 60 mm. In that case, recycled aggregates can be used to replace either coarse aggregates or fine aggregates. However, the aggregate crushed value is usually lower than natural aggregates. The utilization of recycled aggregate made from crushed concrete in paving blocks has been studied by many researchers. Soutsos et al. [6] investigated the use of recycled demolition aggregate in producing concrete paving blocks. It is recommended that, in order to ensure that the strength of recycled concrete block is as same as those blocks that only contain natural aggregates, the replacement level of recycled masonry-derived coarse aggregates shall not exceed 60%. In addition, using those recycled aggregates as fine aggregates, the replacement level shall not go beyond 40%. Recycled rubber particles were also used as fine aggregates in precast concrete paving blocks. Silva [10] studied the properties of concrete tactile paving blocks made with recycled tyre rubber. The study also incorporated GGBS in the blocks as cement replacements. It was found out that using a certain amount of GGBS to replace cement can increase the compressive strength and tensile splitting strength of concrete blocks [17]. Atici and Ersoy [18] evaluated the effect of replacing the cement with GGBS on interlocking paving blocks, and the replacement levels of GGBS used to replace the cement in their research ranged from 10% to 60%. The test results show that the compressive and tensile splitting strength increase with the curing period increasing, and the best strengths are obtained when the GGBS replacement level is ranging from 20% to 60%.

This research aims to investigate the properties' variations of sustainable concrete paving block incorporating different contents of construction wastes. Up to now, recycled materials, such as crumb rubber and crushed tempered glass, are rarely used in industry to replace the natural fine and coarse aggregate in manufacturing concrete paving blocks. Furthermore, the most current research about recycled concrete paving blocks does not cover the block's abrasion resistance and slip resistance. In this research, five different types of waste materials were used to replace the coarse aggregates, fine aggregates or cement in the concrete mixture with a 28-day target mean strength of 30 MPa. For the concrete paving blocks mixed with waste materials, five material properties such as compressive strength, tensile splitting strength, water absorption, abrasion resistance and slip resistance were investigated. Those results were compared with the related standard requirements to obtain the maximum replacement levels of recycled materials used in this study. Furthermore, the test results provided a reference data for the further investigation about the concrete paving block mixed with multiple recycled materials.

Table 1. Requirements of each property for different national codes.

Property	Recommended Value (BS EN-1338: 2003 [15])	Recommended Value (GB 28635-2012 [16])
Dimension tolerance (block thickness < 100 mm):		
- Length	±2.0 mm	±2.0 mm
- Width	±2.0 mm	±2.0 mm
- Thickness	±3.0 mm	≤2.0 mm
Strength performance:		
- Compressive strength		C30: Average strength ≥ 30.0 MPa Any individual strength ≥ 25.0 MPa
- Tensile splitting strength	Average strength ≥ 3.6 MPa Any individual strength ≥ 2.9 MPa	
Weather resistance:		
- water absorption	≤6.0% for Class 2	≤6.5%
Abrasion resistance:		
- Maximum groove	No performance measured for Class 1 23.0 mm for Class 3 20.0 mm for Class 4	≤32.0 mm
Slip resistance	≥60 BPN	≥60 BPN

"BPN" is a unit for slip resistance, which is short for British Pendulum (Tester) Number.

2. Experimental Program

2.1. Materials

The materials used in this series of experiments are listed as follows: natural coarse aggregate, natural fine aggregate, ordinary Portland cement, RCCA (contain crushed concrete, small amount of crushed stone and crushed bricks), RCFA, CR, CG, and GGBS. Figure 1 shows the recycled wastes used in this experiment. Table 2 shows the information of the construction wastes, which contains the size range, the source of each material listed in the table. The detail properties of GGBS provided by the supplier are shown in Table 3.

(a) (b) (c) (d) (e)

Figure 1. Recycled wastes. (**a**) RCCA; (**b**) RFCA; (**c**) crumb rubber; (**d**) crushed glass; (**e**) GGBS.

Table 2. Size range and sources of construction wastes.

Materials	Size Range	Source
Recycled concrete coarse aggregate	5.0–25.0 mm	Crushed concrete
Recycled concrete fine aggregate	0.1–4.0 mm	Crushed concrete
Crumb rubber	1.0–2.0 mm	Recycled tires
Crushed glass	5.0–25.0 mm	Crushed toughened glass

Table 3. Properties of ground granulated blast furnace slag (GGBS).

Properties	Unit	Results	National Code GB/T-18046 [19]
Density	g/cm^3	2.940	$\geq$2.80
Specific surface area	m^2/kg	455.000	400.00–500.00
Mass loss on ignition	%	0.600	$\leq$3.00
MgO percentage	%	9.910	$\leq$14.00
SO_3 percentage	%	1.825	$\leq$4.00
Cl^{-1} percentage	%	0.012	$\leq$0.06
Water percentage	%	0.010	$\leq$1.00
Activity index (7 days)	%	90.000	$\geq$75.00
Activity index (28 days)	%	98.000	$\geq$95.00

2.2. Mix Proportion

In this project, the original mix proportion was designed according to the "mass method" in Chinese standard JGJ55-2011 [20]. The particular mix proportions for this series of experiments are shown in Table 4. The range of replacement levels for each recycled material was determined as follows:

Recycled concrete coarse aggregate, or RCCA for short, is one of the most common recycled waste used for recycled aggregate concrete (RAC) products. According to the previous articles and experiments' results, five replacement percentages of RCCA were selected in this experiment. The concrete paving blocks which were cast by replacing the natural coarse aggregate with different replacement levels of 20%, 40%, 60%, 80%, and 100% [21] by weight of the total coarse aggregate content.

Recycled concrete fine aggregates (RCFA) were also considered as a type of recycled wastes to be used in this experiment. Unlike the RCCA, previous researchers and experimental results [6] indicated that a high replacement level of RCFA would lead to a sharp decline of concrete product's properties. The limitation of the replacement percentage of RCFA was suggested as 30% [22]. Therefore, the replacement levels of RCFA in the experiment were determined as 10%, 20% and 30%.

Both crumb rubber (CR) and crushed glass (CG) were barely used as aggregates in the concrete paving blocks. Therefore, several trial experiments were undertaken to estimate the ranges of replacement levels. Due to the rubber particles having strong rebound resilience [23], the specimens cannot maintain the original dimensions after being taken out from the brick making machine if the rubber content is too high, considering which the replacement levels of CR were selected as 1%, 2%, and 3%. Similarly to RCCA, crushed glass was also used to replace the coarse aggregates in the blocks. According to the previous articles and trial test results [24], the final replacement levels of CG were decided as 10%, 20%, 30% and 40%. GGBS was used to replace the cement with replacement levels selected as 30%, 50% and 70% according to the previous test results [14,25].

In Table 4, "RL" represents "replacement level", "NCA" represents "natural coarse aggregate", and "NFA" represents "natural fine aggregate". The replacement levels of RCCA in the experiment were determined as 20%, 40%, 60%, 80%, and 100%, labeled as RCCB-X, with "X" indicating the replacement level of RCCA in the blocks. The replacement levels of RCFA in the experiment were determined as 10%, 20%, and 30%, labeled as RCFB-X. The replacement levels of CB were selected as 1%, 2%, and 3%, and labeled as CRB-X. The replacement levels of CG were decided as 10%, 20%, 30%, and 40%, and named as CGB-X. The replacement levels of GGBS were selected as 30%, 50%, and 70% and labeled as GGBS-X.

2.3. Specimens Preparation

The specimens were designed as concrete paving blocks with typical dimensions. Figure 2 demonstrates the process of manufacturing the concrete paving blocks in this project. The whole procedures, including preparing the concrete mixture, casting concrete paving blocks and curing, were in accordance with the actual industrial production standards by using a small scale brick-making machine as shown in Figure 3 with maximum pressing force of 80 kN. Before the casting phase, all the materials used in the experiments were completely dried in the oven for 24 h. At the concrete mixture preparation phase, water was added after the coarse aggregates, cement, and sand mixing in the mixer for two minutes, and then the whole materials were mixed for another three minutes. For one batch of the concrete mixture, the total mass was 60 kg for casting 20 concrete paving blocks. For one batch of blocks (20 blocks), five specimens were used to test the compressive strength, five specimens for tensile splitting strength, and the other three for water absorption, slip resistance tests, and abrasion resistance [15]. After casting, all the specimens were cured in a thermostatic chamber for 28 days until testing. The temperature in the chamber remained at $20 \pm 2\ °C$, and the humidity was controlled at 50% by the temperature and humidity controller.

Table 4. Mix proportions (kg/m^3) of concrete paving blocks.

Mix	RL	Water	Cement	NCA	NFA	RCCA	RCFA	Rubber	Glass	GGBS
RCCB-0	0%	152	380	959	959	0	0	0	0	0
RCCB-20	20%	152	380	767	959	192	0	0	0	0
RCCB-40	40%	152	380	575	959	384	0	0	0	0
RCCB-60	60%	152	380	384	959	575	0	0	0	0
RCCB-80	80%	152	380	192	959	767	0	0	0	0
RCCB-100	100%	152	380	0	959	959	0	0	0	0
RCFB-10	10%	152	380	959	863	0	96	0	0	0
RCFB-20	20%	152	380	959	767	0	192	0	0	0
RCFB-30	30%	152	380	959	671	0	288	0	0	0
CGB-10	10%	152	380	863	959	0	0	0	96	0
CGB-20	20%	152	380	767	959	0	0	0	192	0
CGB-30	30%	152	380	671	959	0	0	0	288	0
CGB-40	40%	152	380	575	959	0	0	0	384	0
CRB-1	1%	152	380	959	949	0	0	10	0	0
CRB-2	2%	152	380	959	940	0	0	19	0	0
CRB-3	3%	152	380	959	930	0	0	29	0	0
GGBS-30	30%	152	266	959	959	0	0	0	0	114
GGBS-50	50%	152	190	959	959	0	0	0	0	190
GGBS-70	70%	152	114	959	959	0	0	0	0	266

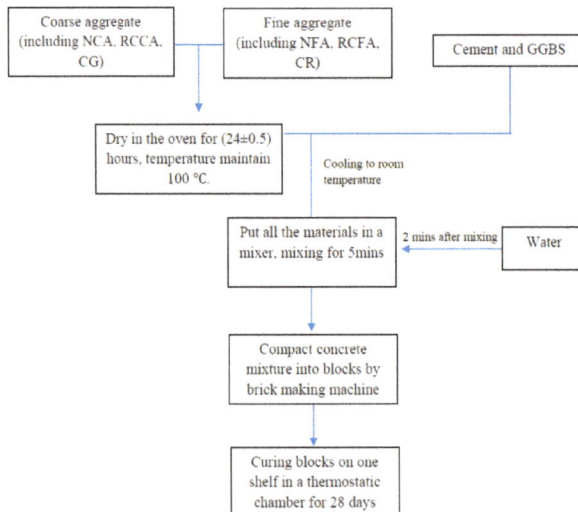

Figure 2. The process of manufacturing the concrete paving blocks.

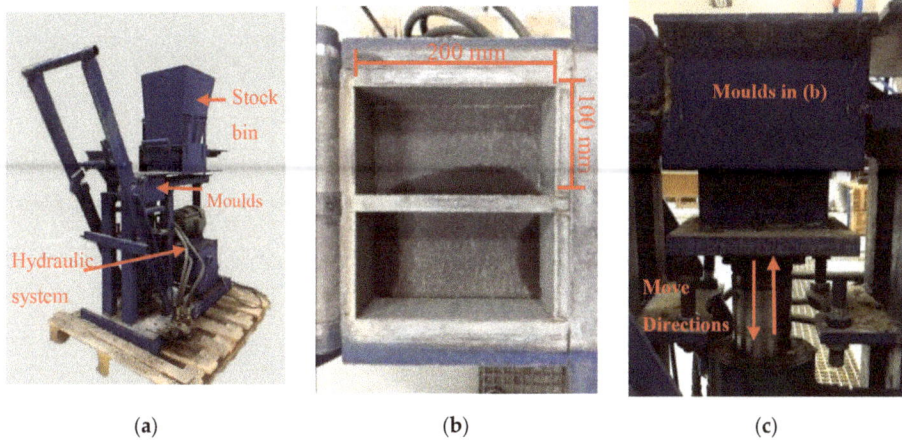

Figure 3. Full view and detailed images of the brick making machine. (**a**) hydraulic brick making machine; (**b**) two moulds for deforming the concrete paving blocks; and (**c**) baseboard for load application.

2.4. Experiments

As the essential component of the concrete mixture, the properties of different types of raw material were tested, including sieving analysis [7] and aggregate water absorption [26].

The blocks' strength tests and coarse aggregates' aggregate crushed value test were conducted on a compression testing machine as shown in Figure 4a. In order to obtain the compressive strength of blocks, the applied load on the pedestal was controlled by force under a constant rate of 3 kN/min, and, for the tensile splitting test, the rate was 0.1 kN/min with reference to BS EN 1338: 2003 [15].

Slip resistance was measured using pendulum friction test equipment as shown in Figure 4b. Specimens were adjusted to ensure the contact surface was horizontal before testing. Water absorption of blocks was also determined in accordance with BS EN 1338: 2003 [15]. Blocks were immersed in water at a temperature of $20 \pm 5\,^{\circ}\mathrm{C}$ for three days before testing.

Abrasion resistance of blocks was tested by a special test machine: CM-B type unglazed brick abrasion testing machine. Figure 4c shows the image of the testing machine. According to BS EN 1338: 2003 [15], the "wide wheel abrasion test" method was selected in this series of experiments.

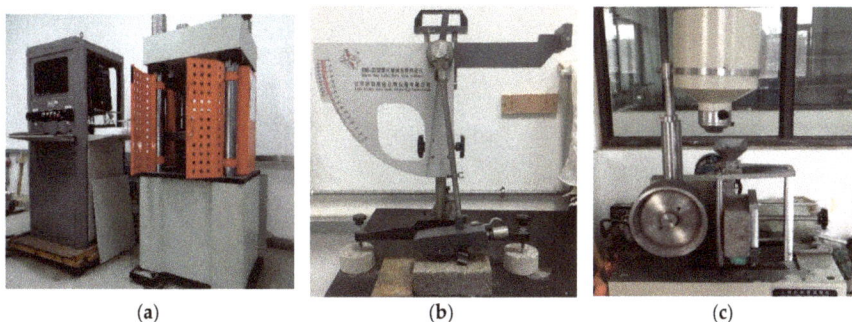

Figure 4. Test equipment in this series of experiments; (**a**) compression testing machine; (**b**) pendulum friction test equipment; (**c**) unglazed brick abrasion testing machine.

3. Results and Discussion

3.1. Aggregates' Properties

3.1.1. Sieve Analysis

The sieving analyses' results of different aggregates used in this experiment are shown in Figures 5 and 6. Based on the comparison of sieve analysis results between NCA and RCCA, it is clear that the RCCA contained a much higher percentage of aggregates whose sizes were less than 5 mm. Most of those aggregates were the small pieces of crushed concrete and dust.

According to Figure 6, comparing the NFA result and RCFA results, it is easy to observe that the RCFA contained higher percentages of particles with a size less than 0.15 mm. Most of those particles were soil particles and dust.

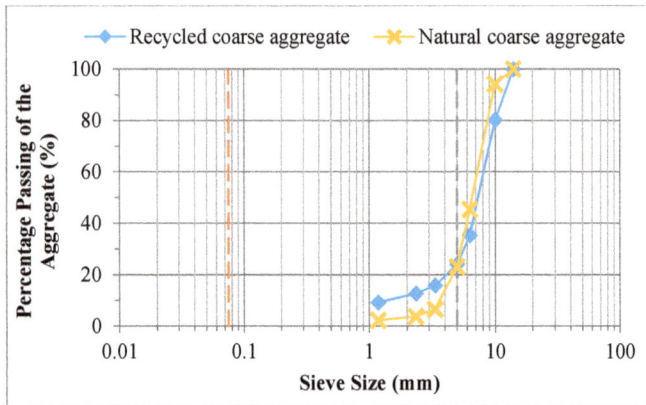

Figure 5. Sieving analysis of NCA and RCCA.

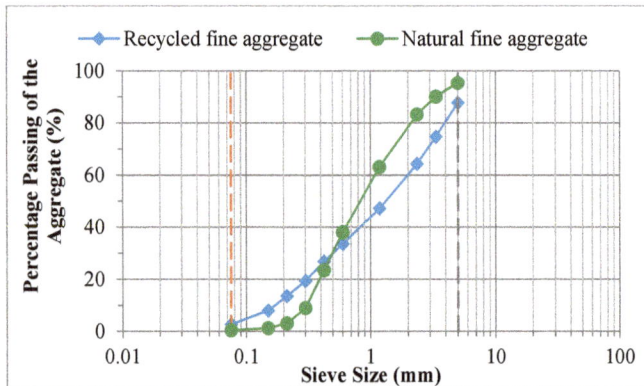

Figure 6. Sieving analysis of NFA and RCFA.

3.1.2. Aggregate Water Absorption

The water absorptions of different coarse aggregates are shown as below in Table 5.

The results confirmed that the RCCA had a higher porosity than the NCA, which led to a stronger capability to absorb water while the crushed glass particles had almost zero water absorption [27].

Table 5. Water absorption of aggregates.

Types of Aggregates	Water Absorption of Aggregates
Natural aggregate	1.53%
Recycled concrete coarse aggregate	6.28%
Crushed glass	0.00%

3.2. Dimensions

All the concrete paving blocks in this series of experiments were checked against the nominal dimension of 200 mm × 100 mm × 60 mm, and the error of each side was less than 2 mm. Figure 7 shows the front and side faces of one concrete paving block.

(a) (b)

Figure 7. Front and side faces of one concrete paving block. (**a**) plan view of a concrete paving block; (**b**) lateral view of a concrete paving block.

3.3. Compressive Strength

Figure 8 shows the relationship between the different replacement levels of materials and the compressive strength of specimens. It is clear that all the replacement materials had an adverse effect on the concrete paving blocks' compressive strength except the crushed glass, which maintains a higher performance in compressive strength with the increase in the percentage of crushed glass. Specimens mixed with RCCA maintained a stable compressive strength until the replacement level reached 40%. The compressive strengths reached 41.07 MPa, 38.04 MPa, and 40.38 MPa when replacement levels were 0%, 20%, and 40%, respectively. After the replacement level exceeding 40%, the compressive strength decreased in an almost linear fashion from 40.38 MPa at 40% to 21.04 MPa. According to BS EN 1338: 2003 [15], the compressive strength of concrete paving blocks shall be equal to or higher than 30 MPa. The specimens with replacement level higher than 60% were unable to meet the specification [16]. According to Figure 8, the specimens mixed with CG had higher compressive strengths than the one mixed with natural coarse aggregate when the replacement levels of crushed glass increased from 10% to 40%. The strength peaked to the highest value, 43.45 MPa, at the replacement level of 20%. Specimens with RCFA and crumb rubber as a replacement of fine aggregates have lower compressive strength compared to specimens without any replacement. When the mass content of crumb rubber in the block increased from 1% to 3%, the compressive strength sharply drops from 31.27 MPa to 21.00 MPa. With regard to RCFA, the reduction of compressive strength was from 32.51 MPa to 23.00 MPa with the replacement level changed from 10% to 30%. The compressive strength of blocks mixed with GGBS kept the same level, nearly 40 MPa when the replacement levels were 0%, 30%, and 50%, respectively. However, the strength sharply decreased to 31.06 MPa when the GGBS content reached 70%.

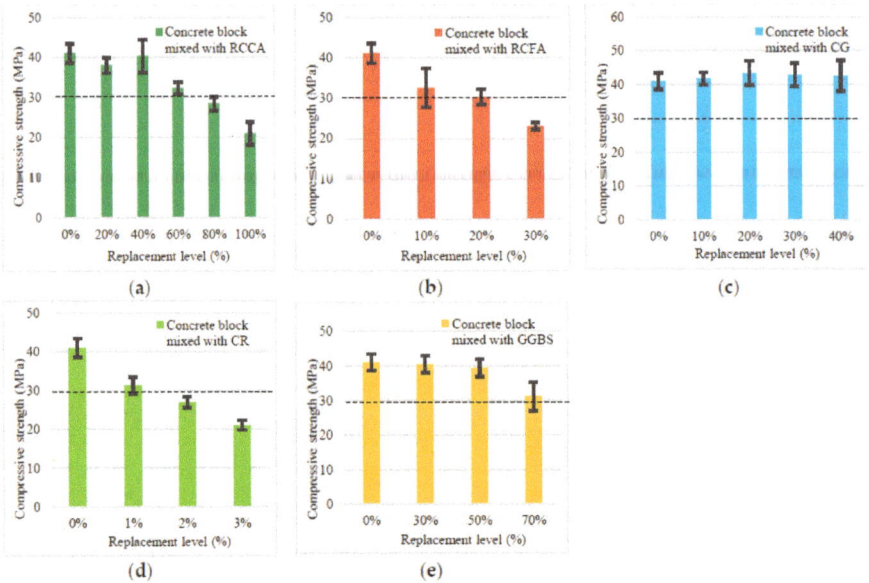

Figure 8. Relationship between the different replacement levels of materials and the compressive strength of specimens.

3.4. Tensile Splitting Strength

Figure 9 demonstrates the relationships between the replacement levels and the tensile splitting strength of concrete paving blocks for various types of waste materials. Specimens mixed with RCCA had an even higher tensile splitting strength of 11.73 MPa at a replacement level of 20%. However, the tensile splitting strength decreased in an almost linear trend from 11.73 MPa to 8.80 MPa when the replacement level increased from 20% to 80%. The lowest splitting tensile strength is 6.37 MPa with 100% RCCA in the specimen. A positive effect on the blocks' tensile splitting strength was observed when crushed glass was used as the replacement for coarse aggregates. The overall tendency was that the tensile splitting strength kept increasing with the replacement level varying from 10% to 40% and reached the highest value of 12.79 MPa at 40%. When the replacement level was 30%, the tensile splitting strength was only 12.45 MPa. However, this value was still higher than the blocks mixed with natural aggregates (11.34 MPa). The tensile splitting strength of the blocks mixed with RCFA declined directly from 10.80 MPa to 7.41 MPa as the replacement level rose from 10% to 30%. Similar to the trend observed for the compressive strength, the tensile splitting strength of specimens containing recycle rubber also remarkably declined with the increase of replacement level. It reached the lowest at 5.53 MPa when the crumb rubber's content is 3%. The tensile splitting strength of blocks mixed with GGBS maintained the same level (12.81 MPa and 13.06 MPa) when the replacement levels rose from 30% to 50%. However, when the percentage of GGBS increased to 70%, the tensile splitting strength was only 9.53 MPa.

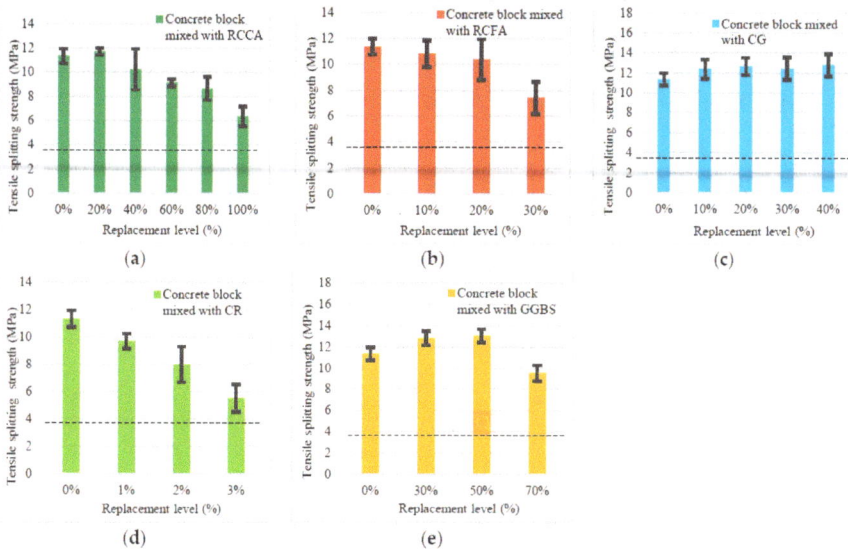

Figure 9. Relationship between the different replacement levels of materials and the tensile splitting strength of specimens.

3.5. Water Absorption

According to Figure 10, it is clear to observe that, for the blocks mixed with RCCA, RCFA, and crumb rubber, the increase in the replacement levels of those materials increases the specimens' water absorption. Compared with samples without any replacement (with 3.89% water absorption), the specimens mixed with RCCA and RCFA maintained similar water absorption of 3.91% and 3.90%, at the replacement levels of 20% and 10%, respectively. The value reached the peak of 7.60% and 4.87%, respectively, while the replacement levels were 100% and 30%. The water absorption of blocks sharply rose with the increase of crumb rubber content in the blocks. With only 3% crumb rubber mixed in the blocks, the water absorption reached 6.02%. Crushed glass is one of the materials that can reduce the water absorption when it was mixed into blocks. The values of water absorption ranged from 3.50%, 3.24%, 3.65% and 3.72% while the replacement levels were 10%, 20%, 30%, and 40%, respectively. All the results were lower than that of the control group at 3.89%. Addition of GGBS also helped reduce the block's water absorption. The water absorptions of samples that contained 30% and 50% percentage of GGBS were 3.58% and 3.77%, respectively. When the replacement level of GGBS peaked to 70%, the water absorption value increased to 4.26%, which is smaller than the standard requirement at 6.0% [15].

3.6. Abrasion Resistance

Figure 11 demonstrated the relationship between materials' replacement levels and the blocks' maximum groove lengths that were obtained from abrasion resistance tests. According to the British standard BS EN 1338: 2003 [15], the higher value of groove length represents a lower performance of block's abrasion resistance. In general, all the samples' test results were higher than 23 mm, which fall short according to the standard requirements for class 3 blocks [15]. The impacts of different replacement levels of materials on the samples' abrasion resistance are shown in Figure 11.

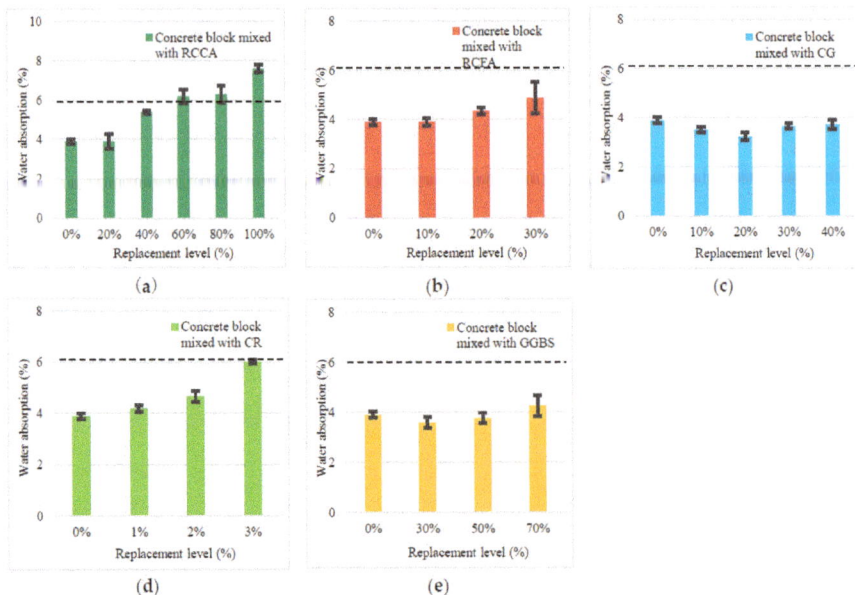

Figure 10. Relationship between the different replacement levels of materials and the water absorption of specimens.

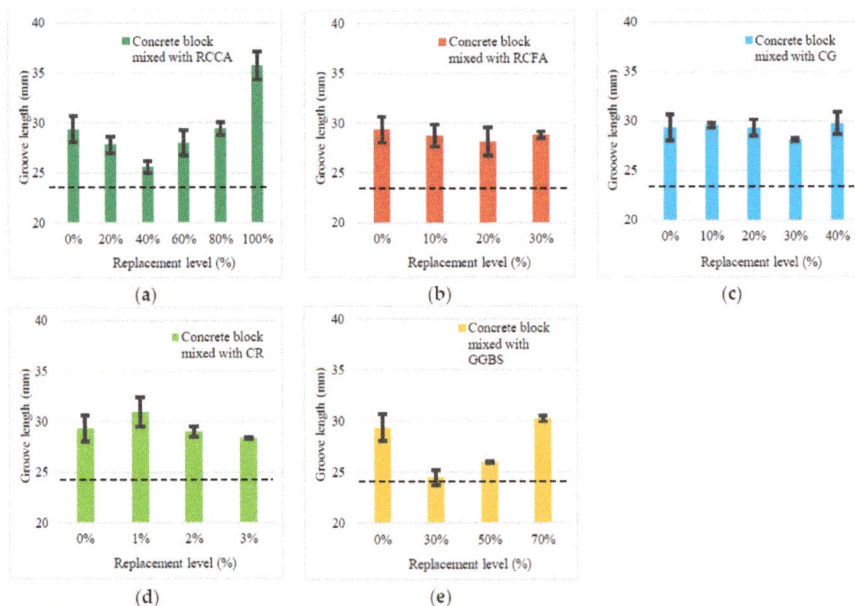

Figure 11. Relationship between the different replacement levels of materials and the maximum groove lengths of specimens.

The measured groove length of the blocks mixed with natural aggregates was 29.34 mm. For the RCCA, when the replacement level changed from 0% to 40%, the length decreased straight from

29.34 mm to 25.58 mm. Then, with increasing content of RCCA, the value of groove length also started to rise and peak at 35.73 mm. A low percentage of RCFA can increase the blocks' abrasion resistance. The value of groove length dropped from 28.74 mm to 28.14 mm when the content of RCFA changed from 10% to 20%. Then, at a 30% replacement level, the length returned to 28.81 mm. To sum up, adding RCFA in the concrete mixture had a relatively positive impact on the blocks' abrasion resistance, and the impact was not significant. For the CG, the values of groove lengths were 29.55 mm, 29.33 mm, 28.12 mm, and 29.77 mm when the replacement levels were 10%, 20%, 30%, and 40%, respectively. The abrasion resistance only had a sharp increase at 30%. The variation of abrasion resistance of the blocks mixed with different percentages of GGBS was significant. At a 30% replacement level, the groove length was only 24.49 mm, which was the lowest in this series of experiments. The value of length rose to 25.98 mm when the replacement level was 50%, and that result was still much lower than 29.34 mm (replacement level: 0%). However, at a 70% replacement level, the result peaked at 30.23 mm. For the blocks mixed with CR, the groove length was very high (30.96 mm) at a 1% replacement level, then declined to 29.02 mm at 2%, and finally reduced to 28.36 mm at 3%.

3.7. Slip Resistance

Figure 12 shows that relationships between the blocks' slip resistance and materials' replacement levels. In general, the blocks mixed with the waste materials had a lower slip resistance than the blocks made with natural aggregates.

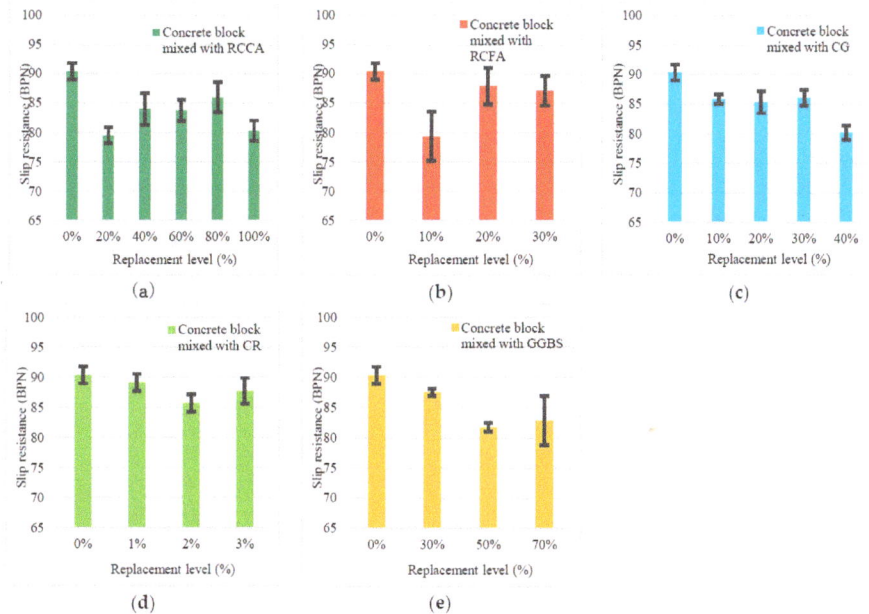

Figure 12. Relationship between the different replacement levels of materials and the slip resistance of specimens.

The slip resistance of blocks mixed with natural aggregates was 90.38 BPN. For the RCCA, the slip resistance sharply decreased to 79.50 BPN when the replacement level was 20%. The slip resistance increased and was maintained at 84.00 BPN, 83.80 BPN, and 86.00 BPN when the replacement levels of RCCA were 40%, 60%, and 80%, respectively. The value decreased to 80.30 BPN with a 100% replacement level of RCCA. The slip resistance of blocks mixed with RCFA also had a significant reduction when the replacement level was 10%. When the replacement levels increased to 20% and

30%, the corresponding slip resistance increased to 87.92 BPN and 87.11 BPN, respectively. For blocks utilizing CG, the values of slip resistance slightly fluctuated from 85.33 BPN to 86.08 BPN when the replacement levels changed from 10% to 30%. When the replacement level increased to 40%, the slip resistance had a sharp decline to 80.20 BPN. The slip resistance of block mixed with CR did not change too much compared with the block mixed with other recycled materials. The values were 89.17 BPN, 85.75 BPN, and 87.75 BPN when the replacement levels increased from 1% to 3%. A reduction of slip resistance can also be observed when the cement replaced by GGBS in the concrete paving blocks. The value decreased from 87.58 BPN to 81.75 BPN when the replacement level of GGBS increased from 30% to 50%. At the 70% replacement level, the slip resistance increased to 82.92 BPN. However, the standard derivation of the average value at this point was higher than the standard derivations of other results in this group.

3.8. Discussion

According to the test results, it is obvious to note that replacing either coarse aggregate or fine aggregate with different waste materials can influence block's properties. Furthermore, different types of waste materials have different effects on various properties. By observing the section of crushed blocks mixed with different recycled coarse aggregates, RCCA and CG, it can be noticed that the coarse aggregate particles maintained their shapes after blocks crushing, which proves that the failure was due to the failure of mortar. By comparing the aggregate properties among NCA, RCCA, and CG, the biggest physical difference between those aggregates is the aggregate water absorption. Since the mix proportion was designed with "Mass method," the aggregates should be totally dry before casting. In that case, the RCCA particles would absorb more free water, and the cement hydration reactions would be more inadequate compared with the reactions in the block mixed with NCA and CG. Hence, increasing the replacement level of RCCA would lead to the strengths decreasing of concrete paving block's. The block's water absorption increases for the same reason. Furthermore, due to fact that the crushed glass particles have a smaller water absorption value than the natural coarse aggregates, replacing the NCA with CG would lead to the strengths increasing and the water absorption decrease of concrete paving blocks.

The same phenomenon was also observed for the blocks mixed with NFA and RCFA. According to the sieve analysis result, the RCFA contained a larger amount of dust than NFA. This dust can absorb free water and reduce the cement hydration reactions in block, and then decrease the block's compressive strength and tensile splitting strength, as well as lead to the increase of the water absorption.

Compared with other recycled aggregates, crumb rubber particles have an extremely small elastic modulus. In that case, when the concrete mixtures were cast into blocks under pressure, the crumb rubber particles deformed and such deformation is recovered after taking the blocks out of moulds. This recovering deformation causes volume expansion (especially on the surfaces of one block), which contributes to the decrease of block's compressive strength and tensile splitting strength, as well as the increase of the water absorption. In addition, there is a potential alkali-silica reaction for blocks mixed with glass. As presented, the block mixed with 40% crushed glass has higher compressive strength and tensile splitting strength than the block mixed without crushed glass at 28 days. An addition batch of blocks mixed with 40% crushed glass are stored in the laboratory for the purpose of long-term investigation.

4. Conclusions

According to the experiment results in the properties of concrete paving blocks mixed with different types of recycled wastes, the following conclusions can be drawn:

- Using RCCA as coarse aggregate for concrete paving blocks can lead to a significant decrease of both compressive strength and tensile splitting strength, and also increase the water absorption of the concrete paving blocks. The abrasion resistance test results indicate that more than 80% RCCA replacing the natural coarse aggregate can have a negative impact on the abrasion resistance. In order to meet the standard requirements, the replacement level of RCCA shall be less than 60%.

- As a recycled waste used to replace the fine aggregate in the blocks, RCFA has an adverse effect on the compressive strength and tensile splitting strength of concrete blocks. Plus, a higher percentage of RCFA also leads to higher water absorption of concrete paving blocks. Nevertheless, adding RCFA in the blocks do not have a negative effect on the block's abrasion resistance. To sum up, in order to meet the standard requirements, the replacement level of RCFA shall be less than 20%.

- Replacing the coarse aggregate with crushed glass (CG) can enhance the strength of concrete paving blocks. With the replacement level ranging from 10% to 40%, all concrete paving blocks mixed with crushed glass had higher strength and lower water absorption than the blocks mixed with natural aggregates. When the replacement level of crushed glass is at 20%, the concrete paving blocks have the highest compressive strength and lowest water absorption. The tensile splitting strength of concrete blocks mixed with crushed glass reached the maximum value when the replacement level was 40%. Furthermore, the highest performance of the block's abrasion resistance was observed when the replacement level of CG was 30%.

- With an increasing amount of crumb rubber in the concrete paving blocks, there is not only a sharp decrease of compressive strength and tensile splitting strength, but also a rapid increase of the water absorption. Adding crumb rubber can also lead a slightly reduction of the block's slip resistance. Although the value of groove length keeps decreasing while the content of CR increasing, the average value of groove length is still very high. To sum up, the crumb rubber is not recommended to replace the fine aggregates of concrete paving blocks.

- Replacing cement with a certain content of GGBS is proved to have a positive effect on the block's properties. Considering the variation of compressive strength, tensile splitting strength and water absorption, 30% and 50% replacement levels of GGBS have an almost equal positive impact on those properties. The blocks with 30% GGBS that replaced cement have a higher performance of abrasion resistance than the one with 50% GGBS. Furthermore, using GGBS to replace 70% cement in the concrete mixture turns out to be harmful to the block's five properties that were tested in this experiment. In general, the replacement level of GGBS shall be equal to or less than 50%.

- This series of experiments and test results provide a large amount of reference data for the further investigation about the concrete paving block mixed with multiple recycled materials. In order to ensure the block mixed with multiple recycled materials could meet the standard requirements, the replacement levels of each recycled materials can be considered as: RCCA 0% to 60%; RCFA: 0% to 20%; CG: 0% to 40%; GGBS: 0% to 50%.

Author Contributions: Conceptualization, C.S.C.; methodology, X.W.; investigation, X.W.; writing—original draft preparation, X.W.; writing—review and editing, C.S.C., J.X.; supervision, C.S.C., J.X.

Funding: This research was funded by Xi'an Jiaotong-Liverpool University.

Acknowledgments: The authors gratefully acknowledge the technical support received from Civil Engineering Department of Xi'an Jiaotong-Liverpool University.

Conflicts of Interest: The authors declare no conflict of interest.

References

1. Xing, Z.; Liu, Y.; Cai, N. Research and application benefit for recycled concrete. *Low Temp. Archit. Technol.* **2011**, *1*, 6.
2. BS ISO 18650-2: 2014. *Publication Building Construction Machinery and Equipment—Concrete Mixers Part 2: Procedure for Examination of Mixing Efficiency*; British Standards Institution: Geneva, Switzerland, 2014.
3. Ravindrarajah, R.S.; Tam, C.T. Properties of concrete made with crushed concrete as coarse aggregate. *Mag. Concr. Res.* **1985**, *37*, 29–38. [CrossRef]

4. Yang, J.; Du, Q.; Bao, Y. Concrete with recycled concrete aggregate and crushed clay bricks. *Constr. Build. Mater.* **2011**, *25*, 1935–1945. [CrossRef]

5. Hansen, T.C.; Narud, H. Strength of recycled concrete made from crushed concrete coarse aggregate. *Concr. Int.* **1983**, *5*, 79–83.

6. Soutsos, M.N.; Tang, K.; Millard, S.G. Use of recycled demolition aggregate in precast products, phase II: Concrete paving blocks. *Constr. Build. Mater.* **2011**, *25*, 3131–3143. [CrossRef]

7. BSI. BS EN 933-11 *Tests for Geometrical Properties of Aggregates*; Austrian Standards Institution: Vienna, Austria, 2012.

8. Shi, C.; Zheng, K. A review on the use of waste glasses in the production of cement and concrete. *Resour. Conserv. Recycl.* **2007**, *52*, 234–247. [CrossRef]

9. Eldin, N.N.; Senouci, A.B. Rubber-tire particles as concrete aggregate. *J. Mater. Civ. Eng.* **1993**, *5*, 478–496. [CrossRef]

10. Da Silva, F.M.; Barbosa, L.A.G.; Lintz, R.C.C.; Jacintho, A.E.P.G.A. Investigation on the properties of concrete tactile paving blocks made with recycled tire rubber. *Constr. Build. Mater.* **2015**, *91*, 71–79. [CrossRef]

11. Issa, C.A.; Salem, G. Utilization of recycled crumb rubber as fine aggregates in concrete mix design. *Constr. Build. Mater.* **2013**, *42*, 48–52. [CrossRef]

12. Bonicelli, A.; Fuentes, L.G.; Ibrahim, K.D.B. *Laboratory Investigation on the Effects of Natural Fine Aggregates and Recycled Waste Tire Rubber in Pervious Concrete to Develop More Sustainable Pavement Materials*; Materials Science & Engineering Conference Series; IOP Publishing: Bristol, UK, 2017.

13. Khatib, Z.K.; Bayomy, F.M. Rubberized Portland cement concrete. *J. Mater. Civ. Eng.* **1999**, *11*, 206–213. [CrossRef]

14. Limbachiya, V.; Ganjian, E.; Claisse, P. Strength, durability and leaching properties of concrete paving blocks incorporating GGBS and SF. *Constr. Build. Mater.* **2016**, *113*, 273–279. [CrossRef]

15. BSI. *BS EN 1338: 2003 Concrete Paving Blocks–Requirements and Test Methods*; British Standards Institution: London, UK, 2003.

16. SAC. *GB 28635-2012 Precast Concrete Paving Units*; Administration of Quality Supervision, Inspection and Quarantine of China: Beijing, China, 2012.

17. Ganjian, E.; Jalull, G.; Sadeghi-Pouya, H. Using waste materials and by-products to produce concrete paving blocks. *Constr. Build. Mater.* **2015**, *77*, 270–275. [CrossRef]

18. Atici, U.; Ersoy, A. Evaluation of destruction specific energy of fly ash and slag admixed concrete interlocking paving blocks (CIPB). *Constr. Build. Mater.* **2008**, *22*, 1507–1514. [CrossRef]

19. SAC. *GB/T 18046-2008. Ground Granulated Blast Furnace Slag Used for Cement and Concrete*; National Cement Standardization Technical Committee(SAC/TC 184): Beijing, China, 2008.

20. Ding, W.; Leng, F.G. *Specification for Mix Proportion Design of Ordinary Concrete (JGJ55-2011)*; China Building Industry Press: Beijing, China, 2011.

21. Corinaldesi, V.; Moriconi, G. Influence of mineral additions on the performance of 100% recycled aggregate concrete. *Constr. Build. Mater.* **2009**, *23*, 2869–2876. [CrossRef]

22. Soutsos, M.N.; Tang, K.; Millard, S.G. Concrete building blocks made with recycled demolition aggregate. *Constr. Build. Mater.* **2011**, *25*, 726–735. [CrossRef]

23. Treloar, L.R.G. *The Physics of Rubber Elasticity*; Oxford University Press: Oxford, MS, USA, 1975.

24. Jin, W.; Meyer, C.; Baxter, S. "Glascrete"-Concrete with Glass Aggregate. *ACI Mater. J.* **2000**, *97*, 208–213.

25. Divsholi, B.S.; Lim, T.Y.D.; Teng, S. Durability Properties and Microstructure of Ground Granulated Blast Furnace Slag Cement Concrete. *Int. J. Concr. Struct. Mater.* **2014**, *8*, 157–164. [CrossRef]

26. BS EN 1097-6. *Tests for Mechanical and Physical Properties of Aggregates. Determination of Particle Density and Water Absorption*; British Standards Institution: London, UK, 2013.

27. Khatib, J. *Sustainability of Construction Materials*; Woodhead Publishing: Sawston, UK; Cambridge, UK, 2016.

![applied sciences logo] *applied sciences*

MDPI

Article

Crushing Performance of Ultra-Lightweight Foam Concrete with Fine Particle Inclusions

Yu Song * and David Lange *

Department of Civil and Environmental Engineering, University of Illinois at Urbana-Champaign, Urbana, IL 61801, USA
* Correspondence: yusong3@illinois.edu (Y.S.); dlange@illinois.edu (D.L.)

Received: 21 January 2019; Accepted: 18 February 2019; Published: 1 March 2019

Featured Application: This study investigates the influence of particle inclusion on the elastic modulus, crushing strength, and 3D foam structure of low-density foam concrete mixtures. The results provide insights into using high-volume fine aggregates or other waste fine particles in low-density foam concrete for its functional applications.

Abstract: Foam concrete is a low-density controlled strength material that can potentially be used for accommodating different types of particles—recycled fine aggregate being an example. The paste matrix of this material has a cellular microstructure, and bulk performance is readily affected by the inclusion of fines. To study the effect of inclusion of fines on mechanical performance and foam structure of foam concrete, a group of 0.55 g/cm^3 foam–sand composite mixtures with high-volume fly ash replacement are investigated. The elastic modulus is measured by a vibrational frequency test. The crushing mechanics are determined by the load-displacement response from a penetration test. The effect of particle inclusion on the foam concrete microstructure is characterized using micro computed tomography. The results indicate that use of fine-graded sand particles at a small dosage simultaneously reduces cement content and enhances the crushing performance, however poor material performance is observed for a high sand content. The cellular structure of the foam–sand composite, and thus its mechanical behavior, can be substantially diminished by larger sand particles, especially when the particle size is larger than the voids in foam.

Keywords: foam concrete; cellular concrete; ceramic foam; modulus; crushing; energy absorbing; CT; foam structure; foam stability

---

1. Introduction

Concrete sustainability has emerged as a key theme across modern concrete studies because of the large environmental impact of cement production [1] and natural aggregate exploitation [2]. All three Rs of environmental protection—reduce, reuse, recycle—can be applied in some ways to concrete infrastructure. Recycling concrete from demolished building structures and pavements as aggregates for new constructions represents a strong potential for concrete recycling. While the use of recycled coarse aggregate is well established [3–6], the fine fractions in recycled concrete aggregates pose greater challenges for material recycling in new construction [7–9]. Due to the high heterogeneity of recycled fine aggregate, there are a wide variety of concrete durability concerns, such as water permeability, long-term shrinkage, and alkali–silica reaction, which makes it difficult to recycle the fines in the production of normal concrete for structural purposes.

However, good potential has been seen for using recycled fine aggregate for making controlled low strength materials (CLSM). New ideas about using fine inclusions in foam concrete, which is classified as a low-density CLSM [10,11], open new possibilities to accommodate high-volume recycled fines for its functional material applications, such as excavatable landfill, thermal and acoustic isolators, and fire resistant and impact absorbing materials [12–15].

Foam concrete is a cellularized cementitious solid made from blending aqueous foam with fresh cement paste. Depending on the volume ratio of the foam, the bulk density of foam concrete typically varies from 0.3 to 1.6 g/cm^3 [16]. Mixtures with density higher than 0.8 g/cm^3 may be applicable for structural applications [12,17,18]. At lower densities, the strength of the foam concrete is greatly reduced, and the material exhibits other beneficial properties, being ultra-lightweight, permeable, and crushable. In this study, attention is focused on the functional potential for the ultra-lightweight foam concrete.

Although the use of fine aggregates is seen in some foam concrete studies, most of the investigated foam mixtures are denser than 0.8 g/cm^3 [12,18,19] and only a few studies focused on the influence of adding sand into low-density foamed paste [2,20,21]. Based on the existing literature, the strength of low-density foam concrete mixtures is sensitive to the inclusion of particles and can be affected by the filler type. It has been reported that foam concrete with fine aggregates shows a greater tendency of shrinkage, and ensuring good curing is critical [12,22]. It is generally understood that greater sand or particle content reduces the foam stability, which usually causes strength reduction in foam concrete [18,23–25]. For most cases, a small amount of inclusion of finer particles results in a higher strength of foam concrete. Fly ash is also considered as a filler in some studies [20,26–28] and a favorable effect has been seen with respects to workability and strength; however, the physical and chemical nature of fly ash can be quite different to typical particle fillers. Some discussions on using other recycled fine aggregates in foam concrete are also found in literature [2,29].

As a cellular solid, the performance of foam concrete is largely governed by its foam structure [17,23,30–32]. Before implementing recycled fine aggregates in foam concrete, it is pertinent to understand how the cellular microstructure is altered with the inclusion of particles, and how this alters mechanical performance. Given the promise of using recycled fines in ultra-lightweight foam concrete for low strength applications, low-density foam mixture design and processing deserve greater attention by researchers. More knowledge about efficiently incorporating particles into the foam structure is needed for advanced use of recycled fines in foam concrete production.

This study focuses on investigating the effect of adding fine particles to the foam concrete crushing behavior and void structure, where the mixture is considered as a foam–particle composite. Fine river sand was used as a standard filler for studying the influence of particle size and dosage on foam concrete mixtures. A high-volume fly ash replacing 40% cement was designed to further advance the engineering sustainability and to improve the mixture uniformity. All the foam–sand mixtures were designed at a 0.55 g/cm^3 bulk density, with three particle sizes (300, 600, and 850 µm) and at three sand-cementitious weight ratios (0.15, 0.3, and 0.5). The elastic modulus of the samples was measured using a vibrational frequency test. For investigating the crushing behavior of the composite material, a penetration test was conducted. In order to clarify the influence of sand inclusion on the foam structure, micro computed tomography (micro-CT) was further used to characterize the 3D morphology of a subgroup of the samples.

2. Mixture Design and Sample Preparation

A controlled sample without sand as well as nine mixtures of foam–sand composite samples of different sand ratio (mass ratio between sand and cementitious materials) and sand size were prepared for this study. The detailed information of the mixture design is given in Table 1, where the mixture naming convention is "sand ratio_sand size". The materials included cement (Essroc Italcementi Group, Type I/II), fly ash (Boral, Class C), water, superplasticizer (Sika, ViscoCrete 2100), an accelerator (Grace, Calcium Chloride 37), a foaming agent (BASF MasterCell 30), and river sand (specific gravity = 2.63). The "300 µm" sand had a size range between 300 to 600 µm, the "600 µm" sand ranged from 600 to 850 µm, and the "850 µm" sand ranged from 850 to 1180 µm.

Table 1. Mixture design and density of the samples.

Mixture	Sand Ratio	Sand Size [μm]	ϱ_{target} [g/cm^3]	$\varrho_{foamed\ paste}$ [g/cm^3]	$\varrho_{measured}$ [g/cm^3]
Control	0	NA	0.55	0.550	0.552
0.15_300 μm	0.15	300	0.55	0.509	0.575
0.15_600 μm	0.15	600	0.55	0.509	0.573
0.15_850 μm	0.15	850	0.55	0.509	0.547
0.3_300 μm	0.3	300	0.55	0.474	0.583
0.3_600 μm	0.3	600	0.55	0.474	0.58
0.3_850 μm	0.3	850	0.55	0.474	0.562
0.5_300 μm	0.5	300	0.55	0.433	0.549
0.5_600 μm	0.5	600	0.55	0.433	0.562
0.5_850 μm	0.5	850	0.55	0.433	0.538

The foam concrete mixture design generally followed ASTM C796 [33], and the entire mixture preparation was divided into four steps: paste mixing, preparation of aqueous foam, foamed paste mixing, and sand inclusion. For preparing the paste, the mixing protocol followed the standard specified in ASTM C305 [34]. As part of the water was used for foaming, the water available for paste mixing was different across the samples. As such, the superplasticizer was used to maintain the same flowability of the fresh paste for the subsequent mixing. The accelerator was dosed as 8% by weight of cement for better stabilizing of the fresh foamed mixture [16]. A standard aqueous foam was prepared separately by blending water with the foaming agent at a 15:1 ratio. After that, the paste and foam were blended together using a high-speed mixer. During this step, different amounts of aqueous foam were progressively added to the mixture for achieving the target bulk density in different samples. The sieved sand particles were then gently poured into the foamed paste, after which the mixing was extended for 30 seconds to ensure an even final mixture.

All the samples were designed to have the same bulk density of 0.55 g/cm^3. The paste in all samples was designed with 0.45 water–cementitious ratio (w/cm) and 40% fly ash replacement (by weight of cement). All liquid phase from the aqueous foam and chemical agents was considered for cement hydration when calculating the w/cm ratio. Considering the mixing process involves only using a portion of water for paste mixing, the cement created in the first step has a low w/cm ratio, down to 0.35. From previous experimental tests, mixing of this lower w/cm paste with the foam could not be blended into a homogenous mixture of foam concrete, even when dosed with a high amount superplasticizer. Therefore, a major portion of the cement was replaced with fly ash to obtain a more flowable paste mixture.

Based on previous experimental work, it was observed the foam had void sizes ranging between 400 to 800 μm. Therefore, the three sand sizes used in Table 1 were selected to differentiate the three cases that particles are smaller, within range, and larger than the voids. Since the density of the river sand is higher than the paste, the effective bulk density of the pure foamed paste decreases when the sand ratio increases (see Table 1).

For each mixture, three prismatic specimens for the vibrational frequency test (50.8 × 50.8 × 203.2 mm) and three cylindrical samples for the penetration test (101.6 mm diameter by 203.2 mm height) were cast. The cylinder specimens were seal-cured in the mold until testing, while the prism specimens were demolded at 7 days and further cured under the sealed condition. The bulk density of each mixture was also measured at this age. Any sample that showed a deviation from the target bulk density larger than 0.05 g/cm^3 was rejected, and the mixture was recast. The bulk density of the accepted mixtures for testing is shown in Table 1.

3. Testing methods

3.1. Measurement of Elastic Modulus Using a Vibrational Frequency Test

The influence of sand inclusion on foam concrete elastic modulus was evaluated using the vibrational frequency test as specified by ASTM C215. Feasibility of using this test for measuring the elastic modulus of foamed cement paste has been confirmed in several studies [15,35]. As the measurement is non-destructive, the three prism specimens of each mixture were repeatedly tested at the age of 7, 14, and 21 days, where the modulus measurement were repeated three times on each specimen during testing. Due to the use of an accelerating admixture, the subsequent increment of the sample modulus was found to be marginal after 21 days and thus negligible. The testing instrument of the elastic modulus measurement included a PCB-352C04 accelerometer, a PCB-482 signal conditioner, and a NI-9171 DAQ for data acquisition. This testing setup yielded a 4 Hz resolution in the frequency spectrum. As suggested by the previous test [35], the sample elastic modulus was interpreted based on the transverse vibration frequency of the prism specimens.

3.2. Measurement of Crushing Behavior Using a Penetration Test

The crushing behavior of the samples was characterized using a penetration test at 7, 14, and 21 days, where one cylindrical specimen was tested per mixture at each age. This test method has been previously used for studying the crushing behavior of foam concrete [15]. For each test, a 47 mm diameter steel rod was vertically displaced into the cylinder specimen at a 75 mm/min rate from the top center, where the loading was controlled using an Instron-4502 testing frame. Due to the 10 kN loading capacity of the load cell used, the maximum indentation stress was 5.7 MPa.

3.3. Micro-CT

The microstructure of the materials used in this study was investigated using an Xradia MicroCT (MicroXCT-200). Due to the low sample density, a low-level X-ray energy setting was used for data acquisition. The micro-CT specimens were extracted from the controlled and composite samples of a 0.5 sand ratio. Each specimen was a small pellet of 5 mm diameter, excised from the undamaged section of the parent cylindrical sample after the 21-day penetration test. For consistency, the specimens were extracted roughly at the same location from the mid-depth of the cylinder. The CT scan captured a volume of 5000-µm diameter cross-section by 5000-µm height, with a voxel size of 5 µm. After testing, the micro-CT scans were processed using image analysis software, ImageJ. After segmenting different phases in the mixtures (air void, hydrated cement paste, and sand), the 3D solid foam structure was reconstructed for each sample, where the model represented a smaller volume of $500 \times 500 \times 500$ voxel^3 ($2500 \times 2500 \times 2500$ µm^3) at the center of the specimens.

4. Results and discussion

4.1. Elastic Modulus of the Samples

The elastic modulus data obtained from the using the vibrational frequency test are summarized in Table 2 and compared in Figure 1. Each mean value in Table 2 represents the result averaged from nine measurements (3 specimens by 3 repeats). Continuous increase of the elastic modulus was seen for all samples from 7 to 21 days, where the average increase was 7.8% at 14 days and only 2.6% at 21 days. Because of the high accelerator dosage used for mixing, the modulus change of the samples from 14 to 21 days was marginal, indicating that cement hydration in the mixtures was almost complete at 21 days. In terms of standard deviation of the measurements, the samples with a larger sand ratio were less consistent than the others at 7 days but more consistent at the later ages.

Table 2. Mean elastic modulus and standard deviation of the samples.

Sample	Mean of Elastic Modulus (and Standard Deviation) [MPa]					
	7 Days		14 Days		21 Days	
Control	1176	(31.1)	1335	(5.5)	1365	(18.5)
0.15_300 µm	1223	(1.8)	1361	(26.0)	1363	(17.3)
0.15_600 µm	1245	(35.8)	1299	(14.7)	1332	(23.4)
0.15_850 µm	997	(30.2)	1071	(4.4)	1126	(7.2)
0.3_300 µm	1081	(24.8)	1166	(23.1)	1201	(23.5)
0.3_600 µm	1076	(20.6)	1161	(25.8)	1202	(34.6)
0.3_850 µm	975	(29.4)	1045	(27.1)	1072	(28.8)
0.5_300 µm	888	(20.1)	922	(24.3)	931	(15.8)
0.5_600 µm	762	(42.9)	821	(43.4)	855	(33.8)
0.5_850 µm	662	(49.1)	709	(8.9)	719	(16.1)

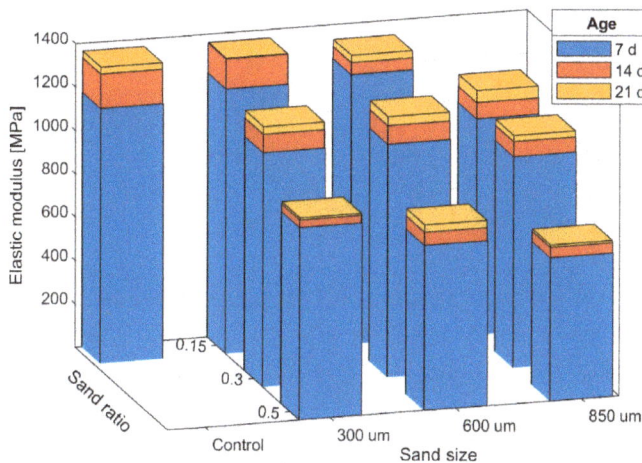

Figure 1. Comparison of the elastic modulus results for all samples.

Figure 1 indicates that elastic modulus decreases for both sand size and sand ratio. The lowest modulus values at all three testing ages are observed with Sample 0.5_850 µm, which had the largest sand ratio and sand size. On the other hand, the modulus of Sample 0.15_300 µm is quite close to the control sample. The influence of sand ratio from 0.15 to 0.5 is more significant than increasing the sand size from 300 to 850 µm in this case. Although the elastic modulus of the fine aggregate is greater than cement paste, the overall trend indicates that the elastic modulus of the composite samples are much reduced at higher sand ratios. However, similar observations have been reported in the literature [12,19,27]. As all samples were designed for the same target bulk density, the effective density of foamed paste is lower in samples with higher sand ratios, as shown in Table 1, meaning that less cement is available to support the cellular structure in foam concrete. This should be the main influence from sand ratio. Interestingly, the 7-day modulus values of Samples 0.15_300 µm and 0.15_600 µm are slightly higher than Control, and continue to stay close to Control at the later ages. This relationship, however, is not observed for Sample 0.15_850 µm. Thus, a small amount of sand inclusion has a minimal influence, or even a slight improvement, on the elastic modulus of the foam–sand composite, but this rule does not hold true when the particle size becomes larger.

The influence of sand size on elastic modulus is not strong in the samples with the 300- and 600-μm sand of 0.15 and 0.3 sand ratios, but an increase in particle size to 850 μm resulted in a substantial reduction for all sand ratios. Seemingly, there is a size threshold between 600 and 850 μm beyond which leads to a marked modulus reduction. It can be noticed that by increasing the sand size, the cellular matrix of the foamed paste loses the ability to support the sand particles within its cellular matrix. Although all three 850 μm samples exhibits lightly lower densities than their counterparts, this difference plays a minor role in the modulus reduction. The lower bulk densities of the 850 μm samples imply that these foam mixtures are less stable during the fresh stage. This issue will be further discussed with the micro-CT evidence later in this paper.

4.2. Loading Behavior

The loading behavior of all the samples was investigated using the penetration test at the same testing ages as the vibrational frequency test. A demonstration of the loading responses of the control sample at the three ages is given in Figure 2, where the indentation stress was calculated by dividing the resistant force by the cross-section area of the cylindrical indenter. A preliminary test has confirmed that the friction between the indenter and sample is so small that such effect is negligible. All the indentation curves of the samples were curtailed on (1) the x axis at 160 mm as further penetration results in boundary effects and (2) the y axis at 5.7 MPa indentation stress as the loading capacity of the testing frame is reached.

The trend of the indentation curve of the samples is consistent with the observation from several other studies on foam concrete [15,36] and other cellular solids [37,38]. As the test initiated, the material behaved in a linear-elastic manner with a steep increase of the indentation stress. After reaching a yielding point shortly at a few millimeters, the increment became substantially milder at a roughly constant rate. Due to the geometry of the indenter, the crushing plateau reported in our previous study was not captured in this test [15]; the material crushing strength, however, can still be quantitatively compared based on the indentation stress at the same indentation depth. In Figure 2, localized material cracking affected the 14- and 21-day loading curves of the controlled sample at around 30 mm, after which the curves returned to their normal trend lines. These events are considered as anomalies that can be generally neglected, and they were rarely observed in the other samples, as cracks could not propagate effectively in the weaker microstructure.

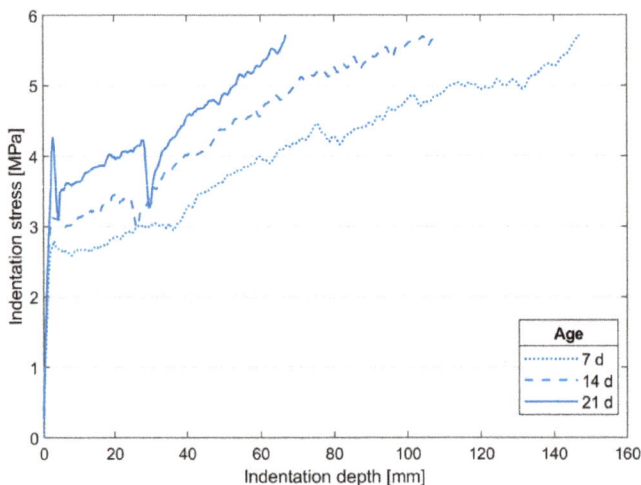

Figure 2. General loading response from the penetration test at 7, 14, and 21 days for the control sample.

The 7-, 14-, and 21-day loading behavior across different samples are compared in Figure 3. In general, the influence from sand ratio and sand size to the crushing strength follows the same trend as observed from the elastic modulus measurement. Comparing the results among the three ages, the strength gain is smaller for the samples with a larger sand ratio. This phenomenon should be mainly caused by the reduced cement usage in the high sand ratio samples. The slope of the load curves after the initial yielding is also relevant to inclusion of sand particles. There is a noticeable rise in the curve slopes of the controlled sample and those of a 0.15 sand ratio. As more sand was added into the mixture, this slope change becomes less pronounce. This observation should be associated with the reduced bulk modulus for samples of higher sand ratio, as the stress build-up underneath the indenter develops at a slower rate in weaker samples. This point is generally supported by data from the elastic modulus measurement.

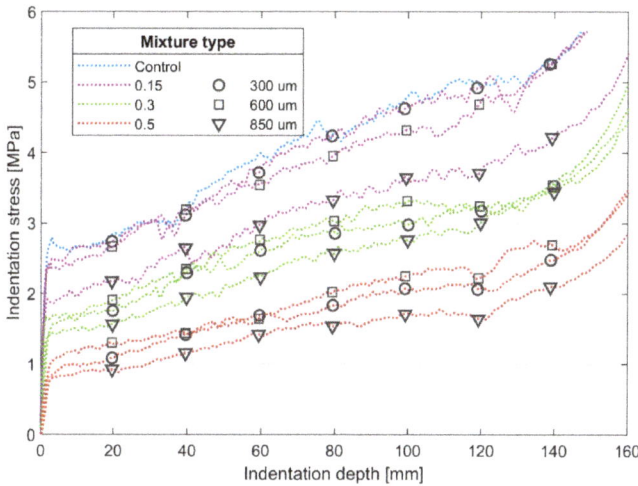

(a)

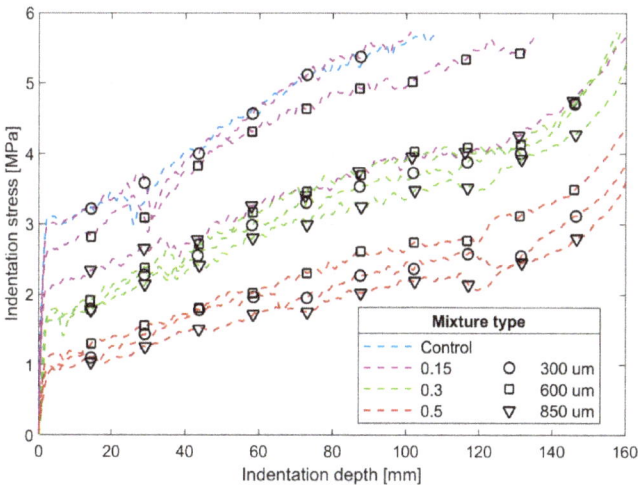

(b)

Figure 3. *Cont.*

247

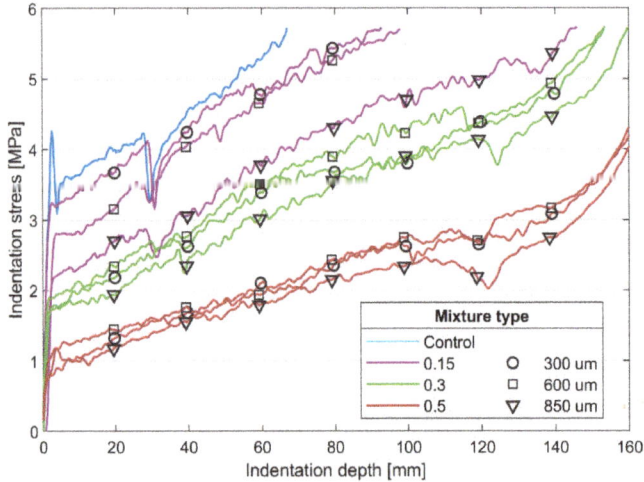

(c)

Figure 3. Loading responses of all the samples at (**a**) 7, (**b**) 14, and (**c**) 21 days.

At a low sand ratio of 0.15, the 7 day loading responses of Samples 0.15_300 μm and 0.15_600 μm are similar to Control, while the indentation stress of Sample 0.15_850 μm is much lower during the entire crushing history. This evidence accords with an earlier study focusing on a higher foam concrete density and with coarser sand [22], which also shows that higher sand size reduces compressive and flexural strength of foam concrete. As a result of the reduced cement ratio, the indentation strength is consistently diminished for samples of the same sand size but higher sand ratio. At higher sand ratios of 0.3 and 0.5, however, it is still evident that the crushing strength is further reduced when larger sand particles are used. Another important observation is that no matter for which sand ratio, the relative strength reduction between the 850 μm and 600 μm samples is larger than that between the 600 μm and 300 μm samples, implying a profound influence from the 850-μm sand.

4.3. Micro-CT Investigation on the Influence of Particle size

The experimental results from both the vibrational frequency test and the penetration test indicate that sand size affects the material performance of the foam–sand composite samples, especially at the large sand ratio of 0.5. It has been known that the mechanical behavior of cellular solid material is primarily determined by the solid property (e.g., solid modulus) and the morphology of the foam structure (e.g., void size distribution) [13,18,23]. In this case, the solid modulus is related to the cement paste and sand; however, this factor remains unchanged under the same sand ratio of 0.5. Therefore, it is more likely that the size of the sand inclusion makes a difference in the foam crushing behavior by affecting the cellular paste matrix.

To study the influence of the inclusion size, the specimens of Samples of 0.5 sand ratio, along with the control sample were further scanned using micro-CT. These samples were selected because they reflected the greatest contrast between the control and composite samples based on the mechanical tests. A sample scan of each sample is shown in Figure 4, along with the corresponding processed image highlighting the different material phases and a 3D reconstruction showing the foam structure. The 3D reconstruction represents a smaller $2500 \times 2500 \times 2500\ \mu m^3$ volume being extracted from the center of each CT scan. Due to the density difference, the sharp contrast made it easy for segmenting between air void and solid phases, while it was more challenging to separate out the sand particles from the cement paste, as the density of C-S-H solid is close to that of the sand [39].

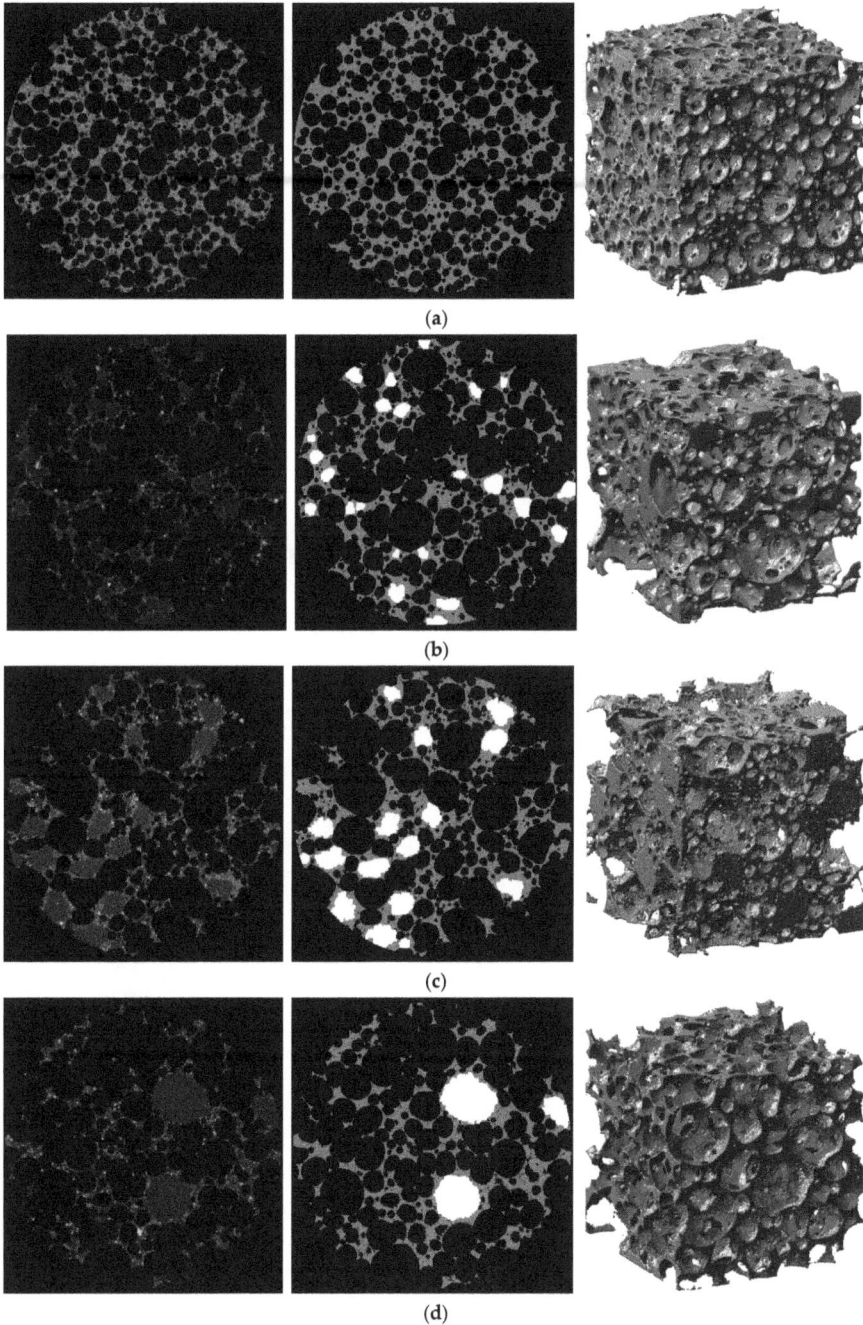

Figure 4. A representative micro-CT scan, corresponding phase segmentation, and 3D reconstruction of samples (**a**) Control, (**b**) 0.5_300 μm, (**c**) 0.5_600 μm, and (**d**) 0.5_850 μm, where the micro-CT scan has a side length of 5000 μm and the 3D reconstruction has a side length of 2500 μm.

To further validate the CT test, the volume fractions of the three phases measured from the 3D reconstruction of each sample were compared with the theoretical values calculated based on the mixture design, as shown in Table 3. In general, the measurement achieves a decent agreement with the theoretical volume for all four samples, especially the void content. Due to the difficulty of separating the paste and sand, the variation on these two phases is greater. With respect to the composite samples, a greater deviation is observed in Sample 0.5_600 μm, but still within a reasonable range.

Table 3. Theoretical volume fractions and measurements from the 3D reconstructions.

Sample	Void Content [%]		Paste Content [%]		Sand Content [%]	
	Calculated	Measured	Calculated	Measured	Calculated	Measured
Control	71.1	70.9	28.9	-	0.0	-
0.5_300 μm	76.8	75.6	18.0	18.9	5.2	5.5
0.5_600 μm	76.8	77.4	18.0	16.8	5.2	5.8
0.5_850 μm	76.8	75.8	18.0	19.3	5.2	4.9

In terms of the spatial distribution of sand, the 3D reconstruction shows that no particle agglomeration in the inspected volumes and these mixtures are generally homogeneous. The segmentation images in Figure 4 confirm the increment on sand size from Sample 0.5_300 μm to 0.5_850 μm. Since the amount of sand inclusion was identical across the three samples, the sand population declines with the particle size. Seemingly, the sand particles in Samples 0.5_300 μm and 0.5_600 μm are more included in the cement paste, but those in Sample 0.5_850 μm are less confined by the cellular matrix.

Regarding the air void size, Sample Control in Figure 4a has the smallest voids among the four samples. This difference is mainly caused by the lower effective density of the foamed paste in the composite samples (see Table 1), and thus less confinement to the air bubbles in fresh cement paste before setting. These larger voids can be partly attributed to the presence of sand as well, as the interaction between the solid sand particles to the soft air bubbles reduces the stability of the foamed paste during mixing [17,23,32]. Another observation is that the sand particles in Samples 0.5_300 μm (Figure 4b) and 0.5_600 μm (Figure 4c) are smaller than the large air voids in the cellular matrix; however, the particle size is evidently larger in Sample 0.5_850 μm (Figure 4d). This difference may be relevant to the greater drop on elastic modulus and crushing strength of samples with the 850 μm sand.

To compare the void systems of the four samples quantitively, their void size distributions were extracted from the 3D reconstructions, as shown in Figure 5. As the air voids that are intersected by the model boundary does not reflect the actual void size, those intersected voids were not counted for the statistical analysis. As a result, the effective model volume was different for each sample. This artifact was compensated by renormalizing the effective model volume back to $2500 \times 2500 \times 2500$ μm^3. A back calculation using the normalized distributions confirmed that the deviation to the actual bulk density of each sample was less than 5%. In the four specimens, voids larger than 800 μm were not observed, which is generally in agreement with the visual inspection. Possibly, air bubbles larger than this size cannot be sufficiently confined by the fresh mixture during mixing.

According to Figure 5a, the log-scaled void population of Sample Control reduces rather linearly as a function of void size. As suggested by the other three samples, the sand inclusion results in fewer small voids and an increase of large voids. This point is better reflected in Figure 5b, as the volume distribution progressively moves to bigger voids with larger sand added. Since a standard aqueous foam was used for casting all mixtures and the only difference was the volume of foam added, ideally, the normalized void size distribution should not shift across different samples. Therefore, the influence on void size distribution must be induced by either the foamed paste or/and the particle inclusion before setting. Figure 5a shows that the overall distributions of Samples 0.5_300 μm and 0.5_600 μm are close to the controlled sample from 100 to 450 μm but consistently higher at larger sizes. The remarkable reduction at 50 μm can be mainly attributed to the diminished stability of small air voids, which have a higher tendency of merging under the reduced paste confinement. Based on the gradual rightward shift of the three distributions in Figure 5b, the larger sand seems to further induces formation of larger voids at the expense of smaller ones.

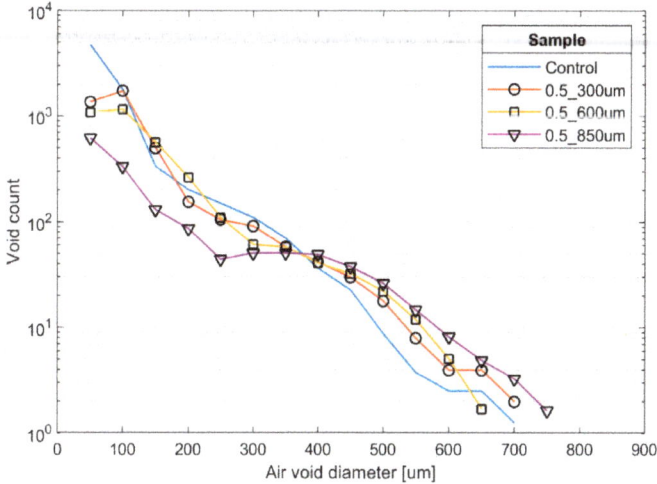

(a)

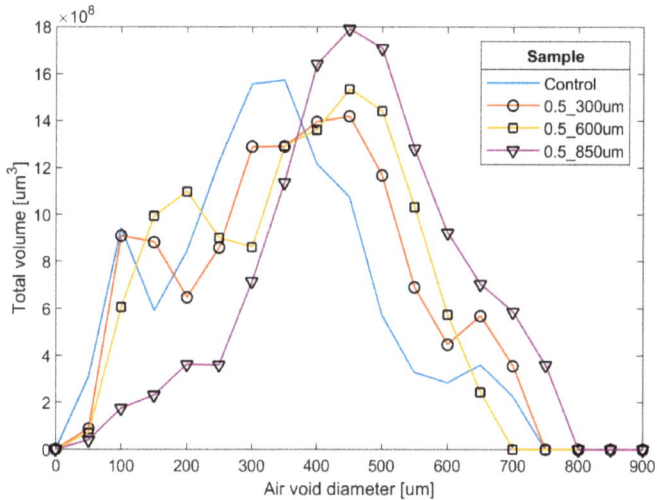

(b)

Figure 5. Statistics of the void (a) population and (b) volume distribution of the micro-CT specimens. These results have been normalized by a unit volume of $2500 \times 2500 \times 2500$ μm^3.

In comparison, the 850-μm sand has a more profound impact on the void distribution. According to Figure 5a, the void population is dramatically reduced before 300 μm and consistently increased beyond 400 μm. In correspondence, its void volume is concentrated at a larger size in Figure 5b. This significant change on the void system can be only reasonably associated with the fact that the 850 μm sand is much larger than the air voids in the sample. Prior studies have reported that the aqueous foam system tends to be more influenced when the inclusion size becomes larger [22,32]. Thus, the void stability in Sample 0.5_850 μm is more diminished than those with smaller sand, leading to the most degenerated foam structure among all four samples. This explanation is also well supported by the experimental results of Sample 0.5_850 μm on elastic modulus as shown in Figure 1, and crushing strength as shown in Figure 3, as well as the other 850-μm sand samples. Therefore, during material design, special attention should be given to control the maximum particle size when any inclusion is used in foam concrete mixtures.

5. Conclusions

This study investigated the influence of sand size and sand content on the mechanical properties of a group of 0.55 g/cm^3 foam concrete mixtures, for which higher sand ratio corresponds to less cement usage and lower density of the foamed paste. The test results of elastic modulus from the vibration frequency test and loading response from the penetration test suggest a strong influence from the material modulus to the crushing behavior of the foam-sand composite. At a small sand ratio of 0.15, the modulus and crushing strength between the controlled and composite samples are very close and even slightly higher when the 300- and 600-μm fine sand were used; the inclusion of the large 850-μm sand, however, caused significantly diminished results for both tests. Under higher sand ratios, the lower effective foamed paste density results in progressively inferior material performance, and the influence from the inclusion size becomes secondary.

In addition to the mechanical tests, the micro-CT scan indicates that the air voids in the composite mixtures are generally enlarged, which is mainly attributed to the reduced paste confinement. Furthermore, increasing the sand size diminishes the stability of small air voids and induces formation of larger voids. This size impact is most profound when the sand particles are larger than the voids.

As a general rule, it is recommended that low-density foam concrete should be designed with a relatively low volume of particles to avoid compromising mechanical performance. Preferably, fine-graded particles should be used to maintain a good foam stability. Due to this direct influence from the sand size to the foam stability, special attention should be paid to the particle dimension and void size of the foam during the mixture design when using particle inclusions in foam concrete.

Author Contributions: Conceptualization, Y.S. and D.L.; Methodology and Data Analysis, Y.S.; Supervision, Project Administration, and Funding Acquisition, D.L.

Funding: The authors would like to acknowledge funding from the University Transportation Center for Research on Concrete Applications for Sustainable Transportation (RE-CAST), O'Hare Modernization Program, and the Chicago Department of Aviation (CDA).

Acknowledgments: We also thank fellow student Jamie Clark for the CT image acquisition and Robbie Damiani for valuable discussion on the test results.

Conflicts of Interest: The authors declare no conflict of interest.

References

1. Chen, C.; Habert, G.; Bouzidi, Y.; Jullien, A. Environmental impact of cement production: Detail of the different processes and cement plant variability evaluation. *J. Clean. Prod.* **2010**, *18*, 478–485. [CrossRef]
2. Lim, S.K.; Tan, C.S.; Li, B.; Ling, T.C.; Hossain, M.U.; Poon, C.S. Utilizing high volumes quarry wastes in the production of lightweight foamed concrete. *Constr. Build. Mater.* **2017**, *151*, 441–448. [CrossRef]
3. Manzi, S.; Mazzotti, C.; Bignozzi, M.C. Short and long-term behavior of structural concrete with recycled concrete aggregate. *Cem. Concr. Compos.* **2013**, *37*, 312–318. [CrossRef]
4. Sagoe-Crentsil, K.K.; Brown, T.; Taylor, A.H. Performance of concrete made with commercially produced coarse recycled concrete aggregate. *Cem. Concr. Res.* **2001**, *31*, 707–712. [CrossRef]
5. Malešev, M.; Radonjanin, V.; Marinković, S. Recycled concrete as aggregate for structural concrete production. *Sustainability* **2010**, *2*, 1204–1225. [CrossRef]

6. Cabral, A.E.B.; Schalch, V.; Molin, D.C.C.D.; Ribeiro, J.L.D. Mechanical properties modeling of recycled aggregate concrete. *Constr. Build. Mater.* **2010**, *24*, 421–430. [CrossRef]

7. Khatib, J.M. Properties of concrete incorporating fine recycled aggregate. *Cem. Concr. Res.* **2005**, *35*, 763–769. [CrossRef]

8. Zega, C.J.; Di Maio, Á.A. Use of recycled fine aggregate in concretes with durable requirements. *Waste Manag.* **2011**, *31*, 2336–2340. [CrossRef] [PubMed]

9. Evangelista, L.; de Brito, J. Durability performance of concrete made with fine recycled concrete aggregates. *Cem. Concr. Compos.* **2010**, *32*, 9–14. [CrossRef]

10. ACI Committee. *229 Report on Controlled Low-Strength Materials*; ACI Committee: Farmington Hills, MI, USA, 2013.

11. ACI Committee. 523 Guide for Cellular Concretes above 50 lb/ft^3 (800 kg/m^3) (ACI 523.3R-93). In *ACI Manual of Concrete Practice*; ACI Committee: Farmington Hills, MI, USA, 2013; ISBN 9780870318856.

12. Ramamurthy, K.; Kunhanandan Nambiar, E.K.; Indu Siva Ranjani, G. A classification of studies on properties of foam concrete. *Cem. Concr. Compos.* **2009**, *31*, 388–396. [CrossRef]

13. Jones, M.R.; McCarthy, A. Behaviour and assessment of foamed concrete for construction applications. In Proceedings of the 2005 International Congress—Global Construction: Ultimate Concrete Opportunities, Dundee, UK, 5–7 July 2005.

14. Wang, J.; Guo, W.; Zhao, R.; Shi, Y.; Zeng, L. Energy-absorbing properties and crushing flow stress equation of lightweight foamed concrete. *J. Civ. Archit. Environ. Eng.* **2013**, *35*, 96–102.

15. Song, Y.; Lange, D. Crushing Behavior and Crushing Strengths of Low-Density Foam Concrete. *Preprints* **2019**, 2019020208. [CrossRef]

16. Wei, S.; Yunsheng, Z.; Jones, M.R. Using the ultrasonic wave transmission method to study the setting behavior of foamed concrete. *Constr. Build. Mater.* **2014**, *51*, 62–74. [CrossRef]

17. Amran, Y.H.M.; Farzadnia, N.; Ali, A.A.A. Properties and applications of foamed concrete; A review. *Constr. Build. Mater.* **2015**, *101*, 990–1005. [CrossRef]

18. Narayanan, N.; Ramamurthy, K. Structure and properties of aerated concrete: A review. *Cem. Concr. Compos.* **2000**, *22*, 321–329. [CrossRef]

19. Jones, M.R.; McCarthy, A. Preliminary views on the potential of foamed concrete as a structural material. *Mag. Concr. Res.* **2005**, *57*, 21–31. [CrossRef]

20. Nambiar, E.K.K.; Ramamurthy, K. Influence of filler type on the properties of foam concrete. *Cem. Concr. Compos.* **2006**, *28*, 475–480. [CrossRef]

21. Nambiar, E.K.K.; Ramamurthy, K. Models relating mixture composition to the density and strength of foam concrete using response surface methodology. *Cem. Concr. Compos.* **2006**, *28*, 752–760. [CrossRef]

22. Lim, S.K.; Tan, C.S.; Zhao, X.; Ling, T.C. Strength and toughness of lightweight foamed concrete with different sand grading. *KSCE J. Civ. Eng.* **2014**, *19*, 2191–2197. [CrossRef]

23. Jones, M.R.; Ozlutas, K.; Zheng, L. Stability and instability of foamed concrete. *Mag. Concr. Res.* **2016**, *68*, 542–549. [CrossRef]

24. Ghorbani, S.; Ghorbani, S.; Tao, Z.; Brito, J.; Tavakkolizadeh, M. Effect of magnetized water on foam stability and compressive strength of foam concrete. *Constr. Build. Mater.* **2019**, *197*, 280–290. [CrossRef]

25. Onprom, P.; Chaimoon, K.; Cheerarot, R. Influence of Bottom Ash Replacements as Fine Aggregate on the Property of Cellular Concrete with Various Foam Contents. *Adv. Mater. Sci. Eng.* **2015**, *2015*, 381704. [CrossRef]

26. Kearsley, E.P.; Wainwright, P.J. The effect of high fly ash content on the compressive strength of foamed concrete. *Cem. Concr. Res.* **2001**, *31*, 105–112. [CrossRef]

27. Jones, M.R.; McCarthy, A. Utilising unprocessed low-lime coal fly ash in foamed concrete. *Fuel* **2005**, *84*, 1398–1409. [CrossRef]

28. She, W.; Du, Y.; Zhao, G.; Feng, P.; Zhang, Y.; Cao, X. Influence of coarse fly ash on the performance of foam concrete and its application in high-speed railway roadbeds. *Constr. Build. Mater.* **2018**, *170*, 153–166. [CrossRef]

29. Jones, R.; Zheng, L.; Yerramala, A.; Rao, K.S. Use of recycled and secondary aggregates in foamed concretes. *Mag. Concr. Res.* **2012**, *64*, 513–525. [CrossRef]

30. Gibson, L.J.; Ashby, M.F. *Cellular Solids*; Cambridge University Press: Cambridge, UK, 1999; ISBN 9781139878326.

31. Ashby, M.F.; Medalist, R.F.M. The mechanical properties of cellular solids. *Metall. Trans. A* **1983**, *14*, 1755–1769. [CrossRef]
32. Nambiar, E.K.K.; Ramamurthy, K. Air-void characterisation of foam concrete. *Cem. Concr. Res.* **2007**, *37*, 221–230. [CrossRef]
33. American Society for Testing and Materials. *ASTM C796—Standard Test Method for Foaming Agents for Use in Producing Cellular Concrete Using Preformed Foam*; ASTM International: West Conshohocken, PA, USA, 2012.
34. ASTM. *ASTM 305. Standard Practice for Mechanical Mixing of Hydraulic Cement Pastes and Mortars of Plastic Consistency*; Annu. B. ASTM Stand Standard; ASTM International: West Conshohocken, PA, USA, 2011.
35. Song, Y.; Lange, D.A. Measuring Young's Modulus of Low-Density Foam Concrete Using Resonant Frequency. *Preprints* **2019**, 2019020207. [CrossRef]
36. Santagata, E.; Bassani, M.; Sacchi, E. Performance of new materials for aircraft arrestor beds. *Transp. Res. Rec.* **2010**, *2177*, 124–131. [CrossRef]
37. Zhou, Q.; Mayer, R.R. Characterization of Aluminum Honeycomb Material Failure in Large Deformation Compression, Shear, and Tearing. *J. Eng. Mater. Technol.* **2002**, *124*, 412–420. [CrossRef]
38. Ramamurty, U.; Kumaran, M.C. Mechanical property extraction through conical indentation of a closed-cell aluminum foam. *Acta Mater.* **2004**, *52*, 181–189. [CrossRef]
39. Allen, A.J.; Thomas, J.J.; Jennings, H.M. Composition and density of nanoscale calcium-silicate-hydrate in cement. *Nat. Mater.* **2007**, *6*, 311. [CrossRef] [PubMed]

*applied
sciences*

MDPI

Article

Effect of Nylon Fiber Addition on the Performance of Recycled Aggregate Concrete

Seungtae Lee

Department of Civil Engineering, Kunsan National University, Kunsan 54150, Korea; stlee@kunsan.ac.kr;
Tel.: +82-10-3666-9561

Received: 6 January 2019; Accepted: 15 February 2019; Published: 22 February 2019

Abstract: The adhered mortars in recycled aggregates (RA) may lower the performance of the concrete, by for instance reducing its strength and durability, and by cracking. In the present study, the effect of nylon fiber (NF) on the permeability as well as on the mechanical properties of concrete incorporating 100% RA was experimentally investigated. Concrete was produced by adding 0, 0.6 and 1.2 kg/m^3 of NF and then cured in water for a predetermined period. Measurements of compressive and split tensile strengths, ultrasonic pulse velocity and total charge passed through concrete were carried out, and the corresponding test results were compared to those of concrete incorporating crushed stone aggregate (CA). In addition, the microstructures of 28-day concretes were examined by using the FE-SEM technique. The test results indicated that recycled coarse aggregate concrete (RAC) showed a lower performance than crushed stone aggregate concrete (CAC) because of the adhered mortars in RA. However, it was obvious that the addition of NF in RAC mixes was much more effective in enhancing the performance of the concretes due to the crack bridging effect from NF. In particular, a high content of NF (1.2 kg/m^3) led to a beneficial effect on concrete properties compared to a low content of NF (0.6 kg/m^3) with respect to mechanical properties and permeability, especially for RAC mixes.

Keywords: recycled coarse aggregate concrete; nylon fiber; mechanical properties; permeability; microstructure

1. Introduction

Recently, a lack of natural aggregates with a high quality has set up the alarm to find alternative uses for recycled aggregate. Environmental and economic benefits led to a higher production and application of recycled aggregate concrete in many countries. A great amount of demolition and construction waste, almost 67,000,000 tons yearly, was generated in South Korea. For the purpose of the wide utilization of recycled aggregate, it is advised to use recycled aggregate up to 30% in concrete construction sites of 24 MPa grade and below [1]. Due to the increased technology in the aggregate production industry, both the quality and quality of recycled aggregates has increasingly improved [2].

However, there are still some doubts on the performance of concrete using recycled aggregate. In general, the performance of recycled aggregate concrete greatly depends on the adhered mortars oriented from parent concrete. Until now, there have been many reports on the properties of concrete made with recycled aggregate [3–11], but most of them resulted in a lower level of concrete strengths. This is mainly due to the residual impurities on the surface of the recycled aggregates, which blocked the strong bond between the cement paste and aggregate. Moreover, it is well known that there are two interfacial zones in recycled aggregate, which negatively affect concrete properties. In fact, this is the reason that the microstructure of recycled aggregate concrete is much more complicated than that of natural aggregate concrete. The old mortar in recycled aggregate includes many micro-cracks, formed during the production of recycled aggregate concrete, and it has a high porosity.

In order to enhance the performance of recycled aggregate concrete, researchers have come up with several advanced techniques. These techniques include the surface treatment [12], the use of mineral admixtures [13,14] and the addition of fibers [4,15,16], which can improve the performance of recycled aggregate concrete.

In particular, the reinforcement of recycled aggregate concrete using fibers leads to reduced micro-cracks in the cement matrix as well as to an increase in material density. There have been numerous studies on the applications of steel fiber [17–23] and polypropylene (PP) fibers [24–26] in concrete. Moreover, the PP fiber also enjoys popularity in the domain of recycled aggregate concrete [4,27]. However, although nylon fiber (NF) shows a rising acceptance in the literatures [28–30], it remains unpopular compared to steel fiber and PP fiber. In comparison with steel and PP fibers, while there is limited literature available on the use of NF, some authors [28,31] reported that the use of NF stepped up the performance after the presence of cracks in concrete, and sustained high stresses. One needs to examine the applicability of NF for the purpose of the enhanced mechanical properties and the reduced micro-cracks in recycle aggregate concrete.

This study is therefore aimed at investigating the usability of NF in recycled aggregate concrete in order to be used in structural concrete, since the use of NF in field concrete is gaining popularity nowadays. In order to achieve this goal, measurements of compressive strength, split tensile strength, ultrasonic pulse velocity, and chloride ion permeability of recycled coarse aggregate concretes with or without NF were carried out, and the corresponding test results were compared to those of concretes incorporating crushed stone aggregate. Additionally, the microstructures of 28-day concretes were examined by using the FE-SEM technique.

2. Experimental Section

2.1. Materials

In this study, ordinary Portland cement conforming to ASTM C150 was used in preparing the concrete specimens. The cement had been produced by a local cement plant in South Korea. The density and specific surface area of the cement used were 3.15 g/cm^3 and 328 m^2/kg, respectively. The mineralogical compounds of the cement were 54.9, 16.6, 10.3, and 9.1% for C_3S, C_2S, C_3A and C_4AF, respectively.

Crushed stone aggregate (CA) and recycled coarse aggregate (RA) were used as coarse aggregates for the concrete production. RA was produced by crushing the waste concrete with a jaw crusher and an impact crusher. To enhance the purity of RA, impurities such as wood, bricks and glass were manually removed. A water jet with high-pressure was also used to remove mud and debris from RA. Both CA and RA have continuous grains from 5 to 25 mm. The main properties of coarse aggregates are presented in Table 1. The size grading of CA and RA is shown in Figure 1. Natural river sand was used as a fine aggregate with a fineness modulus of 2.80, a water absorption of 0.98% and a density of 2.65 g/cm^3.

Table 1. Physical properties of the coarse aggregates.

Properties	Crushed Stone Aggregate (CA)	Recycled Coarse Aggregate (RA)
Density (g/cm^3)	2.64	2.37
F.M.	7.17	7.43
Absorption (%)	0.66	4.31
Abrasion rate (%)	21.7	47.5
Adhered mortar (%) [1]		5.56

[1] Acid-soluble content.

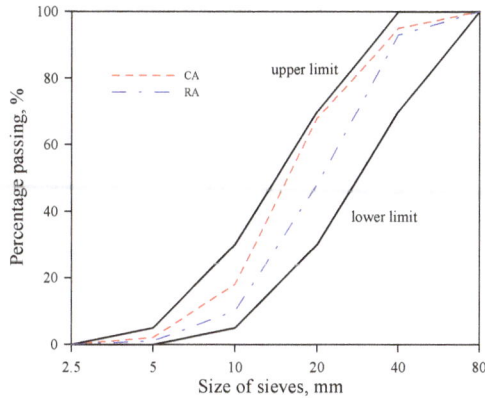

Figure 1. Size grading of coarse aggregates.

The nylon fibers (NF) used in the present study, a picture of which is shown in Figure 2, had been supplied by a local fiber company in South Korea, and they are now commercially available in the domestic market. The properties of NF used in the present study are shown in Table 2. In addition, a polycarboxylate-based superplasticizer (SP) was used to improve the initially low workability of the fresh concrete. The basic properties of SP were provided by a chemical manufacturer in South Korea.

Figure 2. Nylon fibers (NF) used in this study.

Table 2. Physical properties of NF.

Properties	Nylon Fibers (NF)
Diameter (μm)	23
Length (mm)	19
Aspect ratio	826
Density (g/cm^3)	1.16
Tensile strength (MPa)	919
Elastic modulus (GPa)	5.3
Color	white

2.2. Concrete Mix Proportions and Curing

Six concrete mixes of 24 MPa grade concrete have been prepared. For all specimens, the w/c ratio was chosen as 0.50. The desired workability which is the range of 130–170 mm collapse was obtained by means of SP. The RAC was prepared with a 100% replacement of RA with CA. The concrete mix proportions are shown in Table 3. The NF was added at the concentrations of 0, 0.6, and 1.2 kg/m^3 for CAC and RAC mixes, respectively. The SP was initially mixed with water to achieve a uniform dispersion. The concrete mixes were grouped to study the effect of the NF contents. The mixing of concrete was carried out using a manually loaded laboratory mixer. The concrete materials were mixed and then cast into the 100 × 200 mm cylinder molds for compressive and split tensile strength, non-destructive ultrasonic pulse velocity measurements, and a rapid chloride ion penetration test. The concrete specimens were cured under moisture conditions for 24 h, after which they were demolded and moved into a plastic tank for water curing until the time of testing.

Table 3. Mix proportions of concrete.

Concrete Mixes	Mix Proportion (kg/m^3)							Fresh Density [1] (kg/m^3)
	Water	Cement	Sand	CA	RA	NF	SP [3]	
CAC1	170	340	745	1015	-	-	2.45	2345
CAC2	170	340	745	1015	-	0.6 (0.06) [2]	2.45	2350
CAC3	170	340	745	1015	-	1.2 (0.12)	2.45	2386
RAC1	170	340	745	-	912	-	2.72	2278
RAC2	170	340	745	-	912	0.6 (0.06)	2.72	2290
RAC3	170	340	745	-	912	1.2 (1.12)	2.72	2312

[1] Values measured according to ASTM C 138M-17a standard [32]. [2] Fiber volume fraction (v/v %). [3] Superplasticizer.

2.3. Test Methods

The compressive and split tensile strengths of concrete cylinders with dimension of 100 mm in diameter and 200 mm in length after 7, 28, and 91 days of curing were tested according to ASTM C39 [33] and ASTM C469-17 [34], respectively. For evaluating the compressive strength, the cylinders were placed under a compression testing machine of 2000 kN capacity. The test for split tensile strength was also conducted using the same compression testing machine. The mean values of the compressive and split tensile strengths of at least three samples for each concrete mix were taken, and the standard deviation from the test results was calculated. The non-destructive test, like the ultrasonic pulse velocity (UPV), was conducted according to ASTM C597-16 [35] by using PUNDIT LAB, manufactured by PCTE, Australia. The used transducers are 50 mm in diameter, and have a maximum resonant frequency of 54 kHz. A rapid chloride ion penetrability test was used to evaluate the permeability of the concrete specimens. This test was based on the standard test method of ASTM C1202-18 [36]. After having been cured for 28 days, a concrete disc, 50 mm in thickness and 100 mm in diameter, was connected to two chambers: one was filled with 3% NaCl solution and the other with 0.3M NaOH solution to form electrodes, as shown in Figure 3. An electric charge of 60 V was applied to the electrodes for 6 h. The current flowing through the concrete disc and the temperature of the solution in the chambers were recorded with an interval of 30 min. The higher level of the total charge passed represents the higher permeability of concrete, as shown in Table 4.

The microstructures of concrete fractions after the compressive strength tests of 28-day concrete specimens, were investigated using field emission scanning electron microscopy (FE-SEM) equipped with an EDXA Falcon energy system. The concrete samples were gold-coated after drying in a desiccator for 24 h.

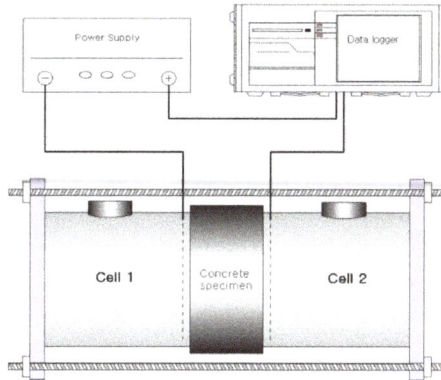

Figure 3. Schematic of the rapid chloride ion penetration test based on ASTM C1202-18 [36].

Table 4. Chloride ion penetrability based on the charge passed [36].

Charge Passed (Coulombs)	Chloride Ion Penetrability
>4000	High
2000–4000	Moderate
1000–2000	Low
100–1000	Very low
<100	Negligible

3. Results and Discussion

3.1. Compressive Stregnth

Table 5 presents the compressive strength values of both CAC and RAC mixes with different NF content. Due to the adhered mortars in recycled aggregates, the compressive strength values of RAC mixes is much lower than CAC mixes. For example, it can be seen that the compressive strength value of the RAC1 mix without NF decreased by 27% at the age of 28 days, compared with that of the CAC1 mix. At 91 days, the CAC1 mix also exhibited a higher compressive strength value, showing 41.2 MPa, than the RAC1 mix (30.1 MPa). It is presumed that the higher absorption and the larger pores, due to the adhered mortar in the RAC1 mix, contributed to a decreased strength in the RAC1 mix.

The compressive strength generally increased when NF was added to the mixes. Among the concrete mixes with RA, the highest compressive strength was obtained from the RAC3 mix with 52.6 MPa at 91 days of curing. For the CAC mixes, the CAC3 mixes exhibited a good development in compressive strength. It was therefore obvious that the addition of NF led to the increase of compressive strength in both the CAC and RAC mixes. It is worth noting that despite the use of RA, the RAC3 mix with 1.2 kg/m^3 of NF showed much higher compressive strength values, ranging from 27~41%, compared with the CAC1 mix with CA, for all stages of the curing.

In order to highlight the effect of RA and NF on the compressive strength of concrete mixtures, the compressive strength ratio (CSR) of concrete mixes at 7, 28 and 91 days was calculated, and the correspondent results were shown in Figure 4a–c, respectively. The CSR results shown in Figure 4 exhibited almost the same trend, regardless of the curing ages. However, it must be noted that when NF was added in the RAC mixes, there was a beneficial effect on the increase in compressive strength, as a similar result was reported by Song et al. [37]. They found an approximately 11.5% increase in the compressive strength of concrete with 0.6 kg/m^3 of NF content, compared to control concrete without NF.

From Figure 4, it was clearly observed that the CSR increased when the NF contents increased. Increases of CSR in the high content NF containing mixes (CAC3 and RAC3) were more remarkable than for the low content NF mixes such as CAC2 and RAC2. At all curing periods, the higher CSR values were obtained from the RAC3 mix with an NF content of 1.2 kg/m^3, which means that, as the NF content increases, the compressive strength development of recycled aggregate concrete becomes more effective. For example, the CSR of the RAC3 concrete was 194% at 7 days, while the value for the CAC3 mix was only 136%. More importantly, the improved behavior of concrete with NF may be the result of the high content of fibers which form a network that acts as a bridge in the cement matrix, resulting in a reduction of micro-cracks [16,27,38].

Table 5. Compressive strength values of concrete mixes with different NF contents (standard deviation in parenthesis).

Mixes	Compressive Strength (MPa)		
	7d.	28d.	91d.
CAC1	25.6 (0.97)	36.5 (1.14)	41.2 (1.12)
CAC2	37.5 (1.42)	45.6 (1.02)	52.4 (0.76)
CAC3	38.2 (0.88)	49.8 (1.65)	56.6 (1.35)
RAC1	18.6 (1.21)	25.4 (2.18)	30.1 (0.90)
RAC2	27.0 (0.45)	39.5 (0.98)	43.6 (2.33)
RAC3	36.2 (0.81)	47.2 (1.47)	52.6 (0.66)

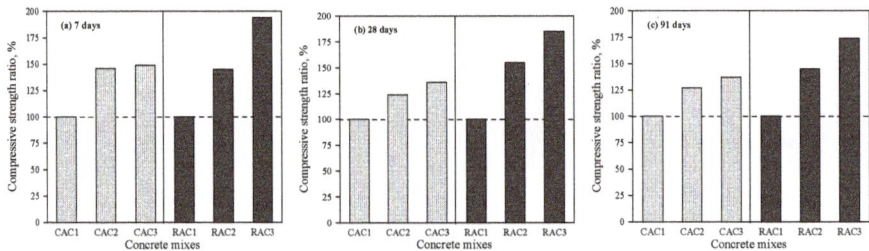

Figure 4. Compressive strength ratios of concrete mixes at: (**a**) 7 days, (**b**) 28 days, and (**c**) 91 days.

3.2. Split Tensile Strenth

The split tensile strength values of the CAC and RAC mixes with different NF contents at 7, 28, and 91 days are listed in Table 6. As expected, the values of the split tensile strength of concrete were significantly dependent on the added NF content. The strength results showed that there was a great increase in the split tensile strength values with the addition of NF up to 1.2 kg/m^3 in both the CAC and RAC mixes. At 91 days, the maximum strength value obtained in the case of the CAC mixes was 6.3 MPa for 1.2 kg/m^3 NF content, an increase of 23.5% over the CAC1 mix. Similarly, for the RAC mixes, the maximum improvement examined in the RAC3 mix with the same NF content, as compared with the unreinforced RAC1 mix, was 80.6%, which has a split tensile strength of 5.6 MPa. These results were in a good agreement with other studies [15,27].

The calculated results of the split tensile strength ratio (SSR) of the concrete mixes at 7, 28 and 91 days were presented in Figure 5a–c, respectively. For both the CAC and RAC mixes, it seems that the addition of NF had much influence on the split tensile strength, especially at the early age of curing. At 7 days, the SSR value for the RAC3 mix was 205% compared to 151% for the CAC3 mix, and similar trends were also observed at 28 and 91 days, as shown in Figure 5b,c.

Within the scope of the present study, it was confirmed that a higher addition of NF led to an increase in both the compressive and split tensile strengths. Furthermore, this was obvious when NF was applied in the RAC mixes compared to the CAC mixes.

Table 6. Split tensile strength values of the concrete mixes with different NF contents (standard deviation in parenthesis).

Mixes	Split Tensile Strength (MPa)		
	7d.	28d.	91d.
CAC1	2.9 (0.14)	4.2 (0.10)	5.1 (0.22)
CAC2	4.2 (0.20)	5.2 (0.24)	6.0 (0.24)
CAC3	4.4 (0.42)	5.6 (0.32)	6.3 (0.33)
RAC1	1.8 (0.12)	2.5 (0.14)	3.1 (0.24)
RAC2	3.1 (0.22)	3.9 (0.11)	4.7 (0.10)
RAC3	3.7 (0.18)	4.7 (0.23)	5.6 (0.19)

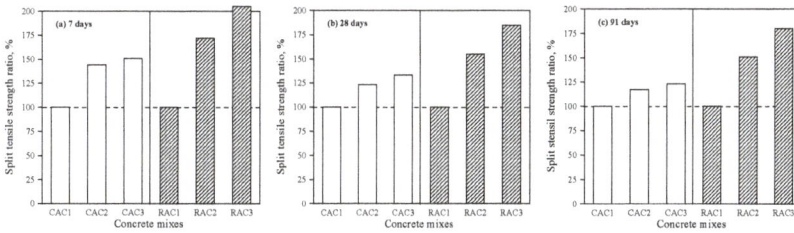

Figure 5. Split tensile strength ratio of concrete mixes at: (**a**) 7 days, (**b**) 28 days, and (**c**) 91 days.

3.3. Ultrasonic Pulse Velocity

The ultrasonic pulse velocity (UPV) measurements have been well known as non-destructive methods to assess the quality of concrete [14,27,39,40]. The UPV values for the CAC and RAC mixes measured at 7, 28, and 91 days are shown in Figure 6. According to the classification criterion for concrete based on ultrasonic pulse measurements by Leslie and Cheeseman [41], the UPV values for concrete mixes observed in this work can be classified as shown in Table 7.

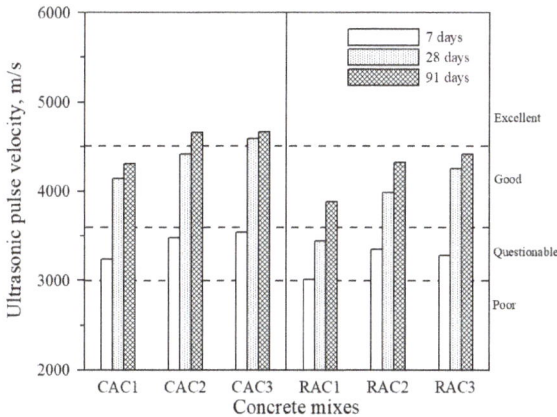

Figure 6. Ultrasonic pulse velocity of concrete mixes.

Table 7. Concrete classification based on the ultrasonic pulse velocity [41].

Pulse Velocity (m/s)	Concrete Classification
V > 4500	Excellent
3600 < V < 4500	Good
3000 < V < 3600	Questionable
2100 < V < 3000	Poor
V < 2100	Very poor

As expected, it was observed from Figure 6 that the UPV values increased with curing time. The increase of UPV with time is a natural development due to the increase of the stiffness via the hydration reaction [42]. Among the concrete mixes made with CA (CAC mixes), the UPV values for the CAC2 mix were almost similar to those for the CAC3 mix, which shows that the content of NF in the CAC mixes does not have a significant influence in the variation of the UPV. However, the CAC2 and CAC3 mixes exhibited a better performance with respect to the UPV compared to the CAC1 mix without NF. This may be partially attributed to the bridge effect of NF which leads to the reduction of micro-cracks in the cement matrix by the addition of NF [29,30,43]. Generally, with an addition of steel fiber, introducing steel fibers of greater length negatively affected the UPV of the concrete specimens [44]. However, in this study, it can be seen that introducing NF positively affects the UPV, although it was relatively long (19 mm in length). This might be attributed to the increase of the materials density and the bridge effect of the fibers, due to incorporation of NF into the concrete mixes. In the case of the RAC mixes, with the further incorporation of NF, the general increases in the UPV were also examined, showing similar trends with regards to the strength properties (Tables 5 and 6) with the increase of NF. At 7 days, the UPV values of the RAC mixes were comparable with the CAC mixes, ranging from 3240 ~ 3540 m/s. This can be attributed to the substantially higher water absorption capacity of RA (4.31%) than CA (0.66%), because of the adhered mortar in RA, resulting in the increased porosity in the RAC mixes. However, relatively small increases in the UPV were observed in the RAC mixes at later ages of curing. Similar to the results of the compressive and split tensile strengths, the UPV results indicated that the CAC mixes showed higher values in the UPV than the RAC mixes, regardless of the aggregate types and NF contents.

The relationship between compressive strength (CS) and split tensile strength (SS), and the UPV obtained from the present study, are presented in Figure 7. It can be seen from this figure that the best fit-curve representing the relationship is given as;

UPV = 2450 Exp (0.012 × CS), determined by a proposed regression model of $R^2 = 0.82$.

UPV = 2526 Exp (0.1016 × SS), determined by a proposed regression model of $R^2 = 0.84$.

High correlation coefficient values (0.82 for UPV-CS curve, and 0.84 for UPV-SS curve) were obtained from the exponential curves, implying that the trend of the UPV values is almost similar to that of both the compressive strength and split tensile strengths [44,45]. Overall, it can be said that the non-destructive UPV measurements are a useful method to determine the mechanical properties of concrete.

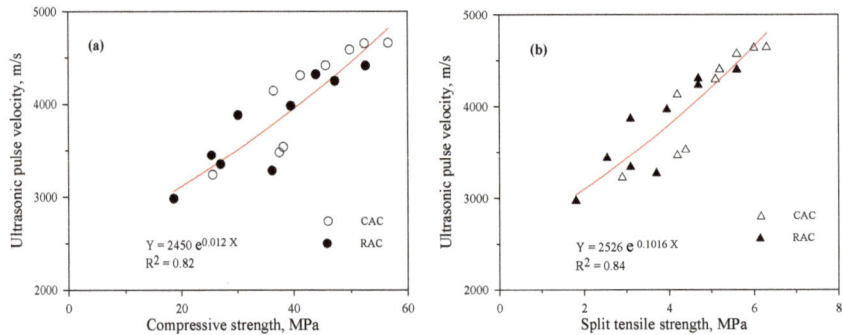

Figure 7. Relationship between the strengths and ultrasonic pulse velocity of concrete mixes: (a) compressive strength, and (b) split tensile strength.

3.4. Rapid Chloride Penetration Test

Based on the ASTM C1202-18 standard [36], the results of the rapid chloride ion penetration test (RCPT) were presented in Figure 8, indicating the total charge that passed through the concrete samples. It can be seen that NF in the CAC mixes reduced the total charge, compared to the CAC1 mix without the addition of NF, and that the reduction of the total charge was more remarkable in the concrete mixes with RA. It was observed that the charge passed for the 28-day RAC mixes were 2872, 2294, and 2050 coulombs for the RAC1, RAC2, and RAC3 mixes, respectively. According to the ASTM C1202-18 criterion, the total charges for all the RAC mixes corresponded to the level of 'Moderate', while they were 'Low' for the CAC mixes, ranging from 1046 to 1424 coulombs (see Table 4). This implies that RA in concrete may impose a high risk to the durability of the concrete structure with respect to steel corrosion. However, the usage of NF in concrete made with RA can mitigate the possibility of steel corrosion oriented from external chlorides due to the reduction of micro-cracks in the cement matrix.

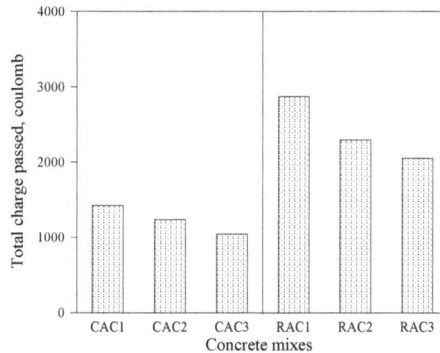

Figure 8. Rapid chloride ion penetration test (RCPT) results of the concrete mixes.

3.5. Microstructure

The results of the microstructural observations of the concrete samples using the FE-SEM technique are shown in Figures 9–14. The analysis was performed on fractions of the CAC and RAC specimens obtained after the compressive strength of concretes, to observe the effect of NF addition. It is well understood that the interfacial transition zone (ITZ) affects the mechanical properties and durability of the concrete [46–49]. In the case of the CAC1 sample (Figure 9), it was found from the image that there were calcium hydroxide (CH) crystals and C-S-H gel, in addition to a small amount of ettringite (AFt) and monosulfate (AFm). Furthermore, the ITZ between the aggregate and bulk cement matrix seems to be dense, indicating a lower porosity and less cracks. The strengthening of the ITZ connection results in the microstructural integrity of the cement matrix. The images for the CAC samples with NF (CAC2 and CAC3) were also examined via the FE-SEM technique, and revealed that C-S-H gel with a dense structure was mainly present through the samples, as shown in Figures 10 and 11.

Comparatively, for the microstructure of the RAC1 samples (Figure 12), there are two interfacial zones; the old ITZ between the aggregate and adhered mortar, and the new ITZ between t recycled aggregate and new mortar. The adhered mortar can be easily differentiated from the new mortar by different degrees of hydration of the cement matrix [27]. It was obvious from the microstructural observation that the crack was formed along this weak interface. This results in both a lower strength and higher permeability in the RAC sample, as shown in Tables 5 and 6, and in Figure 8.

The two interfacial zones were also observed in the microstructure of the RAC2 sample, as shown in Figure 13. Due to the addition of NF, the sample exhibited the cement bulk matrix with a higher density as well as with less pores. With the NF addition of 1.2 kg/m^3 in the RAC mix, as shown in Figure 14, only small cracks were found on the paste of the specimen, which is due to the bridge effect of NF. In addition, the increase in density (see Table 3) of the RAC mixes with NF (RAC2 and RAC3) led to the improvement of mechanical properties such as the strength and UPV as well as to the reduction of permeability. Therefore, it can be confirmed that the enhanced performance of the RAC3 mix compared to the RAC1 mix is due to the reduced micro-cracks via the increase in the fiber content.

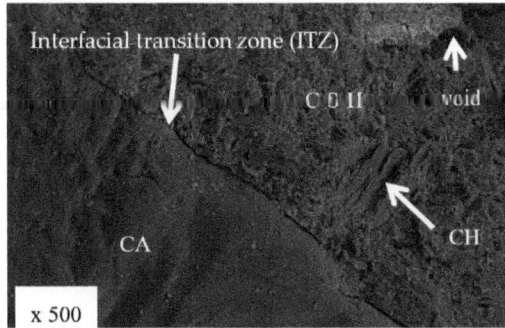

Figure 9. FE-SEM image of CAC1 sample.

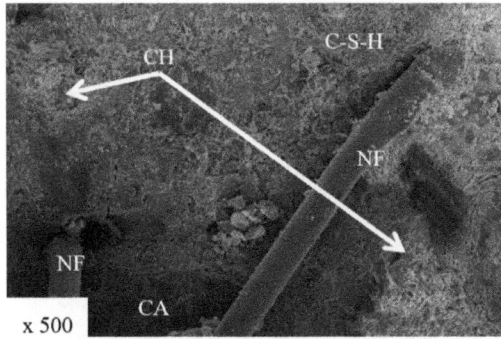

Figure 10. FE-SEM image of CAC2 sample.

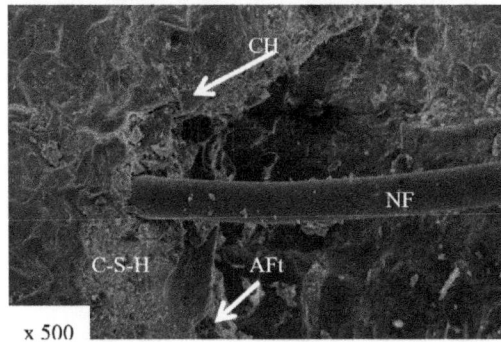

Figure 11. FE-SEM image of CAC3 sample.

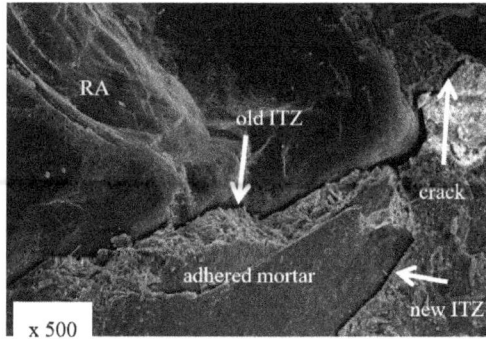

Figure 12. FE-SEM image of RAC1 sample.

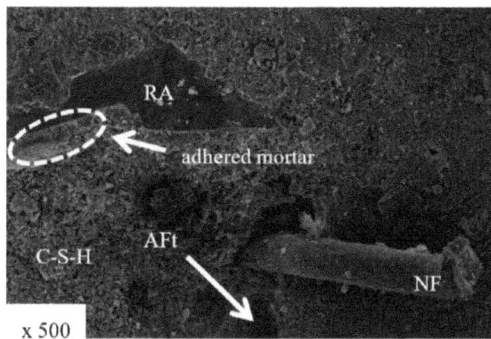

Figure 13. FE-SEM image of RAC2 sample.

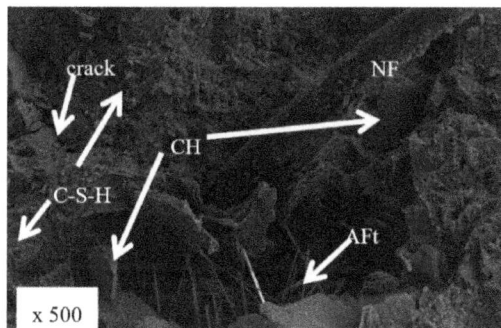

Figure 14. FE-SEM image of RAC3 sample.

4. Conclusions

Experimental works were carried out with the additions of 0, 0.6 and 1.2 kg/m^3 of NF in both CAC and RAC mixes in order to highlight the effect of NF on the mechanical properties of CAC and RAC mixes. The RAC was prepared with a 100% replacement of RA with CA. The main conclusions obtained from the present study are as follows.

(1) Due to the adhered mortar in RA, the compressive strength values of the RAC mixes were significantly lower than those of the CAC mixes. However, we found that the addition of NF led to an increase in compressive strength of both the CAC and RAC mixes. In particular, this trend

was more remarkable in the RAC mixes with a high content of NF. The compressive strength ratio results revealed that there was a beneficial effect of NF on the increase in compressive strength.

(2) As observed in the case of the compressive strength, a similar trend was also examined with the NF content variation for the split tensile strength. Specifically, the test results revealed that there was a significant increase in the split tensile strength, especially with the addition of 1.2 kg/m^3 NF, regardless of concrete types. In the case of the RAC3 mix, we examined an increase of 80.6% in the split tensile strength over the RAC1 mix without NF.

(3) From the test results of the UPV, we observed that in the case of the RAC mixes, there was an increase in the UPV with the further incorporation of NF. This may be due to the increase of the materials density and the bridge effect of NF. Additionally, it seems that the UPV results were closely related to those of both the compressive and split tensile strengths observed, with high correlation coefficient values.

(4) Based on the results of RCPT, we observed that the addition of NF in the CAC mixes reduced the total charge, compared to the control (CAC1) mix without the addition of NF, and the reduction of the total charge was more remarkable in the concrete mixes with RA. This implies that the usage of NF in the RAC mixes can mitigate the possibility of steel corrosion oriented from external chlorides due to the reduction of micro-cracks in the cement matrix.

(5) The microstructural observation of concrete revealed that the micro-cracks propagated along the ITZ between old mortar and aggregate, especially in the RAC mixes. However, for RAC mix with an addition of NF, the NF plays an important role in crack bridging, resulting in a higher strength and lower permeability in concrete.

Within the scope of this study, we can conclude that the addition of NF enhanced the permeability as well as the mechanical properties, especially in concrete incorporating RA. The enhancement is primarily attributed to the bridge effect of NF, which allowed for a higher development of strength and concrete density.

Funding: This research was funded by the Ministry of Land, Infrastructure and Transport (MOLIT) and the Korea Agency for Infrastructure Technology Advancement (KAIA) for a project on "Development of manufacture technology for high-quality recycled aggregate and application of concrete (No. 16C16020280)".

Conflicts of Interest: The author declares no conflicts of interest.

References

1. Ministry of Environment. Construction Waste Recycling Survey Report. 2014. Available online: http://www.me.go.kr/home/web/main.do (accessed on 22 February 2019). (In Korean)
2. Nagataki, S.; Gokce, A.; Saeki, T.; Hisada, M. Assessment of recycling process induced damage sensitivity of recycled concrete aggregate. *Cem. Concr. Res.* **2004**, *34*, 965–971. [CrossRef]
3. Ahmadi, M.; Farzin, S.; Hassani, S.; Motamedi, M. Mechanical properties of the concrete containing recycled fibers and aggregates. *Constr. Build. Mater.* **2017**, *144*, 392–398. [CrossRef]
4. Akca, K.R.; Cakir, O.; Ipek, M. Properties of polypropylene fiber reinforced concrete using recycled aggregates. *Constr. Build. Mater.* **2015**, *98*, 620–630. [CrossRef]
5. Silva, R.V.; de Brito, J.; Dhir, R.K. Properties and composition of recycled aggregates from construction and demolition waste suitable for concrete production. *Constr. Build. Mater.* **2014**, *65*, 201–217. [CrossRef]
6. Lee, S.T.; Swamy, R.N.; Kim, S.S.; Park, Y.G. Durability of mortars made with recycled fine aggregates exposed to sulfate solutions. *J. Mater. Civ. Eng.* **2008**, *20*, 63–70. [CrossRef]
7. Mas, B.; Cladera, A.; Del Olmo, T.; Pitarch, F. Influence of the amount of mixed recycled aggregates on the properties of concrete for non-structural use. *Constr. Build. Mater.* **2012**, *27*, 612–622. [CrossRef]
8. Mefteh, H.; Kebaili, O.; Oucief, H.; Berredjem, L.; Arabi, N. Influence of moisture conditioning of recycled aggregates on the properties of fresh and hardened concrete. *J. Clean. Prod.* **2013**, *54*, 282–288. [CrossRef]
9. Guneyisi, E.; Gesoglu, M.; Algin, Z.; Yazici, H. Effect of surface treatment methods on the properties of self-compacting concrete with recycled aggregates. *Constr. Build. Mater.* **2014**, *64*, 172–183. [CrossRef]

10. Medina, C.; Zhu, W.; Howind, T.; Sanchez, M.I.; Frias, M. Influence of mixed recycled aggregate on the physical-mechanical properties of recycled concrete. *J. Clean. Prod.* **2014**, *68*, 216–225. [CrossRef]

11. Silva, R.V.; de Brito, J.; Dhir, R.K. Establishing a relationship between modulus of elasticity and compressive strength of recycled aggregate concrete. *J. Clean. Prod.* **2016**, *112*, 2171–2186. [CrossRef]

12. Zhang, H.; Ji, T.; Liu, H.; Su, S. Modifying recycled aggregate concrete by aggregate surface treatment using sulphoaluminate cement and basalt powder. *Constr. Build. Mater.* **2018**, *192*, 526–537. [CrossRef]

13. Corinaldesi, V.; Moriconi, G. Influence of mineral additions on the performance of 100% recycled aggregate concrete. *Constr. Build. Mater.* **2009**, *23*, 2869–2876. [CrossRef]

14. Muduli, R.; Bhusan, B.; Mukharjee, B. Effect of incorporation of metakaolin and recycled coarse aggregate on properties of concrete. *J. Clean. Prod.* **2019**, *209*, 398–414. [CrossRef]

15. Ozger, O.B.; Girardi, F.; Giannuzzi, G.M.; Salomoni, V.A.; Majorana, C.E.; Fambri, L.; Baldassino, N. Effect of nylon fibres on mechanical and thermal properties of hardened concrete for energy storage systems. *Mater. Des.* **2013**, *51*, 989–997. [CrossRef]

16. Yap, S.P.; Alengaram, U.J.; Jumaat, M.Z. Enhancement of mechanical properties in polypropylene- and nylon-fibre reinforced oil palm shell concrete. *Mater. Des.* **2013**, *49*, 1034–1041. [CrossRef]

17. Abbass, W.; Khan, I.; Mourad, S. Evaluation of mechanical properties of steel fiber reinforced concrete with different strengths of concrete. *Constr. Build. Mater.* **2018**, *168*, 556–569. [CrossRef]

18. Marcos-Meson, V.; Fischer, G.; Edvardsen, C.; Skovhus, T.L.; Michel, A. Durability of steel fibre reinforced concrete exposed to acid attack—A literature review. *Constr. Build. Mater.* **2019**, *200*, 490–501. [CrossRef]

19. Dinh, H.H.; Parra-Montesinos, G.J.; Wright, J.K. Shear behaviour of steel fibre-reinforced concrete beams without stirrup reinforcement. *ACI Struct. J.* **2010**, *107*, 597–606.

20. Cuenca, E.; Serna, P. Shear behavior of prestressed precast beams made of self-compacting fiber reinforced concrete. *Constr. Build. Mater.* **2013**, *45*, 145–156. [CrossRef]

21. Homma, D.; Mihashi, H.; Nishiwaki, T. Self-healing capacity of fiber reinforced cementitious composites. *J. Adv. Concr. Technol.* **2009**, *7*, 217–228. [CrossRef]

22. Cuenca, E.; Tejedor, A.; Ferrara, L. A methodology to assess crack-sealing effectiveness of crystalline admixtures under repeated cracking-healing cycles. *Constr. Build. Mater.* **2018**, *179*, 619–632. [CrossRef]

23. Caratelli, A.; Meda, A.; Rinaldi, Z.; Romualdi, P. Structural behaviour of precast tunnel segments in fiber reinforced concrete. *Tunn. Undergr. Space Technol.* **2011**, *26*, 284–291. [CrossRef]

24. Navas, F.O.; Navarro-Gregori, J.; Herdocia, G.L.; Serna, P.; Cuenca, E. An experimental study on the shear behaviour of reinforced concrete beams with macro-synthetic fibres. *Constr. Build. Mater.* **2018**, *169*, 888–899. [CrossRef]

25. Conforti, A.; Minelli, F.; Plizzari, G.A. Shear behaviour of prestressed double tees in self-compacting polypropylene fibre reinforced concrete. *Eng. Struct.* **2017**, *146*, 93–104. [CrossRef]

26. Pujadas, P.; Blanco, A.; Cavalaro, S.; Aguado, A. Plastic fibres as the only reinforcement for flat suspended slabs: Experimental investigation and numerical simulation. *Constr. Build. Mater.* **2014**, *57*, 92–104. [CrossRef]

27. Das, C.S.; Dey, T.; Dandapat, R.; Mukharjee, B.B. Performance evaluation of polypropylene fibre reinforced recycled aggregate concrete. *Constr. Build. Mater.* **2018**, *189*, 649–659. [CrossRef]

28. Kurtz, S.; Balaguru, P. Postcrack creep of polymeric fiber-reinforced concrete in flexure. *Cem. Concr. Res.* **2000**, *30*, 183–190. [CrossRef]

29. Spadea, S.; Farina, I.; Carrafiello, A.; Fraternali, F. Recycled nylon fibers as cement mortar reinforcement. *Constr. Build. Mater.* **2015**, *80*, 200–209. [CrossRef]

30. Lee, S.T. Application of nylon fiber for performance improvement of recycled coarse aggregate concrete. *J. Korea Acad. Ind. Coop. Soc.* **2017**, *18*, 785–792. (In Korean)

31. Perez-Pena, M.; Mobasher, B. Mechanical properties of fiber reinforced lightweight concrete composites. *Cem. Concr. Res.* **1994**, *24*, 1121–1132. [CrossRef]

32. *ASTM C138M-17A, Standard Test Method for Density, Yield, and Air Content of Concrete*; ASTM International: West Conshohocken, PA, USA, 2017.

33. *ASTM C39, Standard Test Method for Compressive Strength of Cylindrical Concrete Specimens*; ASTM International: West Conshohocken, PA, USA, 2018.

34. *ASTM C469-17, Standard Test Method for Splitting Tensile Strength of Cylindrical Concrete Specimens*; ASTM International: West Conshohocken, PA, USA, 2017.

35. *ASTM C597-16, Standard Test Method for Pulse Velocity through Concrete;* ASTM International: West Conshohocken, PA, USA, 2016.

36. *ASTM C1202-18, Standard Test Method for Electrical Indication of Concrete's Ability to Resist Chloride Ion Penetration;* ASTM International: West Conshohocken, PA, USA, 2018.

37. Song, P.S.; Hwang, S.; Sheu, B.C. Strength properties of nylon-and polypropylene-fiber reinforced concretes. *Cem. Concr. Res.* **2005**, *35*, 1546–1550. [CrossRef]

38. Bayasi, Z.; Zeng, J. Properties of polypropylene fiber reinforced concrete. *ACI Mater. J.* **1993**, *90*, 605–610.

39. Saint-Pierre, F.; Philibert, A.; Giroux, B.; Rivard, P. Concrete quality designation based on ultrasonic pulse velocity. *Constr. Build. Mater.* **2016**, *125*, 1022–1027. [CrossRef]

40. Solis-Carcano, R.; Moreno, E.I. Evaluation of concrete made with crushed limestone aggregate based on ultrasonic pulse velocity. *Constr. Build. Mater.* **2008**, *22*, 1225–1231. [CrossRef]

41. Leslie, J.R.; Cheeseman, W.J. An ultrasonic method for studying deterioration and cracking in concrete structures. *ACI Mater. J.* **1949**, *46*, 17–36.

42. Shariq, M.; Prasad, J.; Masood, A. Studies in ultrasonic pulse velocity of concrete containing GGBFS. *Constr. Build. Mater.* **2013**, *40*, 944–950. [CrossRef]

43. Jeon, J.K.; You, J.O.; Moon, J.H. Durability evaluation of tunnel lining concrete reinforced with nylon fiber. *J. Korea Concr. Inst.* **2008**, *20*, 487–493. (In Korean)

44. Benaicha, M.; Jalbaud, O.; Alaoui, A.H.; Burtschell, Y. Correlation between the mechanical behavior and the ultrasonic velocity of fiber-reinforced concrete. *Constr. Build. Mater.* **2015**, *101*, 702–709. [CrossRef]

45. Demirboga, R.; Turkmen, I.; Karakoc, M.B. Relationship between ultrasonic velocity and compressive strength for high-volume mineral admixture concrete. *Cem. Concr. Res.* **2004**, *34*, 2329–2336. [CrossRef]

46. Rakshvir, M.; Barai, S.V. Studies on recycled aggregates-based concrete. *Waste. Manag. Res.* **2006**, *24*, 225–233.

47. Thomas, C.; Setien, J.; Polanco, J.A.; Alaejos, P.; de Juan, M.S. Durability of recycled aggregate concrete. *Constr. Build. Mater.* **2007**, *21*, 1054–1065. [CrossRef]

48. Andreu, G.; Miren, E. Experimental analysis of properties of high performance recycled aggregate concrete. *Constr. Build. Mater.* **2014**, *52*, 227–235. [CrossRef]

49. Kwan, W.H.; Ramli, M.; Kam, K.J.; Sulieman, M.Z. Influence of the amount of recycled coarse aggregate in concrete design and durability properties. *Constr. Build. Mater.* **2012**, *26*, 565–573. [CrossRef]

MDPI

St. Alban-Anlage 66

4052 Basel

Switzerland

Tel. +41 61 683 77 34

Fax +41 61 302 89 18

www.mdpi.com

Applied Sciences Editorial Office

E-mail: applsci@mdpi.com

www.mdpi.com/journal/applsci

www.ingramcontent.com/pod-product-compliance
Lightning Source LLC
Chambersburg PA
CBHW051722210326
41597CB00032B/5567